Graphene-Based Sensors

Online at: https://doi.org/10.1088/978-0-7503-5578-0

IOP Series in Sensors and Sensor Systems

The IOP Series in Sensors and Sensor Systems includes books on all aspects of the science and technology of sensors and sensor systems. Spanning fundamentals, fabrication, applications and processing, the series aims to provide a library for instrument and measurement scientists, engineers and technologists in universities and industry.

The series seeks (but is not restricted to) publications in the following topics:

- Advanced materials for sensing
- Biosensors
- Chemical sensors
- Industrial applications
- Internet of Things (IoT)
- Lab-on-a-chip
- Localization and object tracking
- Manufacturing and packaging
- Mechanisms, modelling and simulations
- Microelectromechanical systems/nanoelectromechanical systems
- Micro and nanosensors
- Non-destructive testing
- Optoelectronic and photonic sensors
- Optomechanical sensors
- Physical sensors
- Remote sensors
- Sensing for health, safety and security
- Sensing principles
- Sensing systems
- Sensor arrays
- Sensor devices
- Sensor networks
- Sensor technology and applications
- Signal processing and data analysis
- Smart sensors and monitoring
- Telemetry

Authors are encouraged to take advantage of electronic publication through the use of colour, animations, video and interactive elements to enhance the reader experience.

A full list of titles published in this series can be found here: https://iopscience.iop.org/bookListInfo/iop-series-in-sensors-and-sensor-systems.

Graphene-Based Sensors

Edited by
Jamballi G Manjunatha
Department of Chemistry, Field Marshal KM Cariappa College, Constituent College of Mangalore University, Madikeri, Karnataka, India

IOP Publishing, Bristol, UK

ISBN 978-0-7503-5578-0 (ebook)
ISBN 978-0-7503-5576-6 (print)
ISBN 978-0-7503-5579-7 (myPrint)
ISBN 978-0-7503-5577-3 (mobi)

DOI 10.1088/978-0-7503-5578-0

Version: 20231101

IOP ebooks

British Library Cataloguing-in-Publication Data: A catalogue record for this book is available from the British Library.

Published by IOP Publishing, wholly owned by The Institute of Physics, London

IOP Publishing, No.2 The Distillery, Glassfields, Avon Street, Bristol, BS2 0GR, UK

US Office: IOP Publishing, Inc., 190 North Independence Mall West, Suite 601, Philadelphia, PA 19106, USA

Dedicated to IOPP Book publishing staff.

Contents

Preface

Since the last decade, graphene based materials have attained significant importance in electrochemical and material science. Different forms of graphene like graphene oxide, reduced graphene oxide, graphene quantum dots, pristine graphene, functionalized graphene, graphene nanoplatelets, etc, are used in construction of electrochemical devices. The extensive usage of these materials in electrochemical sensing of significant molecules or ions is due to their exceptional electrical, thermal, electrochemical, mechanical and electronic properties. The global rise in the number of research laboratories devoted to electrochemical sensing devices and applications, demonstrate that this area of research is interdisciplinary in nature.

This book describes graphene-based sensors for real-time detection. This covers the trends and applications of graphene-based electrochemical sensors for diverse applications in biomedical, food, environmental monitoring, etc. The key focus of writing this book is to attract scientific researchers from the field of electrochemistry on graphene-based electrochemical sensors, and to encourage more scientists in this field to develop the science. It will also help to progress the proficiency among researchers from diverse areas like physics, materials science, electronics, biology and forensics, who are expected to come together to do some ground-breaking research in this promising interdisciplinary research area, i.e. electrochemical biosensors.

Dr J G Manjunatha
Editor

Acknowledgements

I gratefully acknowledge my sincere gratitude to my mother Sharadamma Jamballi and to my research colleagues.

Editor biography

Jamballi G Manjunatha

J G Manjunatha is working as an Assistant Professor in Chemistry at FMKMC College, a Constituent College of Mangalore University, Madikeri, India. He received his PhD degree in Chemistry from Kuvempu University and Postdoc from the University of Kebangsaan Malaysia. He has received various awards and published more than 165 research articles in reputed International Journals. An editor for thirteen books (RSC, ACS, IOP, Elsevier and Bentham Science publishers), and special issues (IOP Publishing, *Frontiers in Sensors*, MDPI). He is also an editorial board member for many reputed journals and Editor-in-Chief *Sensing Technology* journal (Taylor and Francis).

List of contributors

Vinayak Adimule
Angadi Institute of Technology and Management (AITM), Belagavi, Karnataka, India

Sanu K Anand
Department of Applied Chemistry, Cochin University of Science and Technology, Kochi, Kerala, India

Nurgul K Bakirhan
Department of Analytical Chemistry, Gülhane Faculty of Pharmacy, University of Health Sciences, Ankara, Turkey

C Balaji
Department of Physics, Periyar University, Salem, India

Merve Yuzbasioglu Baran
Department of Pharmacognosy, Gülhane Faculty of Pharmacy, University of Health Sciences, Ankara, Turkey

S Beena
Department of Chemistry, Amrita School of Physical Sciences, Amritapuri, India

Mohamed Brahmi
Physical Chemistry of Natural Substances and Process Team, Laboratory of Applied Chemistry and Environment (LCAE-CPSUNAP), Department of Chemistry, Faculty of Sciences, University Mohamed Premier, Oujda, Morocco

Thiago C Canevari
LabNaHm: Multifunctional Hybrid Nanomaterials Laboratory. Engineering School, Mackenzie Presbyterian University, São Paulo, SP, Brazil

Sandeep Chandrashekharappa
Department of Medicinal Chemistry, National Institute of Pharmaceutical Education and Research Raebareli, Lucknow, Uttar Pradesh, India

Tanmay Chatterjee
Department of Chemistry (Science and Humanities), PES University, Bengaluru, India

B Chethan
Department of Physics, Indian Institute of Science, Bangalore, Karnataka, India

Bruna Coldibeli
Laboratório de Eletroanalítica e Sensores, Departamento de Química, Centro de Ciências Exatas, Universidade Estadual de Londrina, Londrina, PR, Brazil

Mariana Munik da Silva
LabNaHm: Multifunctional Hybrid Nanomaterials Laboratory. Engineering School, Mackenzie Presbyterian University, São Paulo, SP, Brazil

Octavio Pereira Lopes de Souza
LabNaHm: Multifunctional Hybrid Nanomaterials Laboratory. Engineering School, Mackenzie Presbyterian University, São Paulo, SP, Brazil

Ersin Demir
Afyonkarahisar Health Sciences University, Faculty of Pharmacy, Department of Analytical Chemistry, Afyonkarahisar, Turkey

Aprys Cristina Travasso Sarinho dos Santos Oliveira
LabNaHm: Multifunctional Hybrid Nanomaterials Laboratory. Engineering School, Mackenzie Presbyterian University, São Paulo, SP, Brazil

Yashi Dwivedi
Department of Medicinal Chemistry, National Institute of Pharmaceutical Education and Research Raebareli, Lucknow, Uttar Pradesh, India

Joao Henrique Astolfi Ferreira
LabNaHm: Multifunctional Hybrid Nanomaterials Laboratory. Engineering School, Mackenzie Presbyterian University, São Paulo, SP, Brazil

Gustavo Fix
Laboratório de Eletroanalítica e Sensores, Departamento de Química, Centro de Ciências Exatas, Universidade Estadual de Londrina, Londrina, PR, Brazil

Sonu Gandhi
DBT – National Institute of Animal Biotechnology (NIAB), Hyderabad, Telangana, India
and
RCB – Regional Centre for Biotechnology, Faridabad, Haryana, India

H Ganesan
Sri Ranganathar Institute of Engineering and Technology, Coimbatore, India

M G Gopika
Amrita School of Physical Sciences, Amritapuri, India

A Gowdhaman
Department of Physics, Periyar University, Salem, India

P Karpagavinayagam
PG and Research Department of Chemistry, V.O. Chidambaram College, Thoothukudi, Tamil Nadu, India

Raunak Katiyar
Department of Medicinal Chemistry, National Institute of Pharmaceutical Education and Research Raebareli, Lucknow, Uttar Pradesh, India

B Kavitha
Department of Chemistry, Sri Ranganathar Institute of Engineering and Technology Athipalayam, Coimbatore, India

G Kavitha
S.A. College of Arts & Science, Poonamallee, Avadi Main Road, Thiruverkadu, Chennai, India

Pratik Kolhe
DBT – National Institute of Animal Biotechnology (NIAB), Hyderabad, Telangana, India

K Girish Kumar
Department of Applied Chemistry, Cochin University of Science and Technology, Kochi, Kerala, India

Sevinc Kurbanoglu
Department of Analytical Chemistry, Faculty of Pharmacy, Ankara University, Ankara, Turkey

Laura Chaluleu Lieb
LabNaHm: Multifunctional Hybrid Nanomaterials Laboratory. Engineering School, Mackenzie Presbyterian University, São Paulo, SP, Brazil

J Maalmarugan
Sri Ranganathar Institute of Engineering and Technology, Coimbatore, India

Nidhi Manhas
School of Sciences, Indira Gandhi National Open University, Maidan Garhi, New Delhi, India

Manna Rachel Mathew
Department of Applied Chemistry, Cochin University of Science and Technology, Kochi, Kerala, India

Ana Maria Assed Batistela Medeiros
LabNaHm: Multifunctional Hybrid Nanomaterials Laboratory. Engineering School, Mackenzie Presbyterian University, São Paulo, SP, Brazil

Murat Misir
Kırşehir Ahi Evran University, Faculty of Engineering and Architecture, Kırşehir, Turkey

Luccas Lossano Name
LabNaHm: Multifunctional Hybrid Nanomaterials Laboratory. Engineering School, Mackenzie Presbyterian University, São Paulo, SP, Brazil

Nida Aydogdu Ozdogan
Afyonkarahisar Health Sciences University, Faculty of Pharmacy, Department of Analytical Chemistry, Afyonkarahisar, Turkey
and

Department of Analytical Chemistry, Faculty of Pharmacy, Ankara University, Ankara, Turkey
and
Ankara University, Graduate School of Health Sciences, Ankara, Turkey

S Parameswari
Velammal Institute of Technology, Ponneri, Chennai, India

Drishya Prakashan
DBT – National Institute of Animal Biotechnology (NIAB), Hyderabad, Telangana, India
and
RCB – Regional Centre for Biotechnology, Faridabad, Haryana, India

V Prasad
Department of Physics, Indian Institute of Science, Bangalore, Karnataka, India

R Rajkumar
PG and Research Department of Chemistry, Kamaraj College, Thoothukudi, Tamil Nadu, India

H G Rajprakash
Department of Physics, JNN College of Engineering, Shivamogga, Karnataka, India

R Ramesh
Department of Physics, Periyar University, Salem, India

Titiksha Kumar Sagar
Department of Medicinal Chemistry, National Institute of Pharmaceutical Education and Research Raebareli, Lucknow, Uttar Pradesh, India

Serdar Sanli
Department of Chemistry, Faculty of Science and Literature, Ordu University, Altınordu, Ordu, Turkey

Monima Sarma
Department of Chemistry (Science and Humanities), PES University, Bengaluru, India

Elen Romão Sartori
Laboratório de Eletroanalítica e Sensores, Departamento de Química, Centro de Ciências Exatas, Universidade Estadual de Londrina, Londrina, PR, Brazil

Sıla Ozlem Sener
Department of Pharmacognosy, Gülhane Faculty of Pharmacy, University of Health Sciences, Ankara, Turkey

N Senthilkumar
PG & Research Department of Chemistry, Arignar Anna Government Arts College, Cheyyar, Tiruvannamalai, India

Duygu Sevim
Department of Pharmacognosy, Gülhane Faculty of Pharmacy, University of Health Sciences, Ankara, Turkey

Maitri Shah
DBT – National Institute of Animal Biotechnology (NIAB), Hyderabad, Telangana, India

Kalpana Sharma
Department of Physics, M S Ramaiah Institute of Technology, Bangalore, Karnataka, India

Sreelekshmi
Amrita School of Physical Sciences, Amritapuri, India

M Stanleydhinakar
Department of Physics, Periyar University, Salem, India

A Sunilkumar
Department of Materials Engineering, Indian Institute of Science, Bangalore, Karnataka, India

Abdesselam Tahani
Physical Chemistry of Natural Substances and Process Team, Laboratory of Applied Chemistry and Environment (LCAE-CPSUNAP), Department of Chemistry, Faculty of Sciences, University Mohamed Premier, Oujda, Morocco

S Thomas
International and Inter University Centre for Nanoscience and Nanotechnology, Mahatma Gandhi University, Kottayam, Kerala, India

Daniel Yaezu Tiba
LabNaHm: Multifunctional Hybrid Nanomaterials Laboratory. Engineering School, Mackenzie Presbyterian University, São Paulo, SP, Brazil

C Vedhi
PG and Research Department of Chemistry, V.O. Chidambaram College, Thoothukudi, Tamil Nadu, India

Chapter 1

An overview of graphene properties, types, and role in chemistry

Mariana Munik da Silva, Luccas Lossano Name, Daniel Yaezu Tiba, Octavio Pereira Lopes de Souza, Aprys Cristina Travasso Sarinho dos Santos Oliveira, Joao Henrique Astolfi Ferreira, Laura Chaluleu Lieb, Ana Maria Assed Batistela Medeiros and Thiago C Canevari

In this chapter, a general review of graphene will be carried out, addressing its morphology, properties, synthesis methodology, and characterization.

1.1 Properties

Graphene is one of the carbon allotropes, such as fullerenes, carbon nanotubes, graphite, and diamond. It is a nanomaterial composed exclusively of carbon atom bonds with sp^2 hybridization (C=C), forming a two-dimensional (2D) sheet organized through several hexagonal reticles [1, 2]. Graphene, like other nanomaterials, has several unique characteristics due to the behavior of the atoms of the molecules on a nanoscale. Because of its 2D structure, it has thermal, optical, mechanical, and electrical properties that have attracted great interest for its various applications. The material is known and highlighted for being resistant, light, flexible, and quickly dissipating heat [3, 4]. The electrical properties of graphene are unique to its class of material due to the arrangement of carbon atoms, creating a flat hexagonal 2D structure, such as sheets, having a large contact area, without branches, similar to a honeycomb, allowing the free movement of electrons, thus being able to tune possible characteristics such as conductivity and electrical resistivity, which could lead to energy storage for instance [5, 6]. Graphene is a crystalline material of a 2D, hexagonal structure, resembling a beehive structure comprising carbonic sp^2 hybridization bonds. This carbon crystal has an atom thickness and a fingerprint of a complete form scattered linearly, close to Fermi's level. The atoms in this long chain of carbonic connections show their sp^2 hybridization with the orbitals p_x and p_y, performing σ bonds with the adjacent carbons.

doi:10.1088/978-0-7503-5578-0ch1

This characteristic allows the non-hybridization of the p_z orbital, which is oriented perpendicularly to the plane, forming the valence and conduction bands, symmetrical due to the conformations of the triangular subnetworks formed on the structure, with a singular bandgap. Hybridized orbits include the bands π–π^*, responsible for the free movement of electrons in the 2D plane [5].

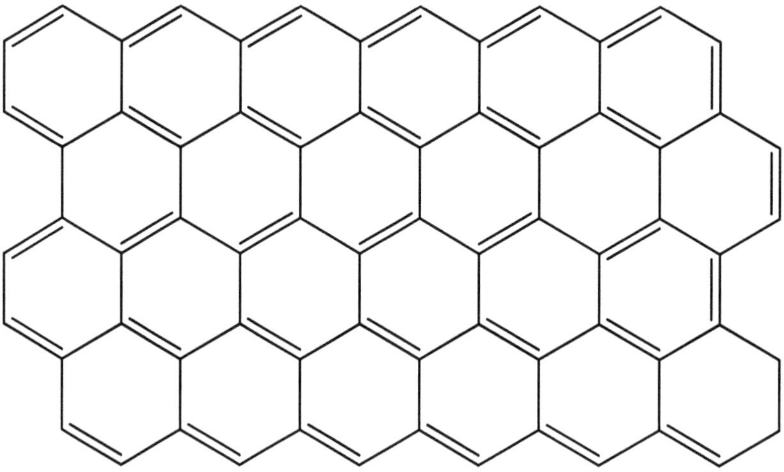

Graphene

Graphene oxide (GO) has a structure composed of bonds between carbon atoms and sp^2 hybridization. Differing structurally from graphene due to several oxygenated functional groups directly influencing the material's properties and characteristics since they directly influence graphene morphology [6].

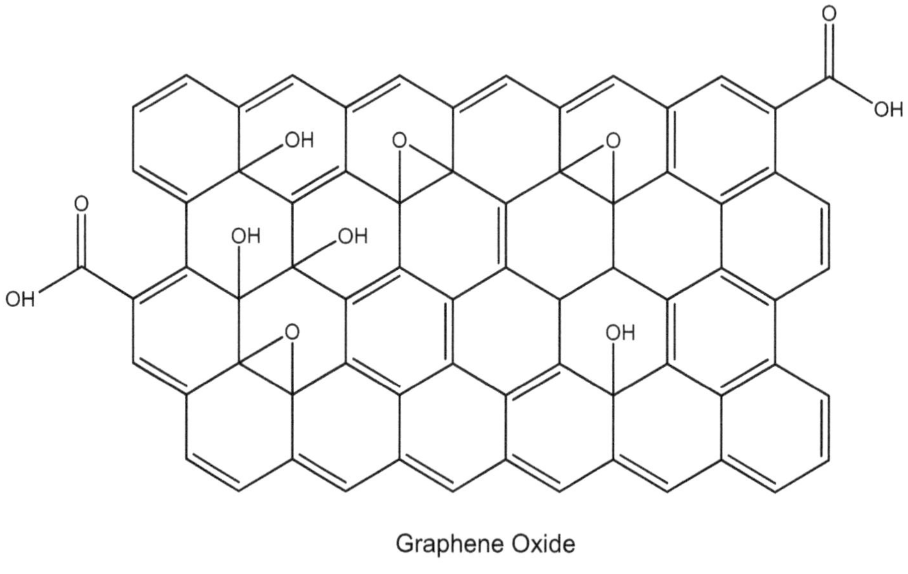

Graphene Oxide

Another compound mentioned is reduced graphene oxide (rGO), obtained from GO. When compared to pristine graphene and GO, the former has intermediate characteristics. rGO has an sp^2 hybridization structure, and unlike pristine graphene, it also has carbon atoms that hybridize in sp^3, although in a lower quantity when compared to GO. These carbon atoms, which are in sp^2 hybridization regions, have three carbon atoms that form adjacent bonds with three other atoms of the same element in a triangular fashion. The valence electron left in the structure occupies a normal position of $2p_z$, creating a bond with a large capacity to be displaced [7].

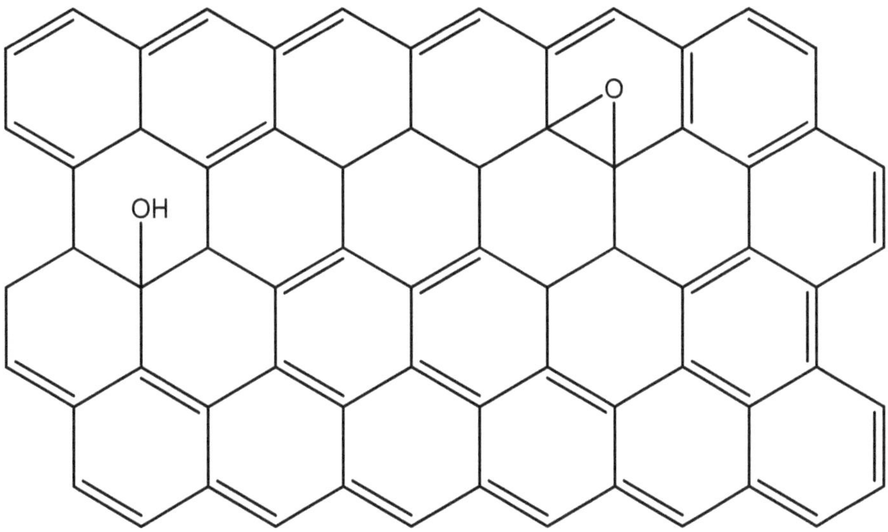

Reduced Graphene Oxide

According to Xu *et al* [8], graphene's flawless or highly pure surface appears highly stable. Therefore, the graphene surface with a high purity level typically interacts with other molecules through physical adsorption by electron transfer as π–π interactions. Functional groups can be introduced through various physical or chemical methods to obtain a more reactive surface of graphene, defects, or surface. One of the most common examples is the chemical doping with boron or nitrogen atoms and functional groups such as carboxyl, carbonyl, and amine, achieving new properties of graphene. Kaplan *et al* [9] reported that the unique electronic structure, frontier orbital electrons in graphene, and free electrons could participate during a chemical reaction. Examples of occurrence of free-electron transfer during methods such as aryl diazonium functionalization and selective oxidation of graphene. Moreover, frontier orbital electrons are involved in edge and nanopore formation and edge functionalization.

Structural defects may appear during growth or processing and can appeal to specific applications. For example, to Liu *et al* [10], there are three defects: point defects as vacancies or interstitial atoms, line defects (lack of continuity of hexagons), and edge defects. Defects are not stationary, and their migration depends on the defect type and increases exponentially with temperature, affecting the product's properties.

The hybridization sp^2 of the carbon atoms allows them to arrange themselves into a diverse number of polygons and form other structures, some of which can create different types of defects such as Stone–Wales, single vacancies, multiple vacancies, one-dimensional (1D) defects, and defects at the edges. The graphene lattice can be reorganized by forming non-hexagonal rings without removing or adding atoms into the structure, as with single and multiple vacancies. Single vacancies undergo a Jahn–Teller distortion leading to the saturation of two of the three dangling bonds toward the missing atom. One bond always remains for geometrical reasons, and two pentagons and two heptagons form in the structure. At the same time, multiple vacancies can be created by the coalescence of two single vacancies or by removing two neighboring atoms, two pentagons, and n octagons appear instead of the honeycomb shape. As reported by Liu *et al* [10], simulations determine that the formation energy of multiple vacancies is in the same order as single vacancies, about 8 eV. Various vacancies are thermodynamically favored over structures with an odd number of missing atoms where an open bond remains.

The 1D defect occurs by tilt boundaries that separate two domains of different lattice orientations with the deflection axis normal to the plane. These lines can be considered lines of reconstructed point defects with or without dangling bonds. Since edges terminate each graphene layer with the atom being either free or suppressed with hydrogen atoms, the most common structures formed are armchair and zigzag orientations. They can appear through local changes through the reconstruction method used or sustained by carbon removal from the edges. For example, removing one carbon atom from a zigzag edge turns it into a pentagon, causing tension in the structure and leading to a more reactive point in the graphene layer. Other reconstructions can combine different pentagons and heptagons arrangements, as the hydrogen atoms and other chemical groups can saturate bonds at the edge, increasing the number of defects.

The defects associated with dangling bonds enhance the reactivity of graphene. Numerous simulations showed that reconstructed vacancies locally change the density of π-electrons, increasing the local reactivity. It's also reported that oxygenated carbon groups can be easily attached to vacancies type-defects. Denis *et al* [11] showed in a study of comparative notes the reactivity of the defects described above. The single vacancy is the most reactive of all the defects since a pentagon ring is formed in the graphene layer. As for the other types of vacancy defects, their reactivity can be like each other depending on the attached group to the vacancy.

Regarding edge defects, the zigzag orientation was more reactive than the armchair orientation since it could present dangling bonds. Also, as a comparative note, the armchair edge gives the most stable form of all other defects.

Furthermore, it is possible to increase or achieve new properties of graphene through methods other than defects and patching. For example, Kaplan *et al* [9] show numerous functionalization possibilities through covalent interactions, exemplified by the interaction between the aryl diazonium salt and the graphene layer, where the reaction mechanism under conditions of low diazonium concentration involves electron transfer from the graphene to aryl diazonium salt, causing the bond between the benzene ring and diazonium group to be broken, resulting in aryl radical group and the release of the nitrogen molecule, leading to a form of a covalent C–C bond between the graphene surface and the aryl group. At the attachment site, the carbon changes from an sp^2 to an sp^3 hybridization to form a bond. The electron transfer depends on the overlap of states between the filled states of the graphene and the empty states of the molecule. The functionalization can be done by applying electrochemical bias or passively without it. As known in previous studies, using electrochemical bias can also change other properties of graphene. Therefore, covalent functionalization is highly explored since the linkage between the material and the graphene is more effective. Thus through covalent functionalization, it is possible to combine other materials with graphene achieving new properties and new hybrid materials. For example, inorganic nanoparticles such as metals and semiconductors have been decorated onto graphene and GO sheets with implications for energy generation, storage, electronics, and biological applications.

Another example is the hybrid materials resulting from graphene and polymers. Many approaches have been extensively studied for the covalent attachment of polymers to graphene employing GO or rGO since there are more functional groups for functionalization. Moreover, the combination can show improvement in mechanical properties.

According to the first Brillouin zone, the electronic structure of graphene has two inequivalent points: K and K′, which are called Dirac points, where two relevant bands occur: valence (π) and conductance (π*) [12]. This happens due to the delocalized π bonds above and below the base plane arising from the sp^2 hybridization [13].

The tight-binding event on the first nearest neighbor interaction provides the dispersion relation of the electrons near these two points (K and K′), which is linear and has a conical shape. As a result, this makes the electronic properties different from other materials [12].

There is also an ambipolar behavior due to the electric field standing out. This behavior shows that positive and negative gate voltages induce significant amounts of electrons (holes). All the induced carriers can reach up to 10^{13} cm^{-2} and are mobile, and graphene has no trapped charges [12].

For undoped graphene, for instance, the π band is filled, which indicates that the Fermi level is at the point where the bands meet. The Fermi level is based on the energy of the most energetic electron in the material at zero Kelvin temperature. When shifting the Fermi level, one can manipulate the carriers and, consequently, the material's properties [14].

At low energies, the charge carriers (electrons and holes) behave like Dirac fermions, defined by quasi-particles without mass. As the dispersion is linear close to the Dirac points, the form of the relativistic Dirac equation is related to the dynamics of the carriers, in which the speed of light is replaced by the Fermi speed (300 times less than the speed of light) [3, 15, 16].

Furthermore, regarding the electrical resistivity of graphene, it is assumed that it is approximately 1.0 μΩ cm, which is slightly lower than that of silver (1.5 μΩ cm) [17]. Other studies were conducted from 2005 to 2007 [3, 18, 19] and showed that graphene's electron mobility ranges from 3×10^3 to 2.7×10^4 cm^{-2} V^{-1} s^{-1}. Although, SLG (single-layer graphene) and MLG (multilayer graphene) have attracted more interest. The electrical conductivity of MLG can increase with the increase of the temperature, which implies a negative coefficient of resistance, unlike SLG, which slightly decreases with the increase of the temperature [17].

Thinking about a planar graphene model without any disorders, the Fermi energy level is placed at the Dirac point, where the conductance and valence bands meet. The dispersion relation around the Dirac point is isotropic and linear. When there is any introduction of disorders into the graphene structure, there is a violation in the electronic homogeneity. Consequently, the latter affects the interatomic bonds' length, lead to the re-hybridization of σ and π orbitals, and may cause the scattering of electron wave, changing the electron trajectory [20].

Moreover, these disorders diverge from the electronic structure in a perfect lattice. The electric properties of graphene may influence the intrinsic ripples by changing the gap, creating polarized carrier puddles, and inducing pseudo-magnetic fields [20].

1.1.1 Rheological properties

The rheology of the nanofluids is based on the shear viscosity, yield stress, and yet the shape of the nanoparticles. These characteristics strongly influence factors such as viscosity and convective heat transfer [21]. However, temperature is the effective parameter for the density of nanofluids [22].

Therefore, according to Mehrali and coworkers, graphene nanofluids demonstrated that the temperature strongly influenced the viscosity, in which with the decrease of the temperature, the density is higher, and with the increase of the temperature will be lower. This happens due to the weakening of the intermolecular adhesion forces [22].

Many other studies have been conducted to understand the rheology of graphene, as in nanofluids or nanocomposite systems.

Dynamic rheological measurements quantify graphene dispersion in a polymer matrix system. The linear viscoelastic behavior (the elastic modulus (G') at low frequency) provides information on the graphene nanosheet distribution in the polymer matrix. Nevertheless, developing a linear plateau level of G' at low frequency indicates the rheological percolation threshold, which is connected to forming an interconnected network of graphene nanosheets. The latter is based on these measurements, which can quantify the dispersion and level of interaction between nanosheets and the polymer chains. On the other hand, the same observations can be considered by measuring the complex viscosity of the polymer nanocomposite systems [23].

When the aspect ratio and surface of the graphene nanosheets are high, instead of a good quality dispersion of nanosheets, the opposite happens due to the stronger interaction between these nanosheets, which is called aggregation [23].

In addition to the fracture strength of graphene, single-layer graphene is considered the strongest material ever tested, with an intrinsic strength of 130 GPa, when graphene is defect-free. However, when referring to the armchair, zigzag directions are 90 and 105 GPa [24, 25].

1.1.2 Thermal conductivity

Thermal conductivity is a characteristic of a material that measures how well it conducts heat, and it is based on Fourier's law. Graphene has exceptionally high thermal conductivity (k), ranging from 2600 to 5300 W m^{-1} K^{-1} at room temperature. On the other hand, when graphene is supported on a substrate, the thermal conductivity is lower, ranging from 370 to 600 W m^{-1} K^{-1}. It is expressed as $k \sim C\nu\Lambda$, where C is the specific heat capacity, ν is the phonon group velocity, and Λ is the phonon mean free path (MFP) [26, 27].

As a result of the 2D structure and because of the strong sp^2 bonding, graphene has a high phonon group velocity given by $\nu \sim \sqrt{C/M}$, where C is the force constant related to the bond strength, M is the atomic mass [26].

Moreover, this high thermal conductivity is explained by the graphene's ideal 2D structure, and the long phonon MFP is phonon MFP (Λ) is given by $\Lambda = \tau\nu$, where τ is phonon relaxation time. Therefore, various defects in the structure reduced the extremely high conductivity of graphene, for instance: dopants, vacancies, and grain boundaries [26].

1.2 Routes of synthesis

Geim and Novoselov made the first top-down approach to graphene synthesis. The mechanical exfoliation (using scotch tape) led to the first process for graphene isolation and, consequently, began a 'gold rush' toward synthesizing graphene on a larger scale. Their method is well known but not beneficial for industrial scale due to the low yield. Nowadays, graphene production is very studied, presenting new ways to produce the material with more efficient methods [27, 28].

The top-down approach involves a dimensional reduction of a starting material throughout any process, such as the exfoliation of graphite powder. In this scenario, bulk graphite (a 3D material) is transformed into a 2D structure with a thickness within the nanometric range. Even though the most known method is the chemical exfoliation of Hummers, there are other approaches for graphene synthesis, including liquid-phase exfoliation (LPE), electrochemical exfoliation, unzipping carbon nanotubes, and arc discharge [27, 29].

The LPE technique normally involves organic solvents and an external energy source. Two common ways to apply this energy are sonication or mechanical mixing. This process uses external sources to break the van der Waals interaction in the graphite layers. LPE has two main steps. The first is the oxidation of the graphite, which increases the interlayer space and creates cavities like bubbles, which reduces the van der Waals forces and allows organic solvent inclusion. The second step involves shear forces, like mechanical stress involving an external source [27, 29, 30].

Typically, graphene flakes are exfoliated by organic solvents. However, one can use others, like ionic liquids or surfactants. The first proposed production by LPE was operated using *N*-methyl pyrrolidone (NMP) and sonication of graphite powder. The product using LPE has better electrical properties than other routes. However, the result tends to be a mixture of single-layer to few-layer graphene flakes, limiting the single-layer amount [27, 29, 30].

Another process is electrochemical exfoliation, where the precursor is a graphite work electrode. The process relies on a single step in which an electric voltage is applied, forcing ionic species to intercalate the graphite, creating gaseous molecules that allow the exfoliation of the layers. The product can either be stored in an organic solvent showing great stability or filtrated by vacuum [27, 29, 31].

The working voltage should be carefully chosen as it may induce alterations in the graphene structure. If it uses negative potential, the graphite works as a cathode, attracting cationic species; the opposite occurs when a positive potential is applied. Graphite will work as an anode, attracting anionic species. This way, the graphene can be modified for a desired application [31].

The arc-discharge method is one of the most traditional graphene synthesis routes. The process consists of an electric-arc chamber, a graphite rod anode, the sacrificial electrode, and a cathode inside a water-cooled vacuum chamber. For the discharge of the graphite rod, one can apply different gas atmospheres, such as He, H_2, Ar, NH_3, CO_2, and their mixtures. These gases directly affect the graphene produced, whether doped or pure. When the constant direct current (DC) or alternating current (AC) passes through the electrodes, a high-temperature plasma is formed, vaporizing the anode and creating graphene [31].

The unzipping of carbon nanotubes can also be presented. The unzipping is made by processes such as a chemical attack, intercalation and exfoliation, plasma etching, and electrochemical. The process involves cutting the carbon nanotube (CNT) in an axial direction into a single graphene sheet. Single-walled and multi-walled carbon nanotubes (SWCNTs and MWCNTs) can be starting materials. In the chemical unzipping, the CNT is treated with H_2SO_4 and then oxidized with $KMnO_4$ [30, 31].

1.2.1 Top-down and bottom-up

One of the most highlighted top-down methods for graphene preparation is the Hummers method, a chemical exfoliation route. This obtains the derivatives such as graphite oxide (Gr–O), GO, and rGO. The modified Hummers method also consists of modifying synthesis parameters, for instance, oxidizing agents and their quantities, temperature, and oxidation time.

Even though it was developed in 1958, the history of Hummers method began almost a hundred years before with Brodie in 1859. Besides Hummers and Brodie, Staudenmeier (1898) and Hofmann (1917) also worked and developed known 'acidic graphite.' Those four methods have a common reaction over the graphite oxidation step, which will be discussed further. However, the reactants used as oxidizing agents, such as HNO_3, H_2SO_4, or a mix of both, added to $KClO_3$, $KMnO_4$, or $NaNO_3$, would make a mixture that released toxic gases or that had a risk of explosion [25, 27, 32, 33].

Chronologically, with the methods of Brodie (1859), Staudenmeier (1898), Hoffman (1917), and Hummers (1958), the developments and modifications among them were related to lowering the dangerousness of the synthesis. Nowadays, the method used as a reference is Hummers–Offerman (or simply the Hummers method), which implies the use of H_2SO_4 with $KMnO_4$ and $NaNO_3$ only [32, 33].

The Hummers synthesis, and its modified methods, consist of three steps: oxidation of graphite structure, in which Gr–O is obtained; the step of mechanical exfoliation, obtaining and finally, the reduction step, getting rGO.

Given those three main steps, it is described as follows:

- **Oxidation step (obtaining Gr–O):** In this step, the oxidizing agents act by inserting oxygenated groups between the graphene layers in the graphite structure so that the expansion between them can occur and, consequently, the weakening of the van der Waals forces that keeps the structure stacked. Among the steps, this one has the biggest significant impact on the final form of the derivatives of graphene, since the higher the grade of oxidation of structure, the higher will be the number of oxygenated moieties inserted (such as hydroxyls, ketones, epoxy, carboxylic acids, and carboxylates), and as a consequence the rupture of sp^2 hybridizations forming sp^3 hybridizations, leading to the formation of 'holes' in the structure, or even a rearrangement of the hexagonal rings forming pentagons and heptagons (Stone–Wales effect) [33, 34].
- **Exfoliation step (obtaining GO):** The exfoliation step separates the GO layers from the macrostructure of Gr–O. This exfoliation occurs in an ultrasound bath which easily detaches the layers because of the weakening of the van der Waals forces between the layers of graphite. Not necessarily 100% of the monolayer structures of GO could be obtained from bilayers, trilayers, and multilayers [33, 34].
- **Reduction step (obtaining rGO):** GO reduction stage consists of removing the oxygenated moieties; in some cases, restoring a few defects caused in the oxidation step could be possible. It is known that the groups and defects inserted during the oxidation step may be propagated until the reduction step, affecting the basal plane of GO and its quality [33, 34].

The best way to keep track of the quality of obtained GO or rGO is by using Raman spectroscopy. On the obtained spectra through this technique, two main bands are observed, the D band centered approximately at 1350 cm^{-1} and the G band centered approximately at 1560 cm^{-1}. The D band is related to the structure defects, rupture of sp^2 hybridization, and edge effect, yet the G band is related to the preservation of sp^2 hybridization. It is possible to relate those two bands by a relation of I_D/I_G, in which the higher the value obtained from this relation, the higher the structure's degradation rate. Meanwhile, values close to zero mean better preservation of the structure [35–37].

Nowadays, the modified Hummers methods adds little towards the safety of synthesis itself but it does preserve the structure of GO and rGO to obtain a final nanomaterial closer to a pristine form. It is worth saying that the quality of the final material is also related to the presence of impurities, such as metals in the graphite precursor, and the size of the flake or powder used [34].

Bottom-up processes for nanomaterials represent the opposite strategy for previously mentioned top-down processes. In this scenario, the objective is a dimensional increase, usually from an atomic or molecular level to larger structures.

Even though there are many routes to produce graphene, it is expected that bottom-up techniques should be more feasible and have higher controllability, which leads to higher quality end-products [28].

Chemical vapor deposition (CVD) is one of the most promising techniques in this scenario. A reactant gas mixture is fed into a reaction chamber under controlled conditions, either starting a reaction or just diffusing throughout the surface of a substrate. This ultimately leads to a thin film's growth in the solid–gas interface, which yields the desired product.

Commonly used substrates include transition metals, such as copper or nickel, or insulating materials, such as silica or alumina. Pretreatment is advised for higher quality graphene and includes electropolishing and annealing. On the other hand, reactant gases usually involve a mixture of methane and hydrogen [38]. Carbon sources, including light hydrocarbons such as acetylene [39], alcohols, volatile organic compounds, plant oils, aromatic compounds, and even polymers and raw carbon materials, could be used. Different sources are directly linked to the desired characteristics of an end-product, such as a heteroatom-bearing structure, or evaluation of opportunities regarding underused raw materials or residues, such as green synthesis approaches of carbon dioxide methanation and conversion to graphene [40].

An alternative method for graphene production is the so-called epitaxial growth from SiC. Epitaxy derives from Greek, in which 'epi' stands for 'above' and 'taxis' stands for 'an ordered manner.' The main benefit of such a process is the attainment of high-quality films that can be as large as the substrate of use. Graphene sheets are grown in a SiC substrate due to the exposition of the material to high temperatures. Silicon undergoes sublimation faster, leaving an organized carbon layer at the surface [41]. One can choose many experiment setups to produce such materials. Some strategies involve ultra-high vacuum chambers or furnaces with an inert argon atmosphere [42].

The quality of the manufactured product is directly related to the surface of the substrate. SiC lattices can be either C or Si-terminated. A growth considering the first case would lead to various orientations [43]. Growth from the Si-face of silicon carbide allows unique tuning possibilities, which lead to graphene homogeneity and large domain size [42].

The last bottom-up approach to be covered in this chapter is chemical synthesis. This strategy aims to provide nanographene through the reaction between building blocks in a reaction media. Some characteristics to be explored by nanographene synthesis include mass production of virtually similar molecules, tunable electronic properties by shape and size, and control of chemical properties by introducing side groups [44]. Notably, the quality of a graphene sheet is directly related to the chosen precursor, as observed using the very same protocol for different building blocks [45].

Most common approaches involve the application of polycyclic aromatic hydrocarbons (PAH) as precursors. In this context, one should understand that PAHs with lateral sizes larger than 1 nm are considered nanographene molecules that belong to the classification of 'nanographene.' Nanographenes comprise many compounds, including nanographene molecules, quasi-1D graphene nanoribbons, and graphene quantum dots. However, all the cited materials show lateral sizes within the 1 to nm [46].

Even though a synthetic route can start from very simple molecules such as benzene or hexane as interfacial reactions catalyzed by Lewis acids leading to graphene sheets [45], many kinds of literature focus on the manipulation of PAHs as 'monomers' for bottom-up strategies for further growth into graphene or graphene-like structures. Cross-coupling reactions like the Suzuki–Miyaura and other reactions, such as [2 + 2 + 2] cyclotrimerization or Diels–Alder, may be used for highly substituted large aromatic molecules, which are further graphitized through Scholl reaction or transition metal cyclodehydrogenation reactions [46].

1.3 Applications

The electrochemical applications of graphene can be separated into two main categories (bio) sensing and energy. This section will briefly show a few applications of graphene and its based materials in these categories.

An electrochemical sensor is a device that can convert chemical information into an electric signal, which can be used to correlate the concentration of a specific sample to an analytical signal [47]. Therefore, it must have main parts: a recognition element, a transducer, and a signal processing unit. In the recognition, the element where chemical reactions occur, the transducer converts the signal generated by the recognition element into a detectable one. In other words, it can transform a chemical signal into an electric signal. Finally, it consists of an electronic circuit that converts the transducer signal into a readable format [48].

It's usual to prepare composites and hybrid materials with GO and rGO, such as mixing polymers and metal nanoparticles. The new material has enhanced properties that become interesting in electrode building for electrochemical detection. Graphene can be both a recognition element and a transducer.

The analytes sensed by graphene and its related materials are broad and can be organic or inorganic species. Inside the organic species biomarker molecules can be detected: dopamine [49], serotonin [50], nicotinamide adenine dinucleotide (NADH/ NAD^+) [51], ascorbic acid [52], uric acid [53], glucose [54], histamine [55], epinephrine [56] and cholesterol [57]. Active pharmaceuticals ingredients, for example, paracetamol [58], diethylstilbestrol [59], allopurinol [60], acetazolamide [61], rizatriptan [62], naproxen [63], pregabalin [64], caffeine [65] and chloramphenicol [66] can be found. And other organic species can be detected: catechol [67], hydroquinone [67], bisphenol-A [68], parathion [69], tripropylamine [70], and tartrazine [71]. Inside the inorganic species heavy metals can be sensed, such as Pb^{2+} [72], Hg^{2+} [73], Cd^{2+} [72], Cu^{2+} [74] and Zn^{2+} [75], hydrazine [76], hydrogen peroxide [77], nitric oxide (NO) [78], nitrogen dioxide (NO_2) [79].

A biosensor is an electrochemical sensor that uses specific biochemical reactions mediated by biostructures [47], such as enzymes, whole cells, antibodies, aptamers, and DNA, for recognition element [80] and then converts this biological response into a signal [48]. It's mainly used when it's impossible to detect certain types of analytes by a chemical sensor, and high selectivity and sensitivity are needed.

The analytes that the biosensors can detect can be viruses, like SARS-CoV-2 [81], bacteria, *Escherichia coli* [82], fungi, *Aphanomyces invadans* [83]), protozoa, *Toxoplasma gondii* [84], DNA [85], cancer cells, leukemia [86], enzymes, alanine aminotransferase [87], organic molecule, glucose [88] and inorganic metal ions, Cd^{2+} [89].

The other electrochemical application of graphene and its based materials is for energy. There are two main applications inside of it: energy storage and energy generation. The devices of energy storage are batteries and (super)capacitors. They can store energy as chemical energy [90] and charge accumulated between poles [91], respectively. The energy generation devices are fuel and photovoltaic cells that can directly convert a combustible's potential energy into electric current [92] and transform light into electric power [93].

The most popular battery nowadays is Li-ion batteries, composed of an anode, cathode, and electrolyte. When discharged, the Li-ions are extracted from the anode, typically made of graphite, passing through the electrolyte and reaching the cathode. The charging reverses the pathway [94]. Graphene can substitute graphite on the anode because of its higher specific capacity for lithium storage [95]. However, a significant side effect is the aggregation of the graphene sheets, decreasing their capacity. To overcome this phenomenon, it's inserted between the layers of another nanomaterial, avoiding aggregating, like Fe_3O_4 [96]. It can also be used as a cathode, combining rGO and $LiFePO_4$ [97].

The (super)capacitors can accumulate large amounts of charge and deliver it quickly, making this device important for every electric and electronic apparatus. Graphene is essential due to its unique properties, high surface area, high conductivity, and high charge transfer, obtaining high-power densities. Although the same issue on Li-ion batteries occurs, graphene tends to stack its sheets as described beforehand. Therefore, one uses another nanomaterial to intercalate to fix it, like $NiCoO_4$ [98].

A fuel cell has three parts, anode, cathode, and electrolyte membrane. At the anode, hydrogen is split into H^+ ions and electrons, generating an electrical current. The protons are conducted into the electrolyte membrane to the cathode, bonded with oxygen-producing water. Graphene can be used in three components, the cathode, and anode. It functions as a metal catalyst supporter, improving conductivity [99, 100] when combined with Nafion, a polymer widely used electrolyte membrane, improving its properties [101, 102].

Photovoltaic cells can transform light into electric current by the semiconductor's electron-pair principle. In addition, graphene can be used as a charge collector at the anode due to its high transparency and conductivity [103].

Other roles of graphene shall be discussed in the following section. Due to functionalization, which is necessary to reduce the van der Waals force between graphene nanosheets, other properties can be added to the material, such as better biocompatibility and higher affinity to a given matrix. One of its main applications is increasing the strength of polymer blends. Furthermore, one can mention other characteristics, such as thermal conductivity tunning and absorption of solar light in different wavelengths, leading to relevant photothermal effects for heat generation [104].

Graphene-based materials have an exciting role in heterogeneous catalysis, mainly as a precursor for graphene-supported or graphene-protected catalysts, even though they can be used as a free nanomaterial for the same purposes. The biggest challenge yet is to activate the inert surface of graphene effectively and practically chemically. The key to this process's feasibility is the exploration of heteroatomic ligands, vacancies, and substrate engineering [105].

Looking closely at a graphene-protected catalyst, it can have its reactivity modulated by the atomic layer of carbon, leading the catalysis to a new form of reaction. These new perspectives are due to graphene's peculiar atomic and electronic structure [105].

Graphene-based materials play an essential role as corrosion inhibitors. The deposited graphene can successfully create a shielding effect much like a labyrinth, in other words, promoting a tortuous diffusion path, retarding the corrosive agents from reaching the substrate surface. One must be aware that the surface must be covered with a material with very few defects. Otherwise, the very same maze-like behavior could be compromised. Additionally, the deposition of materials is a critical step since surface optimization involves low conductivity, and maximum shielding must be reached. Even though the conductivity may be considered corrosion-promoting and an issue, this characteristic can be beneficial when applied in a hybrid system involving zinc. In this scenario, high conduction enhances the sacrificial protection involving the metal [106]. The functionalization of graphene can further improve its potential as a corrosion inhibitor. The new functional groups can lead to better adhesion to the metal surface. Some examples include the silane coupling agents KH550 and KH560 [107].

Nanomaterials such as graphene or its composites can also be used for filtration technology, mainly for water and air. Among the target contaminants to be captured by these matrixes, the following groups can be cited: organic pollutants, heavy metals and ions, and even microorganisms [108]. Other environmental solutions include graphene/metal oxide composites that would work as disinfectants in the water treatment chain with high effectiveness and low cost [109].

Graphene aerogels proved to be a highly effective alternative to absorb air pollutants, potentially solving the rising demands such as CFCs (chlorofluorocarbons) and HFCs (hydrofluorocarbons). Further applications of graphene in air filtration also include doped graphene nanosheets, both pristine and ferrous doped graphene, which show results of absorption of CO, CO_2, SO_2, and H_2S [108].

The material can also be incorporated into natural fiber clothes and infused in the fabric, building inherent properties into wearable goods. For instance, the innovative approach provides a 'green' alternative to thermoplastics. Using graphene to infuse material showed better results, as a wearable good, than thermoplastics and similar synthetic polymers [110]. Developing such technology involves simply dipping the fabric into the solution, including dispersed graphene, and treating afterward to ensure the infusion and coat [111]. This procedure fills up the gaps in the fiber's matrix, developing similar elastomer properties due to the hybrid graphene composites [110].

Furthermore, developing the so-called intelligent textiles is rising, integrating graphene sensors and microelectronics allied to artificial intelligence into the fabric and opening a door into healthcare. The material is capable of sensing and responding to minor changes in the environment, such as chemical, electrical, mechanical, and thermal changes, and, more specifically, to medical applications: changes in heart rate, stress levels, breathing rate, blood pressure, and muscle activity and respond focusing in a physiologically based need to ensure the body's wellness while making the individual aware of its physical condition. The textile's multidimensional structure comprises one or more nanopolymer fibers that perform many functions while maintaining comfort, perspiration, reusability, and longevity [112].

Further into the medical field, graphene has recently been tested to deal with regenerative medicine challenges, such as neural regeneration as a treatment for neurodegenerative diseases. The material can promote stem cell differentiation and growth, sustain these cells and even improve their responses by using electrical stimulation due to its conduction properties, durability, and biocompatibility [113].

Coming from this principle, regenerative tissue engineering is also aiming for bone regeneration, using graphene and its derivates to serve both as a scaffold—so the bone can regrow in the right way—and increase the recovery time by reconstructing the tissue using stem cells—in this case, the nanomaterial would serve as a differentiation promoter. Combining the two usages makes tissue engineering a much more precise and practical science. It can be employed for biomechanics, reducing weight, and increasing prosthetics' strength, flexibility, and durability [114].

1.3.1 Final considerations

Since Geim and Novoselov isolated graphene, considered the basic unit of carbon allotropes, several studies were and are being carried out to understand its properties. As a result, graphene and its different nanoforms are being increasingly inserted in several areas due to the possibility of improving its intrinsic properties. In addition, the formation of graphene-based hybrid nanomaterials makes it possible to further expand their applications in all areas. As a result, new graphene-based nanomaterials will emerge, and unique properties and applications will be discovered with them.

References

[1] Berger C, Song Z, Li T, Li X, Ogbazghi A Y, Feng R and De Heer W A 2004 Ultrathin epitaxial graphite: 2D electron gas properties and a route toward graphene-based nanoelectronics *J. Phys. Chem.* B **108** 19912–6

[2] Novoselov K S, Jiang D, Schedin F, Booth T J, Khotkevich V V, Morozov S V and Geim A K 2005 Two-dimensional atomic crystals *Proc. Natl Acad. Sci.* **102** 10451–3

[3] Geim A K and Novoselov K S 2007 The rise of graphene *Nat. Mater.* **6** 183–91

[4] Geim A K 2009 Graphene: status and prospects *Science* **324** 1530–4

[5] Hass J, De Heer W A and Conrad E H 2008 The growth and morphology of epitaxial multilayer graphene *J. Phys.: Condens. Matter* **20** 323202

[6] Novoselov K S, Geim A K, Morozov S V, Jiang D E, Zhang Y, Dubonos S V and Firsov A A 2004 Electric field effect in atomically thin carbon films *Science* **306** 666–9

[7] Pereira Junior A A M, da Silveira P H P M, Isidorio D K M, Filho A L N and da Veiga Junior V F 2021 Caracterização Química E Morfológica De Óxido *De* Grafeno Reduzido A Partir Do Uso De Naringenina Como Agente Redutor *Encontro nacional de modelagem computacional e encontro de ciência e tecnologia de materiais*

[8] Xu C, Xu B, Gu Y, Xiong Z, Sun J and Zhao X S 2013 Graphene-based electrodes for electrochemical energy storage *Energy Environ. Sci.* **6** 1388–414

[9] Kaplan A, Yuan Z, Benck J D, Rajan A G, Chu X S, Wang Q H and Strano M S 2017 Current and future directions in electron transfer chemistry of graphene *Chem. Soc. Rev.* **46** 4530–71

[10] Liu L, Qing M, Wang Y and Chen S 2015 Defects in graphene: generation, healing, and their effects on the properties of graphene: a review *J. Mater. Sci. Technol.* **31** 599–606

[11] Denis P A and Iribarne F 2013 Comparative study of defect reactivity in graphene *J. Phys. Chem.* C **117** 19048–55

[12] Zhu Y, Murali S, Cai W, Li X, Suk J W, Potts J R and Ruoff R S 2010 Graphene and graphene oxide: synthesis, properties, and applications *Adv. Mater.* **22** 3906–24

[13] Wei W and Qu X 2012 Extraordinary physical properties of functionalized graphene *Small* **8** 2138–51

[14] Yan J, Zhang Y, Kim P and Pinczuk A 2007 Electric field effect tuning of electron-phonon coupling in graphene *Phys. Rev. Lett.* **98** 166802

[15] Novoselov K S, Geim A K, Morozov S V, Jiang D, Katsnelson M I, Grigorieva I V and Firsov A 2005 Two-dimensional gas of massless Dirac fermions in graphene *Nature* **438** 197–200

[16] Zhou S Y, Gweon G H, Graf J, Fedorov A V, Spataru C D, Diehl R D and Lanzara A 2006 First direct observation of Dirac fermions in graphite *Nat. Phys.* **2** 595–9

[17] Fang X Y, Yu X X, Zheng H M, Jin H B, Wang L and Cao M S 2015 Temperature-and thickness-dependent electrical conductivity of few-layer graphene and graphene nanosheets *Phys. Lett.* A **379** 2245–51

[18] Tan Y W, Zhang Y, Stormer H L and Kim P 2007 Temperature dependent electron transport in graphene *Eur. Phys. J. Spec. Top.* **148** 15–8

[19] Berger C, Song Z, Li X, Wu X, Brown N, Naud C and de Heer W A 2006 Electronic confinement and coherence in patterned epitaxial graphene *Science* **312** 1191–6

[20] Yang G, Li L, Lee W B and Ng M C 2018 Structure of graphene and its disorders: a review *Sci. Technol. Adv. Mater.* **19** 613–48

[21] Wang L, Chen H and Witharana S 2013 Rheology of nanofluids: a review *Recent Pat. Nanotechnol.* **7** 232–46

[22] Mehrali M, Sadeghinezhad E, Latibari S T, Kazi S N, Mehrali M, Zubir M N B M and Metselaar H S C 2014 Investigation of thermal conductivity and rheological properties of nanofluids containing graphene nanoplatelets *Nanoscale Res. Lett.* **9** 1–12

[23] Bouhfid R, Essabir H and el kacem Qaiss A 2016 Graphene-based nanocomposites: mechanical, thermal, electrical, and rheological properties *Rheology and Processing of Polymer Nanocomposites* (Wiley) pp 405–30

[24] Wang M C, Yan C, Ma L, Hu N and Chen M W 2012 Effect of defects on fracture strength of graphene sheets *Comput. Mater. Sci.* **54** 236–9

[25] Papageorgiou D G, Kinloch I A and Young R J 2017 Mechanical properties of graphene and graphene-based nanocomposites *Prog. Mater Sci.* **90** 75–127

[26] Pyun K R and Ko S H 2019 Graphene as a material for energy generation and control: recent progress in the control of graphene thermal conductivity by graphene defect engineering *Mater. Today Energy* **12** 431–42

[27] Kim T Y, Park C H and Marzari N 2016 The electronic thermal conductivity of graphene *Nano Lett.* **16** 2439–43

[28] Wang X Y, Narita A and Müllen K 2017 Precision synthesis versus bulk-scale fabrication of graphenes *Nat. Rev. Chem.* **2** 0100

[29] Kumar N, Salehiyan R, Chauke V, Botlhoko O J, Setshedi K, Scriba M and Ray S S 2021 Top-down synthesis of graphene: a comprehensive review *Flat Chem.* **27** 100224

[30] Xu Y, Cao H, Xue Y, Li B and Cai W 2018 Liquid-phase exfoliation of graphene: an overview on exfoliation media, techniques, and challenges *Nanomaterials* **8** 942

[31] Yu P, Lowe S E, Simon G P and Zhong Y L 2015 Electrochemical exfoliation of graphite and production of functional graphene *Curr. Opin. Colloid Interface Sci.* **20** 329–38

[32] Santhiran A, Iyngaran P, Abiman P and Kuganathan N 2021 Graphene synthesis and its recent advances in applications—a review *C* **7** 76

[33] Hummers W S and Offeman R E 1958 Preparation of graphitic oxide *JACS* **80** 1339

[34] Nagaoka D A 2019 Síntese e caracterização de nanocompósitos baseados em derivados de grafeno de alta qualidade e óxido de cobalto bidimensionais *Dissertação (Mestrado em Engenharia de Materiais e Nanotecnologia)* Universidade Presbiteriana MackenzieSão Paulo

[35] Singh R K, Kumar R and Singh D P 2016 Graphene oxide: strategies for synthesis, reduction and frontier applications *RSC Adv.* **6** 64993–5011

[36] Malard L M, Pimenta M A, Dresselhaus G and Dresselhaus M S 2009 Raman spectroscopy in graphene *Phys. Rep.* **473** 51–87

[37] Cançado L G, Jorio A, Ferreira E M, Stavale F, Achete C A, Capaz R B and Ferrari A C 2011 Quantifying defects in graphene via Raman spectroscopy at different excitation energies *Nano Lett.* **11** 3190–6

[38] Sun L, Yuan G, Gao L, Yang J, Chhowalla M, Gharahcheshmeh M H and Liu Z 2021 Chemical vapor deposition *Nat. Rev. Methods Primers* **1** 5

[39] Wang H, Zhou Y, Qi M, Liu C, Xu X and Ren Z 2018 Direct Growth of graphene on fused quartz by atmospheric pressure chemical vapor deposition with acetylene *J. Phys. Chem.* C **123** 2370–7

[40] Yang J, Hu P and Yu G 2019 Design of carbon sources: starting point for chemical vapor deposition of graphene *2D Mater.* **6** 042003

[41] Yazdi G R, Iakimov T and Yakimova R 2016 Epitaxial graphene on SiC: a review of growth and characterization *Crystals* **6** 53
[42] Kumar B, Baraket M, Paillet M, Huntzinger J R, Tiberj A, Jansen A G M and Duclairoir F 2016 Growth protocols and characterization of epitaxial graphene on SiC elaborated in a graphite enclosure *Physica* E **75** 7–14
[43] Tetlow H, de Boer J P, Ford I J, Vvedensky D D, Coraux J and Kantorovich L 2014 Growth of epitaxial graphene: theory and experiment *Phys. Rep.* **542** 195–295
[44] Diez-Perez I, Li Z, Hihath J, Li J, Zhang C, Yang X and Tao N 2010 Gate-controlled electron transport in coronenes as a bottom-up approach towards graphene transistors *Nat. Commun.* **1** 31
[45] Lopes L C, da Silva L C, Vaz B G, Oliveira A R, Oliveira M M, Rocco M L and Zarbin A J 2018 Facile room temperature synthesis of large graphene sheets from simple molecules *Chem. Sci.* **9** 7297–303
[46] Wang X Y, Yao X and Müllen K 2019 Polycyclic aromatic hydrocarbons in the graphene era *Sci. China Chem.* **62** 1099–144
[47] McNaught A D and Wilkinson A 1997 *Compendium of Chemical Terminology* 2nd edn (Wiley–Blackwell)
[48] Velusamy K, Periyasamy S, Kumar P S, Rangasamy G, Pauline J M N, Ramaraju P and Vo D V N 2022 Biosensor for heavy metals detection in wastewater: a review *Food Chem. Toxicol.* **168** 113307
[49] Liu C, Zhang J, Yifeng E, Yue J, Chen L and Li D 2014 One-pot synthesis of graphene-chitosan nanocomposite modified carbon paste electrode for selective determination of dopamine *Electron. J. Biotechnol.* **17** 183–8
[50] Kim S K, Kim D and Jeon S 2012 Electrochemical determination of serotonin on glassy carbon electrode modified with various graphene nanomaterials *Sensors Actuators* B **174** 285–91
[51] Nagarajan R D, Murugan P and Sundramoorthy A K 2020 Selective electrochemical sensing of NADH and NAD+ using graphene/tungstate nanocomposite modified electrode *ChemistrySelect* **5** 14643–51
[52] Ma Y, Zhao M, Cai B, Wang W, Ye Z and Huang J 2014 3D graphene foams decorated by CuO nanoflowers for ultrasensitive ascorbic acid detection *Biosens. Bioelectron.* **59** 384–8
[53] Baig N, Kawde A N and Elgamouz A 2022 A cost-effective disposable graphene-based sensor for sensitive and selective detection of uric acid in human urine *Biosens. Bioelectron.: X* **11** 100205
[54] Lu L M, Li H B, Qu F, Zhang X B, Shen G L and Yu R Q 2011 *In situ* synthesis of palladium nanoparticle–graphene nanohybrids and their application in nonenzymatic glucose biosensors *Biosens. Bioelectron.* **26** 3500–4
[55] Saghatforoush L, Hasanzadeh M and Shadjou N 2014 Polystyrene–graphene oxide modified glassy carbon electrode as a new class of polymeric nanosensors for electrochemical determination of histamine *Chin. Chem. Lett.* **25** 655–8
[56] Sainz R, Del Pozo M, Vilas-Varela M, Castro-Esteban J, Pérez Corral M, Vázquez L and Casero E 2020 Chemically synthesized chevron-like graphene nanoribbons for electrochemical sensors development: determination of epinephrine *Sci. Rep.* **10** 1–11
[57] Agnihotri N, Chowdhury A D and De A 2015 Non-enzymatic electrochemical detection of cholesterol using β-cyclodextrin functionalized graphene *Biosens. Bioelectron.* **63** 212–7

[58] Li J, Liu J, Tan G, Jiang J, Peng S, Deng M and Liu Y 2014 High-sensitivity paracetamol sensor based on Pd/graphene oxide nanocomposite as an enhanced electrochemical sensing platform *Biosens. Bioelectron.* **54** 468–75

[59] Zhang K, Duan X, Zhu X, Hu D, Xu J, Lu L and Dong L 2014 Nanostructured graphene oxide–MWCNTs incorporated poly (3, 4-ethylene-dioxythiophene) with a high surface area for sensitive determination of diethylstilbestrol *Synth. Met.* **195** 36–43

[60] Raj M A and John S A 2014 Electrochemical determination of xanthine oxidase inhibitor drug in urate lowering therapy using graphene nanosheets modified electrode *Electrochim. Acta* **117** 360–6

[61] Shakibaian V and Parvin M H 2012 Determination of acetazolamide by graphene paste electrode *J. Electroanal. Chem.* **683** 119–24

[62] Teradal N L, Narayan P S, Satpati A K and Seetharamappa J 2014 Fabrication of electrochemical sensor based on green reduction of graphene oxide for an antimigraine drug, rizatriptan benzoate *Sensors Actuators* B **196** 596–603

[63] Qian L, Thiruppathi A R, Elmahdy R, van der Zalm J and Chen A 2020 Graphene-oxide-based electrochemical sensors for the sensitive detection of pharmaceutical drug naproxen *Sensors* **20** 1252

[64] Gholivand M B, Jalalvand A R and Goicoechea H C 2014 Developing a novel computationally designed impedimetric pregabalin biosensor *Electrochim. Acta* **133** 123–31

[65] Jose J, Subramanian V, Shaji S and Sreeja P B 2021 An electrochemical sensor for nanomolar detection of caffeine based on nicotinic acid hydrazide anchored on graphene oxide (NAHGO) *Sci. Rep.* **11** 11662

[66] Cardoso A R, Marques A C, Santos L, Carvalho A F, Costa F M, Martins R and Fortunato E 2019 Molecularly-imprinted chloramphenicol sensor with laser-induced graphene electrodes *Biosens. Bioelectron.* **124** 167–75

[67] Erogul S, Bas S Z, Ozmen M and Yildiz S 2015 A new electrochemical sensor based on Fe_3O_4 functionalized graphene oxide-gold nanoparticle composite film for simultaneous determination of catechol and hydroquinone *Electrochim. Acta* **186** 302–13

[68] Zhou W, Sun C, Zhou Y, Yang X and Yang W 2014 A facial electrochemical approach to determinate bisphenol A based on graphene-hypercrosslinked resin MN202 composite *Food Chem.* **158** 81–7

[69] Gong J, Miao X, Wan H and Song D 2012 Facile synthesis of zirconia nanoparticles-decorated graphene hybrid nanosheets for an enzymeless methyl parathion sensor *Sensors Actuators* B **162** 341–7

[70] Yu Y, Zhou M, Shen W, Zhang H, Cao Q and Cui H 2012 Synthesis of electrochemiluminescent graphene oxide functionalized with a ruthenium (II) complex and its use in the detection of tripropylamine *Carbon* **50** 2539–45

[71] Cheng S, Tang D, Zhang Y, Xu L, Liu K, Huang K and Yin Z 2022 Specific and sensitive detection of tartrazine on the electrochemical interface of a molecularly imprinted polydopamine-coated PtCo nanoalloy on graphene oxide *Biosensors* **12** 326

[72] Promphet N, Rattanarat P, Rangkupan R, Chailapakul O and Rodthongkum N 2015 An electrochemical sensor based on graphene/polyaniline/polystyrene nanoporous fibers modified electrode for simultaneous determination of lead and cadmium *Sensors Actuators* B **207** 526–34

[73] Ghanei-Motlagh M, Taher M A, Heydari A, Ghanei-Motlagh R and Gupta V K 2016 A novel voltammetric sensor for sensitive detection of mercury (II) ions using glassy carbon electrode modified with graphene-based ion imprinted polymer *Mater. Sci. Eng.* C **63** 367–75

[74] Wang Y, Zhao S, Li M, Li W, Zhao Y, Qi J and Cui X 2017 Graphene quantum dots decorated graphene as an enhanced sensing platform for sensitive and selective detection of copper (II) *J. Electroanal. Chem.* **797** 113–20

[75] Ruecha N, Rodthongkum N, Cate D M, Volckens J, Chailapakul O and Henry C S 2015 Sensitive electrochemical sensor using a graphene–polyaniline nanocomposite for simultaneous detection of Zn (II), Cd (II), and Pb (II) *Anal. Chim. Acta* **874** 40–8

[76] Rahman H A, Rafi M, Putra B R and Wahyuni W T 2023 Electrochemical sensors based on a composite of electrochemically reduced graphene oxide and PEDOT:PSS for hydrazine detection *ACS Omega* **8** 3258–69

[77] Mathivanan D, Devi K S, Sathiyan G, Tyagi A, da Silva V A O P, Janegitz B C and Gupta R K 2021 Novel polypyrrole-graphene oxide-gold nanocomposite for high performance hydrogen peroxide sensing application *Sensors Actuators* A **328** 112769

[78] Wen W, Chen W, Ren Q Q, Hu X Y, Xiong H Y, Zhang X H and Zhao Y D 2012 A highly sensitive nitric oxide biosensor based on hemoglobin–chitosan/graphene–hexadecyltrimethylammonium bromide nanomatrix *Sensors Actuators* B **166** 444–50

[79] Fei H, Wu G, Cheng W Y, Yan W, Xu H, Zhang D and Wu H C 2019 Enhanced NO_2 sensing at room temperature with graphene via monodisperse polystyrene bead decoration *ACS Omega* **4** 3812–9

[80] Kim Y S, Raston N H A and Gu M B 2016 Aptamer-based nanobiosensors *Biosens. Bioelectron.* **76** 2–19

[81] Seo G, Lee G, Kim M J, Baek S H, Choi M, Ku K B and Kim S I 2020 Rapid detection of COVID-19 causative virus (SARS-CoV-2) in human nasopharyngeal swab specimens using field-effect transistor-based biosensor *ACS Nano* **14** 5135–42

[82] Akbari E, Buntat Z, Afroozeh A, Zeinalinezhad A and Nikoukar A 2015 *Escherichia coli* bacteria detection by using graphene-based biosensor *IET Nanobiotechnol.* **9** 273–9

[83] Qi X, Chen T, Lu D and Chen B 2017 Graphene-Au nanoparticle based electrochemical immunosensor for fish pathogen *Aphanomyces invadans* detection *Fuller. Nanotub. Carbon Nanostructures* **25** 12–6

[84] Jiang S, Hua E, Liang M, Liu B and Xie G 2013 A novel immunosensor for detecting *Toxoplasma gondii*-specific IgM based on goldmag nanoparticles and graphene sheets *Colloids Surf.*, B **101** 481–6

[85] Hwang M T, Heiranian M, Kim Y, You S, Leem J, Taqieddin A and Bashir R 2020 Ultrasensitive detection of nucleic acids using deformed graphene channel field effect biosensors *Nat. Commun.* **11** 1543

[86] Khoshroo A, Hosseinzadeh L, Adib K, Rahimi-Nasrabadi M and Ahmadi F 2021 Earlier diagnoses of acute leukemia by a sandwich type of electrochemical aptasensor based on copper sulfide-graphene composite *Anal. Chim. Acta* **1146** 1–10

[87] Her J L, Pan T M, Lin W Y, Wang K S and Li L J 2013 Label-free detection of alanine aminotransferase using a graphene field-effect biosensor *Sensors Actuators* B **182** 396–400

[88] Wu P, Shao Q, Hu Y, Jin J, Yin Y, Zhang H and Cai C 2010 Direct electrochemistry of glucose oxidase assembled on graphene and application to glucose detection *Electrochim. Acta* **55** 8606–14

[89] Wang X, Gao W, Yan W, Li P, Zou H, Wei Z and Ding K 2018 A novel aptasensor based on graphene/graphite carbon nitride nanocomposites for cadmium detection with high selectivity and sensitivity *ACS Appl. Nano Mater.* **1** 2341–6
[90] Oxford Learner's Dictionaries Battery (Oxford University Press) (https://oxfordlearnersdictionaries.com/definition/english/battery?q=battery)
[91] Simon P and Gogotsi Y 2008 Materials for electrochemical capacitors *Nat. Mater.* **7** 845–54
[92] Steele B C and Heinzel A 2001 Materials for fuel-cell technologies *Nature* **414** 345–52
[93] E.P.I. Association 2013 *EPIA Report* (https://www.earth-policy.org/?/indicators/C47/)
[94] Brownson D A and Banks C E 2014 *The Handbook of Graphene Electrochemistry* (Springer)
[95] Lian P, Zhu X, Liang S, Li Z, Yang W and Wang H 2010 Large reversible capacity of high quality graphene sheets as an anode material for lithium-ion batteries *Electrochim. Acta* **55** 3909–14
[96] Zhou G, Wang D W, Li F, Zhang L, Li N, Wu Z S and Cheng H M 2010 Graphene-wrapped Fe_3O_4 anode material with improved reversible capacity and cyclic stability for lithium ion batteries *Chem. Mater.* **22** 5306–13
[97] Zhou X, Wang F, Zhu Y and Liu Z 2011 Graphene modified $LiFePO_4$ cathode materials for high power lithium ion batteries *J. Mater. Chem.* **21** 3353–8
[98] Xu J, Li L, Gao P, Yu L, Chen Y, Yang P and Yang P 2015 Facile preparation of $NiCo_2O_4$ nanobelt/graphene composite for electrochemical capacitor application *Electrochim. Acta* **166** 206–14
[99] Li Y, Li Y, Zhu E, McLouth T, Chiu C Y, Huang X and Huang Y 2012 Stabilization of high-performance oxygen reduction reaction Pt electrocatalyst supported on reduced graphene oxide/carbon black composite *JACS* **134** 12326–9
[100] Tao Z, Chen W, Yang J, Wang X, Tan Z, Ye J and Zhu Y 2019 Ultrathin transferable platinum and platinum Ruthenium-modified graphene film as an efficient methanol oxidation catalyst *Sci. China Mater.* **62** 273–82
[101] Zarrin H, Higgins D, Jun Y, Chen Z and Fowler M 2011 Functionalized graphene oxide nanocomposite membrane for low humidity and high temperature proton exchange membrane fuel cells *J. Phys. Chem.* C **115** 20774–81
[102] Zhang B, Cao Y, Jiang S, Li Z, He G and Wu H 2016 Enhanced proton conductivity of Nafion nanohybrid membrane incorporated with phosphonic acid functionalized graphene oxide at elevated temperature and low humidity *J. Membr. Sci.* **518** 243–53
[103] Tang Y B, Lee C S, Xu J, Liu Z T, Chen Z H, He Z and Lee S T 2010 Incorporation of graphenes in nanostructured TiO_2 films via molecular grafting for dye-sensitized solar cell application *ACS Nano* **4** 3482–8
[104] Sun X, Huang C, Wang L, Liang L, Cheng Y, Fei W and Li Y 2021 Recent progress in graphene/polymer nanocomposites *Adv. Mater.* **33** 2001105
[105] Yam K M, Guo N, Jiang Z, Li S and Zhang C 2020 Graphene-based heterogeneous catalysis: role of graphene *Catalysts* **10** 53
[106] Zhang R, Yu X, Yang Q, Cui G and Li Z 2021 The role of graphene in anti-corrosion coatings: a review *Constr. Build. Mater.* **294** 123613
[107] Ding R, Chen S, Lv J, Zhang W, Zhao X D, Liu J and Li W H 2019 Study on graphene modified organic anti-corrosion coatings: a comprehensive review *J. Alloys Compd.* **806** 611–35

[108] Karanjikar S R, Sena A S, Manekar P, Mudagi S and Juneja A S 2022 Utilization of graphene and its derivatives for air & water filtration: a review *Mater. Today Proc.* **50** 2007–17
[109] Upadhyay R K, Soin N and Roy S S 2014 Role of graphene/metal oxide composites as photocatalysts, adsorbents and disinfectants in water treatment: a review *RSC Adv.* **4** 3823–51
[110] Suresha B, Hemanth G, Hemanth R and Lalla N P 2020 Role of graphene nanoplatelets and carbon fiber on mechanical properties of PA66/thermoplastic copolyester elastomer composites *Mater. Res. Express* **7** 015325
[111] da Luz F S, Garcia Filho F D C, Del-Rio M T G, Nascimento L F C, Pinheiro W A and Monteiro S N 2020 Graphene-incorporated natural fiber polymer composites: a first overview *Polymers* **12** 1601
[112] Barman J, Tirkey A, Batra S, Paul A A, Panda K, Deka R and Babu P J 2022 The role of nanotechnology based wearable electronic textiles in biomedical and healthcare applications *Mater. Today Commun.* **32** 104055
[113] Ding X, Liu H and Fan Y 2015 Graphene-based materials in regenerative medicine *Adv. Healthcare Mater.* **4** 1451–68
[114] Luong-Van E K, Madanagopal T T and Rosa V 2020 Mechanisms of graphene influence on cell differentiation *Mater. Today Chem.* **16** 100250

Chapter 2

Recent advances in graphene-based electrochemical sensing devices

Monima Sarma and Tanmay Chatterjee

The design of electrochemical sensors and biosensors should be such that it is eco-friendly, and highly selective as well as for specific analytes. In this context, nanomaterials, especially graphene, and its derivatives, have attracted considerable attention due to their distinctive properties. As such, substantial development, in terms of synthesis, characterization, and applications, has been observed for these materials. Furthermore, such types of nanomaterials having controlled physico-chemical properties are employed to modify electrodes to boost their analytical performances. Graphene, being a transparent yet conductive material, has a low cost and is environmentally friendly, therefore, it is a perfect candidate for the fabrication of electrochemical sensors and biosensors. In this chapter, we discuss graphene and its derivatives and many of its recent applications as electrochemical sensors and biosensors.

2.1 Introduction

Graphene is a nanomaterial that encompasses sp^2-bonded carbon in a 2D array giving a honeycomb (hexagonal) structure. In the last few years, graphene-based nanomaterials have grabbed tremendous attention because of their interesting properties, such as high thermal conductivity, mechanical strength, elasticity, tunable photophysical properties, room temperature electron mobility, etc. Graphene, being a transparent yet conductive material, has a low cost and is environmentally friendly, therefore, it is a suitable candidate for the fabrication of sensors and biosensors. The word sensor is derived from the Latin word *sentire* implying 'to feel.' Sensors have become increasingly valuable with the progress of modern civilization. The invention of the Internet of Things (IoT) has lifted the status of sensors to a much higher level and human life, essentially these days, has become dependent on these man-made sensors [1, 2].

doi:10.1088/978-0-7503-5578-0ch2

In contrast to the traditional detection methods, electrochemical sensor technology has been progressively employed in practical applications due to its unique advantages. Electrochemical sensors are electronic gadgets that are usually employed to quantify electroactive analytes via electrochemical processes. The concentration of the analyte is determined by measuring the changes in electrical current upon application of potential. The pertinent device is a three-electrode assembly containing three electrodes, viz., the working (glassy carbon, carbon paste, graphite, etc), counter (platinum wire), and reference (saturated calomel, Ag/AgCl, etc) electrodes, together with a supporting electrolyte [3, 4].

This experimental setup is attached to an electrochemical analyzer with a display system. The redox reaction occurs at the surface of the working electrode, a closed circuit is established with the help of the counter electrode while the accurate application of working electrode potential is ensured with the use of the reference electrode. Though various electrochemical methods like impedimetry, amperometry, or voltammetry (differential pulse voltammetry (DPV), square wave, cyclic, linear sweep) are known, however, DPV is the most well-known for accuracy as it is a highly sensitive technique [5–7].

The low cost and their simplicity make potentiometric sensors the most widely used practical sensors. Such devices are generally classified into three categories: (i) ion-selective electrodes (ISEs); (ii) coated-wire electrodes (CWEs) and (iii) field-effect transistors (FETs) [8, 9].

Amperometry is more commonly used as an analytical technique of high precision and sensitivity, wherein the driving force for the electrocatalytic redox reactions is the applied voltage responsible for generating electrical currents proportional to the analyte concentration. Another strategy is via impedimetric sensors where the cell is stimulated with a minimal alternating potential and subsequently, the steady state system response is monitored [10].

Potentiometry involves a change in electrode potential in the presence of the analyte according to the Nernst equation:

$$E_{\text{cell}} = E^0_{\text{cell}} - \frac{RT}{nF}\ln\frac{a_{\text{ox}}}{a_{\text{red}}} \approx E^0_{\text{cell}} - \frac{RT}{nF}\ln\frac{[\text{ox}]}{[\text{red}]} \approx E^0_{\text{cell}} - \frac{0.0591}{n}\log\frac{[\text{ox}]}{[\text{red}]}$$

where R is the universal gas constant (~8.314 J mol^{-1} K^{-1}), F is Faraday's constant (~96 485 C mol^{-1}), n is the number of electrons that are involved, E_{cell} is the cell potential (formal potential), E^0_{cell} is the standard cell potential, T is the temperature in Kelvin, and those in parenthesis are the oxidized and reduced species concentrations in solution respectively [11].

Due to their simplicity and low cost, potentiometric sensors are the most widely used practical sensors. There have been numerous examples where they have been used in practicality, for example, when these potentiometric sensors are paired with wearable device technology, they can offer significant information about the physical status during athletic performance via sweat performance [12].

Again, based on the type of membrane, the ion-selective electrodes are classified into glass electrodes, liquid electrodes, and solid electrodes. It has been observed that the usage of glass membrane electrodes to find out pH solutions has proven highly efficient; the only downside is that it is confined to aqueous measurements and effective steps should be taken when determining the hydrogen ion concentrations in nonaqueous liquids [13].

In general, the working electrode dictates the sensitivity and selectivity of electrochemical sensors. However, due to poor electron transfer kinetics, bare electrodes do suffer from an efficient determination of analytes. Therefore, nanomaterials can be used to modify the layers to provide better efficacy for these working electrodes.

2.1.1 Objective of the chapter

The main objective of this chapter is to give the reader a general idea of the carbon-based materials viz., graphene which can be used in several applications, more specifically here, sensors and biosensors. In a broader sense, the goal of electrochemical sensing is to modify the bare electrodes with special carbon-based nanomaterials, metal salt-based nanoparticles, biomaterials, etc. As the electron transfer rate, as well as the effective surface area of these modified electrodes, are significantly improved compared to those of the bare electrodes, much focus is given to developing better-performing electrochemical sensors.

2.2 Graphene in electrochemical sensing devices

Amongst all these electrode nanomaterials used so far, the carbon-based nanomaterial, graphene, has been widely valued and is considered one of the most promising candidates in the electrochemical sensors owing to the versatile benefits it offers, such as rapid sensing capacity, lower overpotential, multiple detection ability, etc in real matrices. It also exhibits outstanding physical properties (large surface area, strong mechanical strength, and excellent optical, thermal, and electrical conductivity). Graphene provides great electrochemical performance and the main reasons for this have been attributed to (i) its honeycomb-like two-dimensional structure, which allows easy transfer of analytes, (ii) its enormous surface area which provides enhanced analyte adsorption as well as improved electroactive sites for electrodes. In addition to that, graphene also shows very good conductivity, transparency, and enhanced electrocatalytic activity [14]. Highly sensitive electrochemical sensors and biosensors are crucial not only in environmental protection but also for biomedical applications and clinical diagnosis. Such sensors are used for sensing sugars, cancer biomarkers, antioxidants, toxins in food samples or drinking water, etc. And a crucial role in graphene and graphene-based material sensors and biosensors is played by the electrodes that are used. Lately, there have been various approaches to modifying electrodes. In a recent review, Cinti and Arduini have represented several strategies for the synthesis and applications of graphene-based screen-printed (bio)sensors [15].

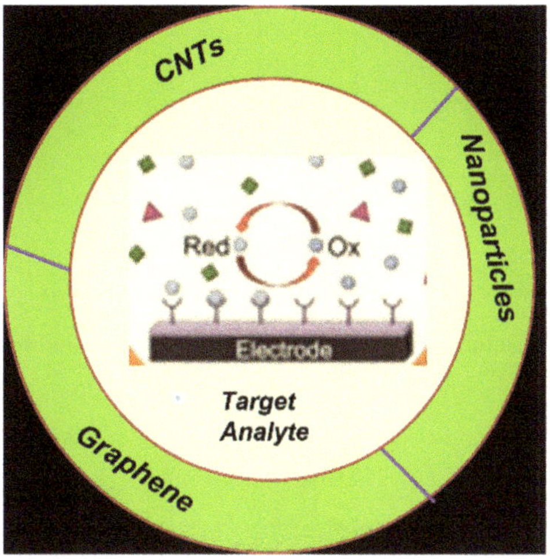

Two core approaches are known for the synthesis of graphene: (i) top-down and (ii) bottom-up methods (figure 2.1). The top-down methods encompass oxidative exfoliation-reduction, arc discharge, liquid-phase exfoliation, electrochemical exfoliation, etc of which the electrochemical exfoliation of graphite is the most accepted as it provides better crystallinity and minimal layers making it a highly suitable candidate for electrodes. The important elements employed for the electrochemical setup for the exfoliation of graphite are the graphite working electrodes, e.g. graphite rod or foil; the counter electrode, e.g. platinum wire, and electrolyte [16]. In electrochemical exfoliation, mechanisms usually depend on the nature of the potential. Anodic exfoliation encompasses the embolism of anionic species in the reaction blend into graphite. A positive charge is created when electrons are withdrawn from the working anode (graphite) when a positive current is passed. This charge boosts the bulky negative ions to be intercalated and this, in turn, increases the interlayer spacing between graphene sheets and supports the resultant exfoliation of sheets. Whereas, in cathodic exfoliation, positively charged ions in solution are attracted toward the negative bias at the working electrode (graphite) and finally these intercalating species open the graphene sheets leading to expansion and exfoliation. This type of technique causes minimal defects and is therefore useful in many energy storage devices, nanocomposites, and electronic devices [17].

The bottom-up approaches like template pathway, chemical vapor deposition (CVD), epitaxial growth, pyrolysis, etc are quite well-known but suffer from low yield and high production costs, with CVD being very common.

CVD gives rise to high-quality, large-area, and single-crystal graphene thereby making them the preferred choice for synthesis. Of late, the cold-wall CVD technique has attracted tremendous attention due to short deposition times, and rapid heating and cooling rates, thereby facilitating the faster growth of graphene. This method renders the entire chamber to be at room temperature so that the gas phase reactions accompanying high-temperature graphene growth can be suppressed to achieve super-clean graphene [18].

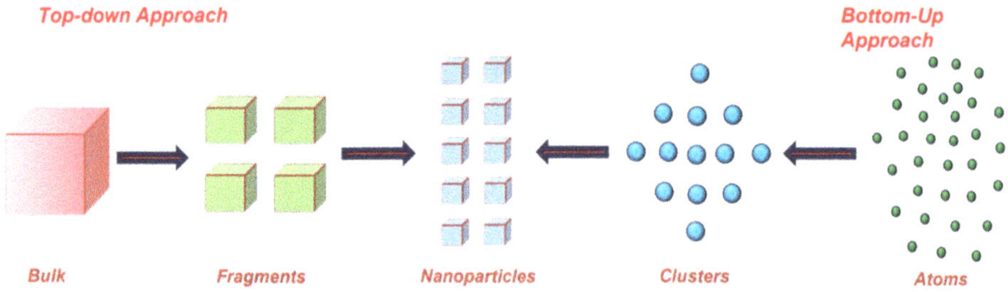

Figure 2.1. Synthesis nanoparticles (viz., graphene) via top-down and bottom-up approaches.

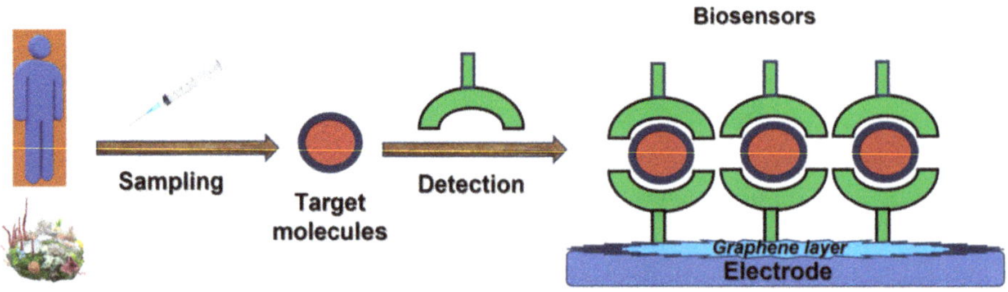

Figure 2.2. Sensing of analyte molecules by receptors in biosensors.

The unique properties of graphene make it suitable for a variety of applications, with great demand from numerous industries. In the next sections, we shall present some details about analytical advances of the graphene-based materials as sensing devices toward cancer biomarkers, heavy metal ions, bisphenol A (BPA), glucose, cholesterol, dopamine (DA), ascorbic acid (AA) uric acid (UA), NADH, gas molecules and psychoactive drugs, etc (figure 2.2).

2.2.1 Cancer biomarkers

Cancer is a dangerous disease that develops in uncontrolled cell growth, initiating the growth of a tumor. These cancerous cells can spread to other body organs through blood vessels, spreading cancer all over the body [19]. According to the World Health Organization (WHO), it has been observed that the mortality rate of people suffering from cancer has been growing over the past few years. Thus, biomarkers play an important role in cancer therapy [20, 21]. Numerous reports have revealed that the checkup of cancer-related biomarkers in the body of an affected patient exhibits the potential for early identification of cancer and its pertinent therapy. Furthermore, with the recent development in biosensor fabrication technologies, tumor-specific biomarkers can also be easily recognized, which opens new prospects to design target-specific sensing platforms for the diagnosis of cancer. Electrochemical technology has become a valuable analytical means for detecting analytes in the contemporary research field. The major gains of this technology are high sensitivity, convenience, and instant detection of infected cells, thus proving it to be a powerful tool for the early diagnosis of cancer cells in clinical medicine. An efficient method for the detection of cancer is based on the usage of metal nanoparticle-graphene-modified electrodes. Graphene/Au nanoparticles (AuNPs) electrochemical responses for various cancer biomarkers were studied for the sensitive detection of tumor markers [22–25].

2.2.2 Heavy metal electrochemical sensors

Heavy metal ions are detrimental to the environment and human health and when they exceed the advisable limit in soil or water, it causes environmental degradation or pollution. Thus, such cases require a simple, fast, and sensitive method for their detection. In contrast to standard detection approaches, electrochemical sensors based on graphene-based materials are highly advantageous due to high sensitivity and selectivity, and rapid detection. Several review papers have been published in recent years showing the recent progress of graphene-based nanomaterials in the detection of heavy metals [26–28]. For example, Qiu *et al* reviewed the optical sensing behavior of metal ions, such as Cu^{2+}, Pb^{2+}, Hg^{2+}, Cd^{2+}, Ag^{+}, which utilized graphene-based nanomaterials. The efficiency of such sensors was essentially enhanced when graphene-based nanomaterials were used. The heavy metals that enter and get stacked in the environment can be further absorbed by plants and animals living in these contaminated areas leading to bioaccumulation. These ultimately biomagnify within the human body through the food chain. Thus, even trace amounts of toxic heavy metals, e.g. cadmium, lead, arsenic, or mercury may have damaging effects on the environment as well as on human health [29].

Among the heavy metals, Hg^{2+}, Cd^{2+}, and Pb^{2+} present severe hazards such as impairment to the immune system of mankind, reproductive toxicity, disorders in the respiratory organs, etc. In recent times, various modified electrodes were employed for separate or simultaneous detection of analytes. For the sensing of the Cd^{2+} ion, Thinakaran *et al* [30] reported a modified glassy carbon electrode that employed reduced graphene oxide (rGO), carboxymethyl cellulose, and glutathione for the electrochemical detection of the pertinent ion. Furthermore, Luo and his group developed a highly sensitive fluorescence sensor that was highly selective for Pb^{2+} detection using graphene quantum dots (GQDs) and gold nanoparticles [31].

2.2.3 Cholesterol and glucose electrochemical (bio)sensors

Variations in cholesterol and glucose levels in the body may pose severe health issues like heart attacks, diabetes, etc. Therefore, their detection and control are crucial [32], and in this regard, biosensors can go a long way in helping doctors and patients by providing real-time information [33–35]. Recently, an amperometric biosensor based on reduced GO/poly(3-aminobenzoic acid) nanocomposite film on a screen-printed carbon electrode was reported by Ounnunkad *et al* for the detection of both glucose and cholesterol, displaying high sensitivity and selectivity. The metabolites could be detected even when present at very low concentrations, thus exhibiting their potential for real sample analysis in clinical diagnosis [36]. Furthermore, effort should also be made for low-cost production as well as to miniaturize the devices for their practical applicability.

2.2.4 Bisphenol A electrochemical detection

BPA (4,4′-(propane-2,2-diyl)diphenol), which is a pure organic compound is extensively used for the fabrication of polycarbonate plastics, and epoxy resins. However, if leached from plastics, it can have adverse effects on the environment and health. It can mimic the role of hormones and finally show negative effects on the reproductive system. Thus, it is vital to develop an effective, sensitive, selective, and rapid sensor for the determination of BPA. When noble metal nanoparticles are combined with graphene, the obtained nanocomposites show outstanding sensing behavior toward BPA [37, 38]. For example, electrodes modified with AuPd nanoparticles loaded on graphene nanosheets [38], or Au nanoparticles loaded on rGO–multiwall carbon nanotubes [39] could even detect very low levels of BPA.

Noble metal nanoparticles are widely used in electrochemical methods because of their versatile and exceptional properties, such as enormous surface area, good electrical conductivity, etc. By using rGO/poly-L-lysine nanocomposites-modified glassy carbon electrode, an effective and selective electrochemical sensor was reported by Zhang *et al* for BPA detection. The high electrocatalytic activity to oxidation of BPA was established by employing DPV. The modified electrode was finally scrutinized for BPA sensing in drinking water which exhibited suitable results. Furthermore, a hydrothermal method was also employed to develop a novel nanocomposite material formed via Mo, Cu, and Se nanoparticles adorned with rGO. This nanocomposite was used for the first time, for detecting for BPA detection [40]. The excellent electrochemical response of the sensor opens new prospects for using electrochemical sensors with hybrid materials. A highly sensitive BPA sensor was also attained via a one-step electrochemical reduction of Cu_2O and graphene oxide which displayed a very low detection limit in real water samples [41]. In addition to that, a good BPA detection limit was also observed for an electrochemical sensor based on porous graphene functionalized black phosphorus composite [42]. Attempts are still being made to develop faster, and more sensitive electrochemical methods for selective detection of BPA detection.

2.2.5 Dopamine, ascorbic acid, and uric acid

The three biological molecules, DA, AA, and UA, play a significant role in human metabolism. As such, abnormal variations in these compounds in the human body may lead to different types of diseases like Parkinson’s disease, gout, cancer, etc. Hence, rapid, selective, and effective electrochemical sensors are particularly required to quantitatively detect these molecules simultaneously [43].

2.2.6 NADH-based electrochemical detection

β-nicotinamide adenine dinucleotide (NAD^+) and its reduced form (NADH) has received considerable interest in the biosensing area as the oxidation of NADH serves as the anodic signal and regenerates the NAD+ cofactor, which can detect biomolecules, such as lactate, glucose, etc [44]. However, this generates large overvoltage for NADH oxidation and also surface fouling. Such issues could be addressed if nanomaterials such as graphene were used. The electrochemical behavior of NADH on graphene-modified electrodes, i.e. chemically reduced graphene oxides, CR-GO was studied by Tang *et al* and it exhibited remarkable electron transfer rates [45]. In another report, Liu *et al* [48] developed a high-performing graphene-modified electrode toward the oxidation of NADH by increasing the graphene dispersity through functionalizing graphene non-covalently with methylene green (MG). The higher activity of functionalized graphene-decorated electrodes compared to the bare electrodes toward NADH oxidation was further confirmed by Lin *et al* [46]. Another report by Pumera *et al* suggested that when graphene is decorated with moieties containing oxygen groups, it can lead to enhanced activity of graphene [47].

2.2.7 Graphene-based gas sensing devices

Breath is a crucial matrix for identifying the diseased or healthy state of an individual through the analysis of volatile organic compounds (VOCs) or gas molecules, generated within the body. These compounds traverse through blood in the body and are eventually exhaled, and then analyzed. Research shows that compounds, viz., isoprene, acetone, and methanol are generally expelled while exhaling. Nevertheless, these three components show abnormally high or low concentrations in diseased states as compared to healthy ones [49, 50]. Owing to the enormous sensitivity of graphene toward gas molecules, a gas-based sensor was fabricated, which could be used to sense ethanol, NH_3, NO_2, and O_2 [51]. The thin layer of graphene is accountable for the adsorption or desorption of these tiny gas molecules on its surface, thereby resulting in variation of conductance. The high surface area of graphene in addition to its 2D structure makes it a suitable candidate for its applicability as a gas sensor in the medical field and is thereby drawing enormous attention [52].

2.2.8 Electrochemical detection of psychoactive drugs

Some recent studies have shown that graphene-modified electrochemical sensors have been used for sensing some psychoactive drugs [53]. As such, there are great prospects for further improvement and application of such nanomaterial-based sensors for real-time diagnosis. One of these is sensing of opioid drugs such as cocaine, morphine, ketamine, tramadol, etc. Opioid drugs are commonly employed as analgesics in pain treatment. However, these psychoactive drugs come with neurotoxic side effects when overdosed and can also create addiction problems. Such opioid-induced neuro-toxicities result in a multifactorial syndrome that could ultimately lead to death. Therefore, nanomaterial-based biosensors are excellent tools to uncover opioid intake and constantly monitor the drugs with high sensitivity and selectivity even at very low detection limits, for example, cocaine, morphine, tramadol, codeine, ketamine, etc [54].

Cocaine is a central nervous system (CNS) stimulant that needs to be consumed only in limited amounts to cure specific illnesses in the human body. Cocaine is the second most ingested banned drug in Europe. As such, graphene-based sensors can be a boon to detecting cocaine intake in the human body. Recently, a core–shell Au@Ag NP-based colorimetric biosensor was reported by Mao *et al*, for the detection of cocaine which exhibited good selectivity and sensitivity [55]. Another study employed nanoporous anodic alumina for detecting cocaine. Herein, the alumina support was functionalized with ssDNA and Rhodamine B. This nanosystem exhibited a limit of detection (LOD) value of 5×10^{-7} M [56]. In addition to this, recently an NP-based potentiometric sensor was used for cocaine detection. These nanoparticles were obtained via polymerization reaction in H_2O and UV-initiated in an organic solvent, all of which exhibited a very strong affinity for cocaine [57]. Furthermore, an aptasensor system produced with captured DNA (coupled with carbon dots) on an AuNP-based electrode with an ITO surface exhibited an excellent detection range (10–70 pm) and a LOD of 0.26 pm [58]. Another novel fluorescent nanoprobe-based on fluoro-graphene-plasmonic nanohybrid aptamer-based was also developed recently for cocaine detection, which displayed an outstanding LOD limit of 4.6 nM [59]. Another innovative method is to use the label-free approach. In a recent report, Thioflavin (T), a fluorescence indicator coupled with an anti-cocaine aptamer, was used to augment the fluorescence signal. This label-free methodology proved to be highly sensitive to cocaine, with a LOD of 250 nM [60]. Furthermore, employing molecularly imprinted polymers on the palladium-decorated graphene surface, a cocaine-selective electrode was developed. To quantify cocaine, the square-wave voltammetric technique was used for this sensor. The sensor was used to measure drug levels of real street samples of spiked cocaine in saliva, and river water which exhibited quite close results to the standard spiked levels [61]. Additionally, a 3D-printed graphene–polylactic acid electrode was also developed to measure cocaine concentrations in different samples with good sensitivity [62].

Although *morphine* is the gold standard of analgesics administered to lessen acute pain, it is prohibited from athletic doping. To detect traces of morphine, GQDs have received much attention. These carbon-based nanomaterials have numerous benefits, such as: (i) they are highly photoluminescent; (ii) chemically inert and biocompatible; and (iii) can be easily functionalized with biological molecules [63–65]. Fluorescent GQDs have been employed to determine the most widely used narcotic substances, viz., morphine and methamphetamine with very low LOD and rapid reaction time [66]. Similarly, alumina NPs can be efficiently used to trace morphine in real-time samples. These NPs exhibited high electron conductivity, and their linear response range and LOD were found to be 0.1–550 and 0.03 μmol l^{-1}, respectively [67].

Tramadol is a pain medication and belongs to the family of opium alkaloids. Tramadol comforts the patient by inhibiting the reuptake of norepinephrine and serotonin thereby augmenting their release [68–70]. It also inhibits heart and lung function as it acts as a depressant for the CNS. This painkiller is extensively employed to treat visceral, neuropathic, and somatic pain [71–73]. People who utilize exceptionally high dosages of tramadol over the prescribed dose can suffer

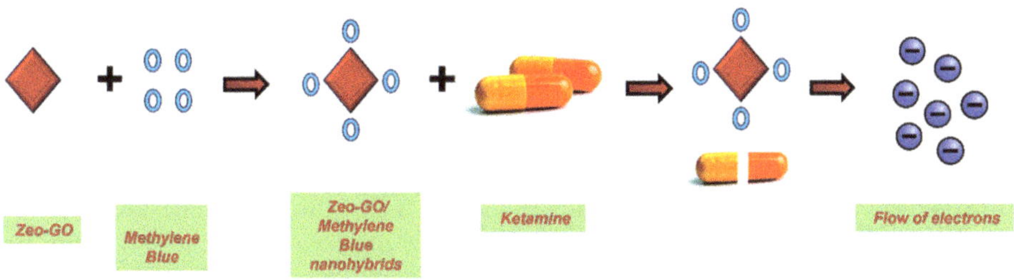

Figure 2.3. The sensing mechanism of ketamine with ZeO–GO/methylene blue nanohybrids.

from vomiting, nausea, cardiopulmonary arrest, and finally death [74–76]. To develop an advanced tramadol sensor, the pencil graphite electrode (PGE) surface was modified with CuO-NPs and polypyrrole (PPy) and this sensor was found to yield outstanding results, as it could be employed for tramadol recognition in drug samples [77]. Another effective electrochemical sensor based on cobalt oxide/ionic liquid crystal/carbon nanotubes decorated-CPE in the presence of SDS surfactant was developed for rapid and sensitive measurement of nalbuphine and tramadol, both of which displayed remarkable performances in urine samples [78].

Another analgesic drug extensively used in the medical field is *ketamine*, but due to its increased usage also for recreational purposes, it has become a key concern (figure 2.3).

Owing to its tasteless and colorless nature and because it is fragrant-free, it aids in criminal activities, when added to drinks [79, 80]. Therefore, to check drug abuse, portable and miniaturized nanomaterial-based sensors are very handy in such situations. In such a context, a graphene-based electrochemical sensor (a microfluidic paper-based analytical device) fabricated from zeolite nanoflakes and graphene oxide (ZeO–GO) with methylene blue was very instrumental in detecting this drug in alcoholic and non-alcoholic drinks with LOD of 1 nM, and was reported to be very selective [81].

Nonetheless, the real challenge remains in the fact the sensor needs to be consistent, highly sensitive and selective, and highly economical such that it can be functional for practical applicability.

2.2.9 Flexible electrochemical sensors

Flexible electrochemical sensors have become the talk of the town and are emerging as an advanced technology to resolve real-world challenges. Flexible electrode-based devices are versatile and handy, occupy less space, have high conductivity, high stability, and mechanical strength, and are biocompatible. Thus, of late, graphene-paper-based electrochemical sensors have demonstrated suitability for crucial applications due to their flexibility and lightweight [82].

Graphene paper, which can be derived from graphene or GO-based is finding widespread usage in scientific research because of its tunable conductivity and versatile functionality and thus, is a good candidate for new-generation flexible materials for electrochemical sensors.

The inherent planar structure of graphene sheets endows them with the flexibility of controlling the alignment of these nanosheets into 2D architectures. Finally, these 2D layered papers or films can be loaded with a variety of functionalized materials, thereby bestowing them with manifold functionalities, essential for specific unique applications [83–87]. A variety of techniques are used to gather the layered structure of graphene papers. These techniques include CVD, solution-processed assembly, electrophoretic deposition (EPD), etc. It has been observed that the graphene films acquired by CVD methods exhibited high transparency, flexibility, and electrical conductivity [88] desirable for various electronic devices. Self-assembly is another very popular methodology for the economical formulation of thin films, and this has also been employed for constructing multilayer graphene papers with varying architecture and composition [89, 90]. Interactions such as weak hydrogen bonding interactions, electrostatic interactions, and covalent bonding are generally responsible in such assemblies. Besides the above-mentioned methods, some other approaches such as spray coating, solution casting, and spin coating are also employed to fabricate graphene papers or films [91, 92]. The EPD process is another valuable approach for depositing graphene on conductive substrates. This process is a widely used technique because of its low cost and easy handling [82].

Synthesized graphene papers have a high potential to be used in energy devices and sensors. Thus, structural characterization is of utmost importance for comprehending the connection between their nanostructures and efficiencies. As such, some sophisticated techniques such as, Fourier transform infrared spectroscopy (FTIR), atomic force microscopy (AFM), scanning electron microscopy (SEM), transmission electron microscopy (TEM), energy dispersive x-ray spectroscopy (EDX), etc are extensively used to characterize materials based on graphene papers.

In recent times, functionalized graphene papers have come into the limelight because they can offer notable benefits owing to their hybrid structure, specific functionality, and enriched physicochemical properties. Conventionally, a variety of materials, viz., polymers, metallic and metal oxide materials, and supramolecular units are employed to functionalize graphene papers for numerous applications.

Papers based on polymer-functionalized graphene are extensively used for sensors, storage, and energy conversion. Stabilization of such polymers with graphene paper can be achieved via two approaches, viz., covalent and non-covalent attachments. Lately, a sequence of polyvinylidene fluoride-modified graphene monolayer papers with a gradient rGO/GO structure was reported by Mu *et al*, which displayed interesting results. The GO region embedded in the graphene paper exhibited a good response to variations in humidity, light, and temperature. Expansion or shrinkage of the GO sheets was observed on its exposure to the different environmental conditions [93]. Another similar example was observed when graphene paper was functionalized with polydopamine. Such responses permit the easy fabrication of artificial muscles and electric generators [94].

Metallic compounds, especially transition metal (TM) oxides or hydroxides are important materials, and owing to their versatile functionality, they can be utilized in diverse applications, such as sensors, energy storage and conversion, catalysis, etc [95]. The MO–graphene papers are fabricated by mixing the GO with the pertinent metal salts and then subjected to a vacuum filtration process [96]. Another technique of preparing large-scale nanostructure-modified graphene paper is self-assembly, which is a simple yet efficient technique. For instance, a homogeneous mixture of AuNPs and a GO suspension exhibited self-assembly at the air/liquid interface to form a multi-layered GO-AuNP composite film [97]. Furthermore, N-doped graphene-SnO_2 papers were fabricated using 7,7,8,8-tetracyanoquinodimethane anion which could act as both a complexing agent and as a nitrogen source. Such doped MO–graphene papers were found to exhibit a large capacity and outstanding cycling stability for electrochemical energy storage [98]. Thus, 2D graphene papers with the added advantages of flexibility, economics, and simpler and safer synthesis when functionalized with polymers, nanostructured metals, and biomolecules are ideal and favorable candidates for (bio)sensing applications.

2.3 Conclusion

Research on electrochemical sensors involves the usage of diverse types of materials ranging from metal nanoparticles, carbon-based nanomaterials to biomaterials. With the development of human civilization, the need for flexible, portable, and lightweight devices attracts more attention. Breaking down the macro-sized materials into nanomaterials imparts remarkable physicochemical properties and the electron transfer rate between the substrate and the electrodes is greatly increased, being directly affected by the application of these modifiers. Consequently, the coupling of nanomaterials with electrochemical techniques boosted the performance of the electrodes and the sensing ability of the electrochemical sensors. Carbon-based nanomaterials such as graphene, carbon nanotubes, etc are highly promising in sensing applications due to some of their exciting properties, such as charge transport, large surface area, charge transport, etc. The application of graphene-based materials as (bio)sensors for the recognition of small biomolecules such as DA, glucose, cholesterol, heavy metal ions, etc, and as cancer biomarkers have yielded remarkable results in complex real samples. Though there are a few hurdles associated with biosensors, one being their shorter lifetimes, this can be circumvented using molecularly imprinted sensors. Such strategies can be beneficial as functional groups anchored to the graphene surface can interact with the large drug molecules with ease, i.e. resulting in faster and more sensitive drug sensors. The enormous surface area and high electrical conductivity of graphene-based electrochemical sensors provide outstanding opportunities for future development in this field. Furthermore, recent advances suggest that materials based on graphene papers could play a substantial role in developing flexible electrochemical sensors. However, these are accompanied by certain challenges, such as electronic conductivity, biocompatibility, etc. Thus, further development of graphene papers via the integration of biocompatible and conducting polymers is highly encouraged.

To sum up, the properties of these carbon-based nanomaterials such as graphene are greatly explored and there is still a plethora of opportunities to delve into these materials.

Conflict of interest

The authors declare no conflict of interest.

References

[1] Schroeder V, Savagatrup S, He M, Lin S and Swager T M 2019 Carbon nanotube chemical sensors *Chem. Rev.* **119** 599–663

[2] Kassal P, Steinberg M D and Steinberg I M 2018 Wireless chemical sensors and biosensors: a review *Sensors Actuators* B **266** 228–45

[3] Hu J, Sang G, Zeng N, Lv C and Xu C 2022 Amperometric sensor for the detection of hydrogen stable isotopes based on Pt nanoparticles confined within single-walled carbon nanotubes (SWNTs) *Sensors Actuators* B **356** 131344

[4] Stasyuk N Y, Gayda G Z, Zakalskiy A E, Fayura L R, Zakalska O M, Sibirny A A, Nisnevitch M and Gonchar M V 2022 Amperometric biosensors for L-arginine and creatinine assay based on recombinant deiminases and ammonium-sensitive Cu/Zn(Hg)S nanoparticles *Talanta* **238** 122996

[5] Zhang C and Du X 2020 Electrochemical sensors based on carbon nanomaterial used in diagnosing metabolic disease *Front. Chem.* **8** 651

[6] Brett A M O and Brett C M 1994 *Electrochemistry Principles, Methods, and Applications* (Coimbra: Oxford University Press)

[7] Holler F J, Skoog D A and Crouch S R 2018 *Principles of Instrumental Analysis* 7th edn (Boston, MA: Cengage Learning)

[8] Baranwal J, Barse B, Gatto G, Broncova G and Kumar A 2022 Electrochemical sensors and their applications: a review *Chemosensors* **10** 363

[9] Shao Y, Wang J, Wu H, Liu J, Aksay I A and Lin Y 2010 Graphene-based electrochemical sensors and biosensors: a review *Electroanalysis* **22** 1027–36

[10] Rahman M, Kumar P, Park D-S and Shim Y-B 2008 Electrochemical sensors based on organic conjugated polymers *Sensors* **8** 118–41

[11] Sarma M 2022 *Carbon Nanomaterial-based Electrochemical Sensors for Biomedical Applications* (Bristol: IOP Publishing) ch 16 p 15

[12] Cuartero M, Parrilla M and Crespo G 2019 Wearable potentiometric sensors for medical applications *Sensors* **19** 363

[13] Cho G, Azzouzi S, Zucchi G and Lebental B 2021 Electrical, and electrochemical sensors based on carbon nanotubes for the monitoring of chemicals in water—a review *Sensors* **22** 218

[14] Beaver K, Dantanarayana A and Minteer S D 2021 Materials approaches for improving electrochemical sensor performance *J. Phys. Chem.* B **125** 11820–34

Prinith N S and Manjunatha J G 2023 Voltammetric analysis of hazardous azo-dye indigo carmine at simple and cost-effective carbon paste sensor modified with L-phenylalanine *Sens. Technol.*

Manjunatha J G 2019 Electrochemical polymerised graphene paste electrode and application to catechol sensing *Open Chem. Eng. J.* **13** 81

Pushpanjali P A, Manjunatha J G, Tigari G and Fattepur S 2020 Poly(niacin) based carbon nanotube sensor for the sensitive and selective voltammetric detection of vanillin with caffeine *Anal. Bioanal. Electrochem.* **12** 553

[15] Cinti S and Arduini F 2017 Graphene-based screen-printed electrochemical (bio)sensors and their applications: efforts and criticisms *Biosens. Bioelectron.* **89** 107–22
[16] Yu P, Lowe S E, Simon G P and Zhong Y L 2015 Electrochemical exfoliation of graphite and production of functional graphene *Curr. Opin. Colloid Interface Sci.* **20** 329–38
[17] Yan Y, Nashath F Z, Chen S, Manickam S, Lim S S, Zhao H, Lester E, Wu T and Pang C H 2020 Synthesis of graphene: potential carbon precursors and approaches *Nanotechnol. Rev.* **9** 1284–314
[18] Jia K, Ci H, Zhang J, Sun Z, Ma Z and Zhu Y 2020 Super clean growth of graphene using a cold-wall chemical vapor deposition approach *Angew. Chem. Int. Ed.* **59** 17214–8
[19] Shekari Z, Zare H R and Falahati A 2021 Dual assaying of breast cancer biomarkers by using a sandwich–type electrochemical aptasensor based on a gold-nanoparticles–3D graphene hydrogel nanocomposite and redox probes labeled aptamers *Sensors Actuators* B **332** 129515
[20] Sankara Aditya Jayanthi V S P K, Das A B and Saxena U 2017 Recent advances in biosensor development for the detection of cancer biomarkers *Biosens. Bioelectron.* **91** 15
[21] Selvolini G and Marrazza G 2017 MIP-based sensors: promising new tools for cancer biomarker determination *Sensors* **17** 718
[22] Verma M and Manne U 2006 Genetic and epigenetic biomarkers in cancer diagnosis and identifying high-risk populations *Crit. Rev. Oncol.* **60** 9–18
[23] Xu S, Chang Y, Wu Z, Li Y, Yuan R and Chai Y 2020 One DNA circle capture probe with multiple target recognition domains for simultaneous electrochemical detection of miRNA-21 and miRNA-155 *Biosens. Bioelectron.* **149** 111848
[24] Tallapragada S R, Layek K, Mukherjee R, Mistry K K and Ghosh M 2017 Development of screen-printed electrode based immunosensor for the detection of HER2 antigen in human serum samples *Bioelectrochemistry* **118** 25–30
[25] Sonuç M N and Sezgintürk M K 2014 Ultrasensitive electrochemical detection of cancer-associated biomarker HER3 based on the anti-HER3 biosensor *Talanta* **120** 355–61
[26] Lu Y, Liang X, Niyungeko C, Zhou J, Xu J and Tian G 2018 A review of the identification and detection of heavy metal ions in the environment by voltammetry *Talanta* **178** 324
[27] Liu X, Yao Y, Ying Y and Ping J 2019 Recent advances in nanomaterial-enabled screen-printed electrochemical sensors for heavy metal detection *TrAC—Trends Anal. Chem.* **115** 187
[28] Ullah N, Mansha M, Khan I and Qurashi A 2018 Nanomaterial-based optical chemical sensors for the detection of heavy metals in water: recent advances and challenges *TrAC—Trends Anal. Chem.* **100** 155
[29] Zhang L, Peng D, Liang R P and Qiu J D 2018 Graphene-based optical nanosensors for the detection of heavy metal ions *TrAC—Trends Anal. Chem.* **102** 280
[30] Priya T, Dhanalakshmi N, Thennarasu S and Thinakaran N 2018 Ultra-sensitive detection of Cd (II) using reduced graphene oxide/carboxymethyl cellulose/ glutathione modified electrode *Carbohydr. Polym.* **197** 366
[31] Niu X, Zhong Y, Chen R, Wang F, Liu Y and Luo D 2018 A 'turn-on' fluorescence sensor for Pb^{2+} detection based on graphene quantum dots and gold nanoparticles *Sensors Actuators* B **255** 1577
[32] Hong K N, Fuster V, Rosenson R S, Rosendorff C and Bhatt D L 2017 How low to go with glucose, cholesterol, and blood pressure in primary prevention of CVD *J. Am. Coll. Cardiol.* **70** 2171

[33] Gahlaut A, Hooda V, Dhull V and Hooda V 2018 Recent approaches to ameliorate selectivity and sensitivity of enzyme-based cholesterol biosensors: a review *Artif. Cells Nanomed. Biotechnol.* **46** 472

[34] Narwal V, Deswal R, Batra B, Kalra V, Hooda R, Sharma M and Rana J S 2019 Cholesterol biosensors: a review *Steroids* **143** 6

[35] Zhang C, Zhang Z, Yang Q and Chen W 2018 Graphene-based electrochemical glucose sensors: fabrication and sensing properties *Electroanalysis* **30** 2504

[36] Phetsang S, Jakmunee J, Mungkornasawakul P, Laocharoensuk R and Ounnunkad K 2019 Sensitive amperometric biosensors for detection of glucose and cholesterol using a platinum/reduced graphene oxide/poly(3-aminobenzoic acid) film-modified screen-printed carbon electrode *Bioelectrochemistry* **127** 125

[37] Mahmoudi E, Hajian A, Rezaei M, Afkhami A, Amine A and Bagheri H 2019 A novel platform based on graphene nanoribbons/protein capped Au-Cu bimetallic nanoclusters: application to the sensitive electrochemical determination of bisphenol A *Microchem. J.* **145** 242

[38] Su B, Shao H, Li N, Chen X, Cai Z and Chen X 2017 A sensitive bisphenol A voltammetric sensor relying on AuPd nanoparticles/graphene composites modified glassy carbon electrode *Talanta* **166** 126

[39] Yu H, Feng X, Chen X X, Qiao J L, Gao X L, Xu N and Gao L J 2017 Electrochemical determination of bisphenol A on a glassy carbon electrode modified with gold nanoparticles loaded on reduced graphene oxide-multi walled carbon nanotubes composite *Chin. J. Anal. Chem.* **45** 713

[40] Duan Y, Li S, Lei S, Qiao J, Zou L and Ye B 2018 Highly sensitive determination of bisphenol A based on MoCuSe nanoparticles decorated reduced graphene oxide modified electrode *J. Electroanal. Chem.* **827** 137

[41] Shi R, Liang J, Zhao Z, Liu A and Tian Y 2017 An electrochemical bisphenol A sensor based on one-step electrochemical reduction of cuprous oxide wrapped graphene oxide nanoparticles modified electrode *Talanta* **169** 37

[42] Cai J, Sun B, Li W, Gou X, Gou Y, Li D and Hu F 2019 Novel nanomaterial of porous graphene functionalized black phosphorus as an electrochemical sensor platform for bisphenol A detection *J. Electroanal. Chem.* **835** 1

[43] Wei Y, Liu Y, Xu Z, Wang S, Chen B, Zhang D and Fang Y 2020 Simultaneous detection of ascorbic acid, dopamine, and uric acid using a novel electrochemical sensor based on palladium nanoparticles/reduced graphene oxide nanocomposite *Int. J. Anal. Chem.* **2020** 13

[44] Wang J 2005 Carbon-nanotube based electrochemical biosensors: a review *Electroanalysis* **17** 7

[45] Tang L H, Wang Y, Li Y M, Feng H B, Lu J and Li J H 2009 Preparation, structure, and electrochemical properties of reduced graphene sheet films *Adv. Funct. Mater.* **19** 2782

[46] Lin W J, Liao C S, Jhang J H and Tsai Y C 2009 Graphene modified basal and edge plane pyrolytic graphite electrodes for electrocatalytic oxidation of hydrogen peroxide and β-nicotinamide adenine dinucleotide *Electrochem. Commun.* **11** 2153

[47] Pumera M, Scipioni R, Iwai H, Ohno T, Miyahara Y and Boero M 2009 A mechanism of adsorption of β-nicotinamide adenine dinucleotide on graphene sheets: experiment and theory *Chem. Eur. J.* **15** 10851

[48] Liu H, Gao J, Xue M Q, Zhu N, Zhang M N and Cao T B 2009 Processing of graphene for electrochemical application: noncovalently functionalize graphene sheets with water-soluble electroactive methylene green *Langmuir* **25** 12006

[49] Bajtarevic A, Ager C, Pienz M, Klieber M, Schwarz K and Ligor M 2009 Noninvasive detection of lung cancer by analysis of exhaled breath *BMC Cancer* **9** 348
[50] Zhou X, Xue Z, Chen X, Huang C, Bai W and Lu Z 2020 Nanomaterial-based gas sensors used for breath diagnosis *J. Mater. Chem.* B **8** 3231–48
[51] Kuretake T, Kawahara S, Motooka M and Uno S 2017 An electrochemical gas biosensor based on enzymes immobilized on chromatography paper for ethanol vapor detection *Sensors* **17** 281
[52] Shahdeo D, Roberts A, Abbineni N and Gandhi S 2020 Graphene-based sensors *Comprehensive Analytical Chemistry* (Elsevier) vol 91 ch 8 pp 175–99 ch 8
[53] Chen Y, Ren R, Pu H, Chang J, Mao S and Chen J 2017 Field-effect transistor biosensors with two-dimensional black phosphorus nanosheets *Biosens. Bioelectron.* **89** 505–10
[54] Razlansari M, Ulucan-Karnak F, Kahrizi M, Mirinejad S, Sargazi S, Mishra S, Rahdar A and Díez-Pascual A M 2022 Nanobiosensors for detection of opioids: a review of latest advancements *Eur. J. Pharm. Biopharm.* **179** 79–94
[55] Mao K, Yang Z, Li J, Zhou X, Li X and Hu J 2017 A novel colorimetric biosensor based on non-aggregated Au@ Ag core–shell nanoparticles for methamphetamine and cocaine detection *Talanta* **175** 338–46
[56] Ribes A, Xifré-Pérez E, Aznar E, Sancenón F, Pardo T, Marsal L F and Martínez-Máñez R 2016 Molecular gated nanoporous anodic alumina for the detection of cocaine *Sci. Rep.* **6** 1–9
[57] Smolinska-Kempisty K, Ahmad O S, Guerreiro A, Karim K, Piletska E and Piletsky S 2017 New potentiometric sensor based on molecularly imprinted nanoparticles for cocaine detection *Biosens. Bioelectron.* **96** 49–54
[58] Azizi S, Gholivand M B, Amiri M, Manouchehri I and Moradian R 2022 Carbon dots-thionine modified aptamer-based biosensor for highly sensitive cocaine detection *J. Electroanal. Chem.* **907** 116062
[59] Meng J, Tang X, Zhou B, Xie Q and Yang L 2017 Designing of ordered two-dimensional gold nanoparticles film for cocaine detection in human urine using surface-enhanced Raman spectroscopy *Talanta* **164** 693–9
[60] Wu Z, Zhou H, Han Q, Lin X, Han D and Li X 2020 A cost-effective fluorescence biosensor for cocaine based on a 'mix-and-detect' strategy *Analyst* **145** 4664–70
[61] Florea A, Cowen T, Piletsky S and De Wael K 2019 Electrochemical sensing of cocaine in real samples based on electrodeposited biomimetic affinity ligands *Analyst* **144** 4639–46
[62] Rocha R G, Ribeiro J S, Santana M H, Richter E M and Muñoz R A 2021 3D-printing for forensic chemistry: voltammetric determination of cocaine on additively manufactured graphene–polylactic acid electrodes *Anal. Methods* **13** 1788–94
[63] Zhu X, Liu Y, Li P, Nie Z and Li J 2016 Applications of graphene and its derivatives in intracellular biosensing and bioimaging *Analyst* **141** 4541–53
[64] Li X, Zhang S, Kulinich S A, Liu Y and Zeng H 2014 Engineering surface states of carbon dots to achieve controllable luminescence for solid-luminescent composites and sensitive Be^{2+} detection *Sci. Rep.* **4** 1
[65] Zhu S, Meng Q, Wang L, Zhang J, Song Y, Jin H, Zhang K, Sun H, Wang H and Yang B 2013 Highly photoluminescent carbon dots for multicolor patterning, sensors, and bioimaging *Angew. Chem.* **125** 4045–9

[66] Masteri-Farahani M and Askari F 2019 Design and photophysical insights on graphene quantum dots for use as nanosensor in differentiating methamphetamine and morphine in solution *Spectrochim. Acta* A **206** 448–53

[67] Arabali V and Sadeghi R 2013 Surface properties of nano-Al_2O_3 film and its application in the preparation of morphine electrochemical sensor *Ionics* **19** 1775–82

[68] Jahromi Z, Mirzaei E, Savardashtaki A, Afzali M and Afzali Z 2020 A rapid and selective electrochemical sensor based on electrospun carbon nanofibers for tramadol detection *Microchem. J.* **157** 104942

[69] Hassamal S, Miotto K, Dale W and Danovitch I 2018 Tramadol: Understanding the risk of serotonin syndrome and seizures *Am. J. Med.* **131** 1–6

[70] Diouf A, Aghoutane Y, Burhan H, Sen F, Bouchikhi B and El Bari N 2021 Tramadol sensing in non-invasive Biological fluids using a voltammetric electronic tongue and an electrochemical sensor based on biomimetic recognition *Int. J. Pharm.* **593** 120114

[71] Soleimani M, Afshar M G, Shafaat A and Crespo G A 2013 High-selective tramadol sensor based on modified molecularly imprinted polymer carbon paste electrode with multiwalled carbon nanotubes *Electroanalysis* **25** 1159–68

[72] Bloor M, Paech M and Kaye R 2012 Tramadol in pregnancy and lactation *Int. J. Obstet. Anesth.* **21** 163–7

[73] Cobianchi M R 2012 Study of codeine-paracetamol combination treatment compared with tramadol-paracetamol in the control of moderate-to-severe low back pain *Minerva Med.* **103** 177–82

[74] Hallinan R, Ray J, Byrne A, Agho K and Attia J 2006 Therapeutic thresholds in methadone maintenance treatment: a receiver operating characteristic analysis *Drug Alcohol Depend.* **81** 129–36

[75] Xia W, Liu G, Shao Z, Xu E, Yuan H, Liu J and Gao L 2020 Toxicology of tramadol following chronic exposure based on the metabolomics of the cerebrum in mice *Sci. Rep.* **10** 11130

[76] Al Samarrai S Y, Abdoon F M and Hashim K K 2019 A simple method to determine tramadol using a coated-wire electrode as a detector in the flow injection analysis *Microchem. J.* **146** 588–91

[77] Arabali V, Malekmohammadi S and Karimi F 2020 Surface amplification of pencil graphite electrode using CuO nanoparticle/polypyrrole nanocomposite; a powerful electrochemical strategy for the determination of tramadol *Microchem. J.* **158** 105179

[78] Atta N F, Galal A and Hassan S H 2019 Ultrasensitive determination of nalbuphine and tramadol narcotic analgesic drugs for postoperative pain relief using nano-cobalt oxide/ionic liquid crystal/carbon nanotubes-based electrochemical sensor *J. Electroanal. Chem.* **839** 48–58

[79] Demoranville L T and Verkouteren J R 2013 Measurement of drug-facilitated sexual assault agents in simulated sweat by ion mobility spectrometry *Talanta* **106** 375–80

[80] Yang Y, Zhai S, Liu C, Wang X and Tu Y 2019 Disposable immunosensor based on electrochemiluminescence for ultrasensitive detection of ketamine in human hair *ACS Omega* **4** 801–9

[81] Narang J, Malhotra N, Singhal C, Mathur A, Chakraborty D, Anil A, Ingle A and Pundir C S 2017 Point of care with microfluidic paper-based device integrated with nano zeolite–graphene oxide nanoflakes for electrochemical sensing of ketamine *Biosens. Bioelectron.* **88** 249

[82] Zhang M, Halder A, Cao X, Hou C and Chi Q 2017 *Graphene-Paper Based Electrochemical Sensors* (TechOpen) vol 3 p 33
[83] Wu D, Zhang F, Liang H and Feng X 2012 Nanocomposites and macroscopic materials: assembly of chemically modified graphene sheets *Chem. Soc. Rev.* **41** 6160
[84] Xu Y and Shi G 2011 Assembly of chemically modified graphene: methods and applications *J. Mat. Chem.* **21** 3311
[85] Joshi R K, Alwarappan S, Yoshimura M, Sahajwalla V and Nishina Y 2015 Graphene oxide: the new membrane material *Appl. Mater. Today.* **1** 1–12
[86] Liu G, Jin W and Xu N 2015 Graphene-based membranes *Chem. Soc. Rev.* **44** 5016–30
[87] Cong H-P, Chen J-F and Yu S-H 2014 Graphene-based macroscopic assemblies and architectures: an emerging material system *Chem. Soc. Rev.* **43** 7295–325
[88] Lee D H, Kim J E, Han T H, Hwang W J, Jeon S W, Choi S Y, Hong S H, Lee W J, Ruoff R S and Kim S O 2010 Versatile carbon hybrid films composed of vertical carbon nanotubes crown on mechanically compliant graphene films *Adv. Mater.* **22** 1247–52
[89] Shao Y, El-Kady M F, Wang L J, Zhang Q, Li Y, Wang H, Mousavi M F and Kaner R B 2015 Graphene-based materials for flexible supercapacitors *Chem. Soc. Rev.* **44** 3639–65
[90] Choi W, Choi J, Bang J and Lee J H 2013 Layer-by-layer assembly of graphene oxide nanosheets on polyamide membranes for durable reverse-osmosis applications *ACS Appl. Mater. Interfaces* **5** 12510–9
[91] Dai M-K, Lian J-T, Lin T-Y and Chen Y-F 2013 High-performance transparent and flexible inorganic thin film transistors: a facile integration of graphene nanosheets and amorphous InGaZnO *J. Mater. Chem.* C **1** 5064–71
[92] Bae S H, Kahya O, Sharma B K, Kwon J, Cho H J, Özyilmaz B and Ahn J H 2013 Graphene-P(VDFTrFE) multilayer film for flexible applications *ACS Nano* **7** 3130–8
[93] Mu J, Hou C, Zhu B, Wang H, Li Y and Zhang Q 2015 A multi-responsive water-driven actuator with instant and powerful performance for versatile applications *Sci. Rep.* **5** 9503
[94] Mu J, Hou C, Wang H, Li Y, Zhang Q and Zhu M 2015 Origami-inspired active graphene-based paper for programmable instant self-folding walking devices *Sci. Adv.* **1** e1500533
[95] Zhang M, Hou C, Halder A and Chi Q 2017 Ultralight, flexible and semi-transparent metal oxide papers for photoelectrochemical water splitting *ACS Appl. Mater. Interfaces* **9** 3922–30
[96] Sumboja A, Foo C Y, Wang X and Lee P S 2013 Large areal mass, flexible and free-standing reduced graphene oxide/manganese dioxide paper for asymmetric supercapacitor device *Adv. Mater.* **25** 2809–15
[97] Liu F, Piao Y, Choi K S and Seo T S 2012 Fabrication of free-standing graphene composite films as electrochemical biosensors *Carbon* **50** 123–33
[98] Wang X *et al* 2012 N-doped graphene-SnO_2 sandwich paper for high-performance lithium-ion batteries *Adv. Funct. Mater.* **22** 2682–90

IOP Publishing

Graphene-Based Sensors

Jamballi G Manjunatha

Chapter 3

Graphene-based sensors for electrochemical detection of textile dye

Bruna Coldibeli, Gustavo Fix and Elen Romão Sartori

This review provides a comprehensive overview of how graphene has been used in the advancement of voltammetric sensors for the detection of textile dyes in water samples over the past decade. It highlights the linear range, limit of detection, interfering compounds investigated, and the different types of water samples analyzed by each of the graphene-based sensors developed.

3.1 Introduction

Water pollution is a major environmental problem that is constantly increasing and negatively affecting the lives of all living beings. It is characterized by changes in the physical, chemical, or biological properties of water that can cause serious problems for the environment and human health [1]. The greatest pollution of water resources is due to anthropogenic activities that directly or indirectly introduce pollutants such as pesticides, synthetic fertilizers, dyes, heavy metals, personal care products, detergents, pharmaceuticals, etc, into water bodies [2, 3].

The disposal into water resources of industrial and domestic wastewater that is not properly treated is a major source of anthropogenic pollution. Despite its great global economic importance, the textile industry is now one of the most polluting industries and is associated with numerous environmental impacts, because of the large amounts of potable water and numerous chemicals consumed in the manufacturing process [3–6]. The composition of wastewater generated in the textile industry is extremely heterogeneous and contains large amounts of highly toxic chemicals, such as heavy metals and non-biodegradable organic compounds, especially textile dyes, which may account for about 10%–45% of the wastewater resulting from the textile manufacturing [4, 7, 8].

Briefly, dyes are synthetic or natural organic compounds that are hydro or oil-soluble and are characterized by the selective absorption of certain wavelengths

doi:10.1088/978-0-7503-5578-0ch3

of visible light according to their chemical structure (chromophore group). Thus, the observed color results from the fraction of light that is not absorbed and reflected (complementary color) [9, 10]. They are used in a variety of applications, including plastics, textiles, food, cosmetics, rubber, paper, and other industries. Considering that the annual world production of synthetic dyes is about 7×10^7 tons, the textile industry is the largest consumer accounting for its consumption of ten thousand tons of these dyes [8, 11].

Dyes are known and used by humans since antiquity, but they were obtained from natural sources such as plants, insects/animals, and minerals, until William Henry Perkin accidentally discovered the first synthetic dye Mauveine, in 1856 while trying to find a way to synthesize quinine (a drug used to treat the tropical disease Malaria). This discovery was followed by a major change in the dyeing process with the development of new products and their industrial production, driven mainly by the demands of the textile industry [9, 11–13].

Dye molecules are composed of the chromophore groups, which are responsible for color production, auxochromes, which complement the chromophore, make the molecule water soluble and provide greater affinity of the dye for the fibers, and the matrix, which corresponds to the rest of the molecule [9, 10, 14]. The most common chromophores are azo (–N=N–), nitro ($-NO_2$), nitroso (–N=O), thiocarbonyl (–C=S), carbonyl (–C=O), and the alkenes (–C=C–), while auxochromes include –COOH, $-SO_3H$, –OH, $-NH_3$, etc [9, 10].

Currently, numerous synthetic textile dyes are commercially available and are classified according to the color index (CI) international system based on their chromophore structure, color, and application method. The CI gives a generic name and constitution number for each dye. For example, the commercial dye known as Congo red is named Direct Red 28 with CI number 22120 according to the CI classification [9, 10]. Azo dyes represent the largest class of synthetic dyes with a wide variety of patterns, including mono-, di-, tri-, and polyazo molecules, i.e. with one, two, three, or more azo structures. Due to their easy preparation with low cost, high stability and color variety, they are the most commonly used dyes in the textile industry [10, 11, 14, 15]. The chemical structures of some textile dyes are shown in figure 3.1. More detailed information about the classes of synthetic textile dyes, their properties, and applications can be found in the published literature [9].

During the textile dyeing process, a significant amount of the dye is not properly fixed in the textile fiber and is released into the water bodies as wastewater [8, 9]. Once the textile dyes are improperly discharged, they not only affect the esthetics of the water bodies, but also impair the penetration of sunlight into the water, resulting in a decrease in the rate of photosynthesis and the amount of dissolved oxygen. They are also persistent pollutants that cause harmful effects on the food chain, as they are bioaccumulative and have carcinogenic, mutagenic, and toxic properties [8, 16].

Given the serious problems associated with the presence of textile dyes in water bodies, it is of great interest to determine these species in natural aquatic resources, in water intended for human consumption, and in industrial effluents treated before disposal in nature, in order to evaluate the efficiency of treatment and to prevent water pollution and any associated harm to living organisms. Several analytical

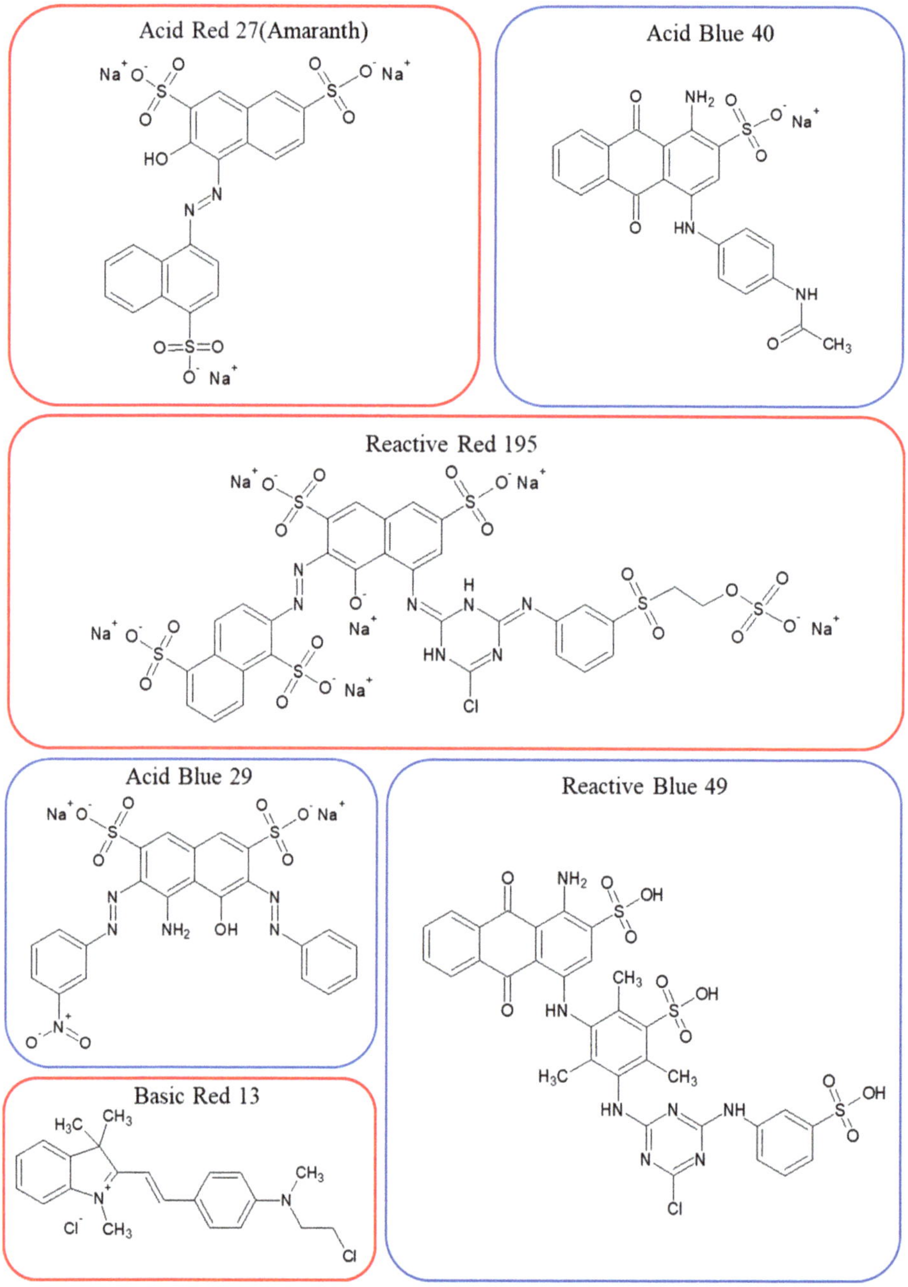

Figure 3.1. Chemical structure of some textile dyes.

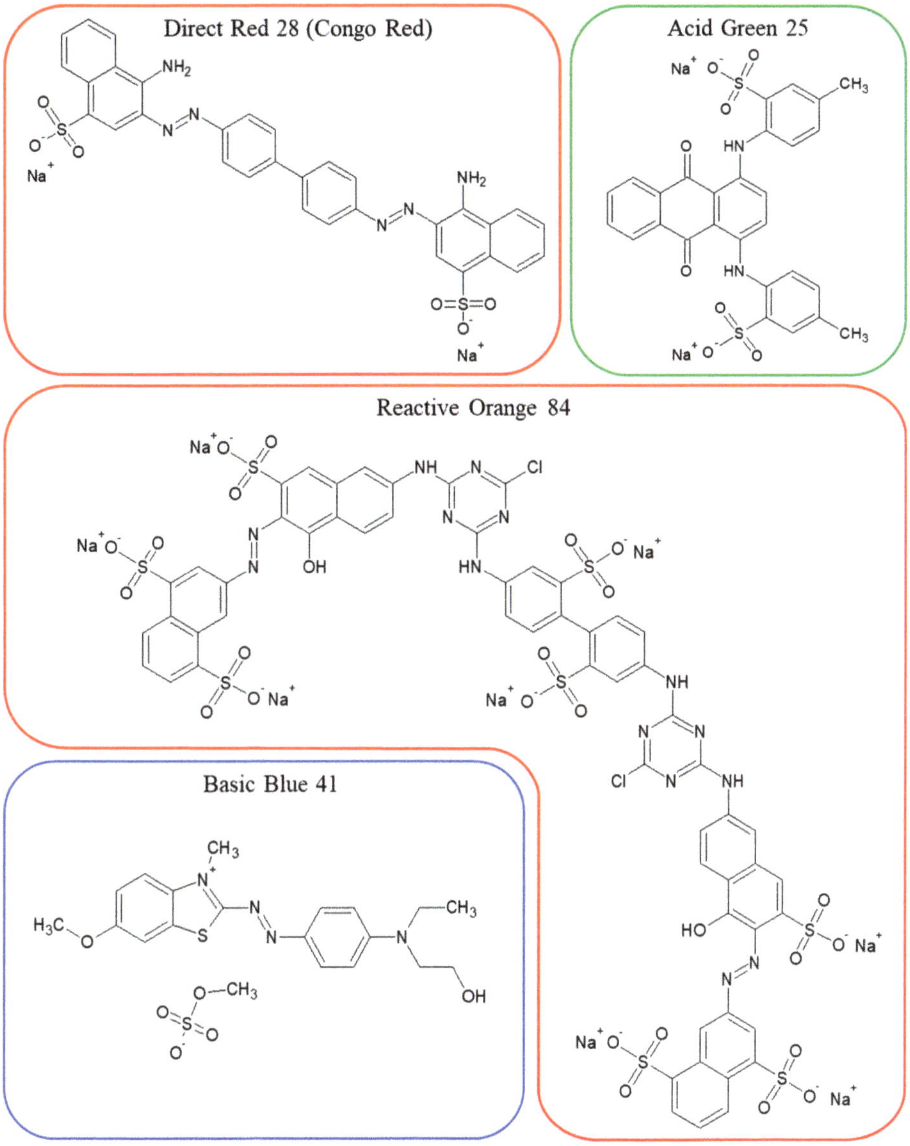

Figure 3.1. (Continued.)

methods have been proposed, mainly based on chromatographic [17–20] and spectrophotometric [21–25] techniques, to quantify the concentrations of dyes in water samples, whether they are natural water, drinking water, or residual water. Although these techniques offer excellent analytical features, in some cases the methods developed have characteristics that are not advantageous from a practical and environmental point of view. These include tedious sample preparation, use of large amounts of toxic organic solvents, long analysis time, expensive and non-portable instrumentation, and high cost. In this regard, to overcome these drawbacks, electroanalytical techniques, such as voltammetry, can be considered as an alternative for the determination of textile dyes in environmental water samples, providing simplicity, cost efficiency, relatively low time analysis, and portability, with comparable analytical performance [26–30].

3.2 Voltammetry

Voltammetry is an electroanalytical technique in which a potential is applied to an electrode to monitor the current flowing through an electrochemical cell [31]. Some examples of voltammetric techniques are cyclic voltammetry (CV), square-wave voltammetry (SWV), and differential pulse voltammetry (DPV) [32]. In these techniques, the potential is varied and the current resulting from the electrolysis of the chemical species in a supporting electrolyte is measured, resulting in a plot of $I = f(E)$ [33]. This graph is called a voltammogram, and the magnitude of the redox current is related to the concentration of the chemical species. Each molecule can have a different redox potential, and these techniques can be conveniently used to determine two or more molecules simultaneously [34–36].

An electrochemical cell consists of three electrodes: working, reference, and auxiliary electrodes [31, 33]. These electrodes are inserted into an electrochemical cell containing the supporting electrolyte, which is usually a buffer, acid, or saline solution. This cell is coupled to a potentiostat for data acquisition, which is interfaced with a computer to display voltammograms (figure 3.2). The redox reaction occurs at the surface of the working electrode when a potential difference is applied between this electrode and the reference electrode. The auxiliary electrode ensures that the current does not flow through the reference electrode, since the latter must maintain a constant potential during the measurements [33]. Generally, the reference electrode consists of an Ag/AgCl or a saturated calomel electrode, the working electrodes are usually made of chemically stable materials, and the commonly used auxiliary electrode is a platinum plate or coiled platinum wire with a surface area significantly larger than that of the working electrode.

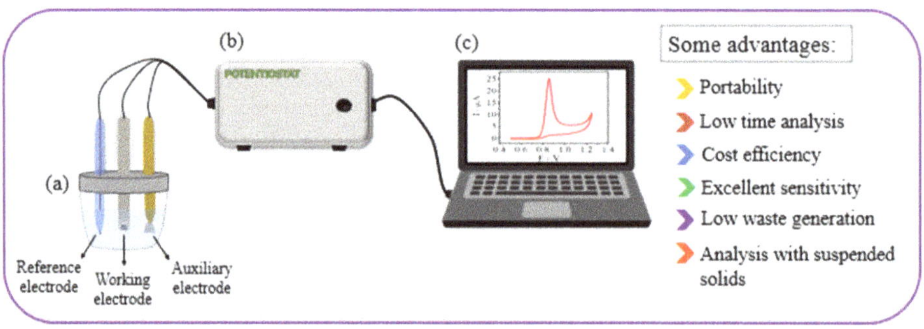

Figure 3.2. Schematic representation of the electrochemical system used in the voltammetric measurements: (a) electrochemical cell, (b) potentiostat, and (c) computer.

The most popular materials used as working electrodes in voltammetry are gold, platinum, mercury, glassy carbon, graphite, and doped diamond films. Several of these cited materials have been used as working electrodes for the determination of textile dyes in environmental samples. Guaratini *et al* used a dropping mercury electrode and the hanging mercury drop electrode for the polarographic determination of Procion Red HE-3B 9 and Procion Green HE-4BD [37]. Both dyes were quantified in Britton–Robinson buffer solution (pH 4 and 3, respectively) by differential pulse polarography in the concentration range of 0.1–10 μmol l^{-1}, with LOD values of 0.1 μmol l^{-1}. Silva *et al* used a cathodically pretreated boron-doped diamond electrode and SWV for the determination of indigo carmine dye in water samples [38]. They obtained a linear concentration range from 0.5 to 84.1 μmol l^{-1}, with LOD of 0.058 μmol l^{-1}, in a 0.3 mol l^{-1} H_2SO_4 solution. Radi *et al* employed a glassy carbon electrode (GCE) and DPV technique for the determination of Reactive Black 5 [39]. This electrode showed a linear concentration range from 0.6 to 1.5 μmol l^{-1}, with LOD of 0.4 μmol l^{-1}, in the phosphate buffer solution (pH 4.2). Later, Radi *et al* employed a GCE for the development of a DPV method for the analytical determination of Reactive Red 231 in river water, tap water, and synthetic effluent samples [40]. This electrode presented a linear relationship from 0.25 to 200 μmol l^{-1}, with LOD of 0.20 μmol l^{-1}, in a phosphate buffer solution (pH 3.77).

To ensure higher sensitivity and selectivity of analytical methods developed for voltammetric determination of textile dyes in environmental matrices using GCE, several authors have used graphene to modify the surface of this electrode or mixed with graphite powder [41]. The use of graphene to modify the surface of carbon-based electrodes has progressed over the years due to its high conductivity, biocompatibility, and low-cost raw materials [42, 43]. The most commonly used method for producing graphene is the chemical oxidation of natural graphite to obtain graphene oxide (GO). Strong oxidizing agents such as sulfuric acid, potassium permanganate, sodium nitrate, and hydrogen peroxide are used. An additional electrochemical, chemical, mechanical, or even thermal step can be used to convert GO to reduced graphene oxide (rGO) [41].

3.3 Determination of textile dyes in environmental matrices employing graphene-based sensors

To date, little has been reported on the use of graphene in the development of voltammetric sensors for the determination of textile dyes in environmental samples. Figure 3.3 illustrates a summary of the types of graphene-based sensors developed over the past decade for the voltammetric determination of textile dyes in environmental samples. It can be seen that the scheme presents four configurations of sensors based on graphene. There are GCE modified with aqueous suspensions of GO or rGO and GCE modified with graphene and also other materials, such as silver nanoparticles (AgNPs), gold nanourchins (GNUs), or nanocomposite: dioxide of ruthenium (RuO_2) and carboxylated multiwalled carbon nanotubes (MWCNTs). There are also graphite paste electrodes modified with graphene, $CoFe_{12}O_{19}$

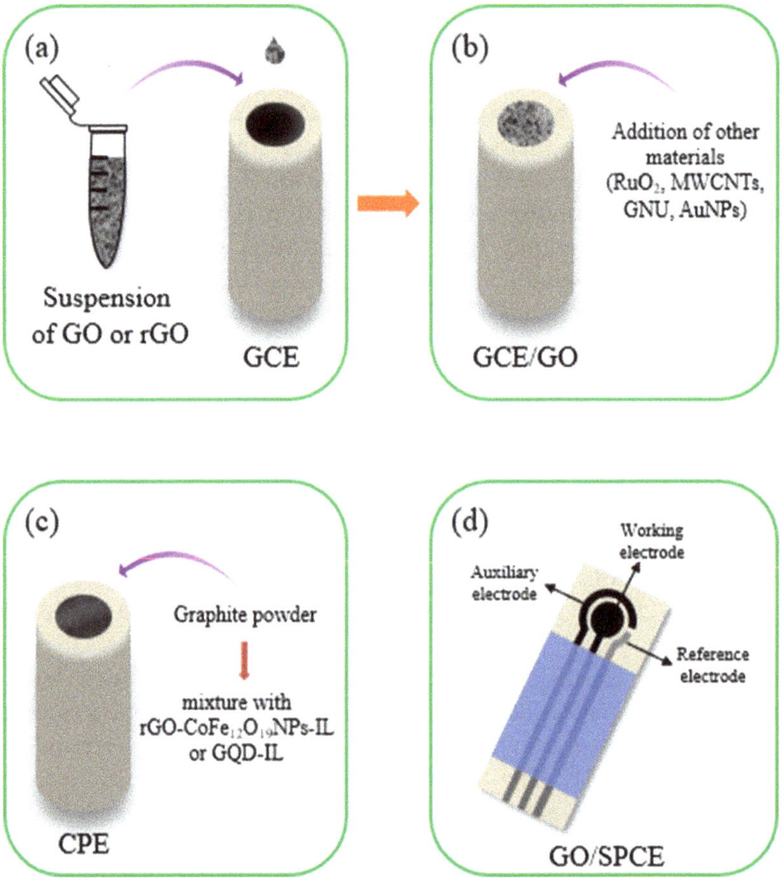

Figure 3.3. Schematic representation of types of graphene-based sensors for the voltammetric determination of textile dyes in environmental matrices reported in the literature over the past ten years: (a) GCE modified with GO or rGO; (b) GCE modified with graphene and other materials, such AgNPs, or GNU, or nanocomposite: RuO_2 and MWCNTs; (c) CPE modified with graphene, $CoFe_{12}O_{19}$NPs, and 1-ethyl-3-methylimidazolium ionic liquid or GQDs and *N*-hexyl-3-methylimidazolium hexafluorophosphate ionic liquid; (d) SPCE modified with GO.

nanoparticles, and 1-ethyl-3-methylimidazolium ionic liquid or graphene quantum dots (GQDs) and *N*-hexyl-3-methylimidazolium hexafluorophosphate ionic liquid and screen-printed carbon electrodes (SPCEs) modified with GO. All these examples of graphene-based sensors were developed for the voltammetric determination of textile dyes in environmental samples over the past ten years. Table 3.1 presents the linear range, LOD value, investigated interferents, and types of analyzed environmental sample. Each of these sensors presented in this table is described below.

Dorabi and Shabani-Nooshabadi modified a carbon paste electrode (CPE) with rGO, $CoFe_{12}O_{19}$ nanoparticles, and 1-ethyl-3-methylimidazolium ionic liquid for analysis of the textile dye Acid Green 25 in hair dyes and water samples [44]. The rGO was synthesized by Hummers method. To obtain this sensor, rGO,

Table 3.1. Analytical features of graphene-based sensors for the voltammetric determination of textile dye in environmental samples.

Textile dye	Sensor	Technique	Linear range (μmol l^{-1})	LOD (μmol l^{-1})	Investigated interferents	Environmental sample types	Reference
Acid Green 25	rGO–$CoFe_{12}O_{19}$NPs-IL/CPE	DPV	0.08–900	0.067	Brilliant blue, Acid Red 33, Acid Yellow 23, Acid Blue 9, Acid Black 1, Red 2G, I^-, K^+, Na^+, PO_4^{3-}, and NO_3^-	Tap water and textile industry wastewater	[44]
Acid Red 27 (Amaranth)	GQD/CPE	DPV	0.1–400	0.03	Glucose, L-threonine, alanine, glycine, Na^+, guanine, ascorbic acid, Cl^-, tryptophan, tyrosine, Sudan I, vitamin B_9, and vitamin B_6	Tap, well, and river water	[45]
Acid Red 27 (Amaranth)	N-rGO/GCE	DPV	0.1–600	0.03	—	Tap water	[46]
Acid Red 27 (Amaranth)	PLcys-GO/GCE	DPV	0.080–1.0	0.009	NO_2^-, SO_4^{2-}, Cl^-, Cr^{6+}, Cr^{3+}, Cd^{2+}, bisphenol, and Sudan I	Confectionery wastewater	[47]
Direct Red 28 (Congo red)	GO/GCE	DPV	0.01–0.2	0.24	Ammonium sulphate, barium chloride, potassium nitrate, copper acetate, ammonium chloride, sodium chloride, potassium chloride, and copper sulphate	Soil and water	[48]
Reactive Blue 49	GO/GCE	DPSV	0.000 28–0.000 79	0.000 23	—	—	[49]
Reactive Red 195	GO/GCE	DPSV	0.0438–0.307	0.0263	—	Industrial waste samples	[50]
Basic Red 13	GO–AgNPs/GCE	DPV	0.5–100	0.2	Na^+, Ca^{2+}; Zn^{2+}, Cl^-, $C_2O_4^{2-}$, NO_3^-, CO_3^{2-}, HSO_4^-, SO_4^{2-}, PO_4^-, Fe^{2+}, Fe^{3+}, CH_3COO^-, Basic Violet 10, Reactive Yellow 28, Reactive Orange 16, Disperse Red 13, and Basic Red 76	Wastewater collected from a textile plant	[51]
Reactive Orange 84	GO–GNU/GCE	DPV	1.0–1200	0.03	—	Solution of the degradation process of the dye	[52]

Acid Blue 40 and Acid Blue 29	rGO–RuO_2–MWCNTs/GCE	SWV	0.05–50 (acid blue 40) and 0.5–75 (acid blue 29)	0.01 (acid blue 40) and 0.1 (acid blue 29)	SO_4^{2-}, CO_3^{2-}, F^-, Br^-, NO_3^-, Mg^{2+}, NH_4^+, Ca^{2+}; Bi^{2+}, Ce^{2+}, Cr^{3+}, Mn^{2+}, Ba^{2+}, Fe^{2+}, Pb^{2+}, Pd^{2+}, Ru^{3+}, Cu^{2+}, Co^{2+}, and Mordant Yellow 10	Wastewater from cloth dyeing	[53]
Basic Blue 41	GO/SPCE	SWV	0.03–2.01	0.005	—	Tap water and wastewater collected from a treatment plant drinking water	[54]

Abbreviations: rGO–$CoFe_{12}O_{19}$/CPE: carbon paste electrode modified with reduced graphene oxide; GQD/CPE: carbon paste electrode modified with graphene quantum dots and ionic liquid; N-rGO/GCE: glassy carbon electrode modified with N-doped reduced graphene oxide; PLcys-GO/GCE: glassy carbon electrode modified with graphene and electropolymerized L-cysteine; GO/GCE: glassy carbon electrode modified with graphene oxide nanoparticles; GO–AgNPs/GCE: glassy carbon electrode modified with silver nanoparticles and graphene; GO–GNU/GCE: glassy carbon electrode modified with graphene oxide gold nanourchins; rGO–RuO_2–MWCNTs: GCE modified with reduced graphene oxide, RuO_2, and carboxylated multiwalled carbon nanotubes; SPCE/GO: screen printed electrode modified with graphene oxide; DPV: differential pulse voltammetry; DPSV: differential pulse stripping voltammetry; SWV: square-wave voltammetry.

$CoFe_{12}O_{19}$NPs, and ionic liquid were mixed with graphite powder and paraffin oil. This modification of the CPE enhanced the conductivity and the surface area of the electrode and thus increased the oxidation current of the Acid Green 25 compared to the unmodified CPE. Water samples such as tap water, and textile industry wastewater were analyzed directly without any pretreatment and the hair dye sample was dissolved in water containing ethanol, filtered, and analyzed in a phosphate buffer solution (pH 7.0). The authors reported that no important interferences were found in the analysis of Acid Green 25 on the surface of rGO–$CoFe_{12}O_{19}$NPs-IL/CPE. The authors did not provide information about the stability and quantities of measurements that this sensor performs.

Akbari modified a CPE with GQDs and *N*-hexyl-3-methylimidazolium hexafluorophosphate ionic liquid for the voltammetric determination of Acid Red 27 (Amaranth) in tap water, well water, and river water [45]. The carbon paste was obtained by mixing GQD and graphite powder using ionic liquid and paraffin oil to obtain a uniformly wetted paste. The GQDs were prepared with straightly pyrolysing citric acid. The use of graphene and this ionic liquid provided an improvement in conductivity of the sensor, large surface area, fast electron transfer, and good dispersing properties; consequently, an increase in the peak current of the textile dye; it also reduces fouling caused by Acid Red 27 or its oxidation product. This sensor was stored under ambient conditions in the buffer solution and the tests carried out for three weeks showed that the oxidation currents had a decrease of 2.3% in relation to the initial value. Several potentially interfering substances in the determination of Acid Red 27 showed tolerance limits of a relative error of ±5%, indicating selectivity for Acid Red 27. The samples were analyzed directly in phosphate buffer solution (pH 7.0) and acceptable recoveries values of Acid Red 27 were obtained with reproducible results.

Moradpour *et al* modified a GCE with N-doped rGO (N-rGO) for the voltammetric determination of Acid Red 27 (Amaranth) in tap water [46]. The N-rGO sheets were prepared by a hydrothermal method. The suspension of this nanomaterial was prepared in deionized water and drop-cast on the GCE surface, then it was allowed to dry completely under ambient conditions. This sensor exhibited better electrocatalytic behavior than the unmodified GCE, showing a high oxidation current at a lower potential. Satisfactory recovery percentages of 97.5–103% were obtained in the determination of this dye in water sample, in which the sample was analyzed directly in a phosphate buffer solution (pH 7.0). The authors did not provide information about the stability and quantities of measurements that this sensor performs.

Gao *et al* modified a GCE using commercial graphene and L-cysteine for the voltammetric determination of Acid Red 27 (Amaranth) in confectionery wastewater [47]. Graphene was dripped on the surface of the electrode, and after complete drying, electropolymerization of the L-cysteine film was performed. The sensor showed selectivity for Acid Red 27 determination in the presence of several tested compounds. The treatment of the sample consisted of centrifugation, and the clear liquid phase was used to determine Acid Red 27 in phosphate buffer solution (pH 2.0). When stored for two weeks at room temperature, peak currents remained at about 96.8% of their initial value.

Shetti *et al* modified a GCE with GO for the voltammetric determination of textile dye Direct Red 28 (Congo red) in soil and water samples [48]. The GO suspension was prepared by dispersing commercial GO in ethanol. This suspension was placed on the surface of the GCE and activation was performed by CV between 0.0 and 1.2 V in phosphate buffer solution (pH 8.0). GO provided a higher voltammetric response and also shifted the anodic peak to a lower potential when compared with bare GCE. This sensor was stored in a bottle containing no air for seven days and, in this condition, showed a response loss of only 4% for a Direct Red 28 concentration of 1.0 mmol l^{-1}. Different metal ions were evaluated as potentially interfering compounds, and the results showed that there was interference from ammonium sulfate, barium chloride, potassium nitrate, and copper acetate on the voltammetric response of Direct Red 28 using the GO/GCE. The authors reported that the oxidation of the Congo red dye is similar to the oxidation of the hydroxyl–substituted azo dye, with the involvement of one electron and one proton. The soil was dispersed in water, centrifuged, filtered, and analyzed in phosphate buffer pH 8.0. The water sample was analyzed directly without any additional pretreatment.

Revathi and Kumar modified a GCE with commercial graphene for the voltammetric determination of Reactive Blue 49 in textile industries effluents [49]. The suspension of this nanostructured material was prepared in water and placed on the surface of the GCE. Employing this sensor, the textile dye is first accumulated by adsorption on the surface of the sensor and then the accumulated molecules are stripped out giving specific signals. In this case, an accumulation potential of 500 mV and an accumulation time of 80 s were used. The authors suggest the application of the developed method in the determination of Reactive Blue 49 in effluents from textile industries.

Revathi *et al* modified the surface of a GCE by dripping a graphene dispersion and used the sensor in the development of a voltammetric method to detect Reactive Red 195 in industrial waste samples [50]. The authors employed commercial graphene and reported an excellent signal of the Reactive Red 195 on the graphene-modified surface of GCE when compared to the unmodified GCE due to the increased surface area powered by the nanostructured material.

Jafari *et al* modified a GCE with silver nanoparticles (AgNPs) and GO for the voltammetric determination of Basic Red 13 in wastewater collected from a textile plant [51]. The GCE surface was modified with commercial GO suspension and then immersed in a $AgNO_3$ solution for electrochemical deposition of the nanoparticles applying eight continuous cycles of potential by CV. With the combination of GO–AgNPs in the composition, it was possible to increase the surface area of the modified electrode. The authors identified the degradation products of Basic Red 13 by gas chromatography coupled to the mass spectrum technique and described that its effectiveness decreases significantly after mechanical damage. Regarding interfering species, its surface is subject to contamination of Fe(II) and Fe(III) ions, and its use in highly contaminated environments with these species should be avoided. Another alternative made by these authors was to modify the surface of GCE with

GO and gold nanourchins (GNU) for the electrochemical detection of Reactive Orange 84 in wastewater [52]. The authors synthesized GO by the modified Hummers method. To obtain this sensor, first the GO suspension was dropped on the GCE surface and, after drying, the GNU suspension was added and then slowly dried. The use of GNU in the modification allowed the increase of the effective surface area of the electrode and its sensitivity to Reactive Orange 84. The authors applied this sensor to determine the concentration of Reactive Orange 84 remaining after the degradation process; however, they did not investigate the effect of possible interferences.

Khosropour *et al* modified a GCE with RuO_2, commercial rGO, and MWCNTs for the simultaneous electrochemical detection of Acid Blue 40 and Acid Blue 29 in wastewater samples [53]. The suspension with these materials was prepared in ethanol and deposited on the GCE surface. The modified electrode showed a high ability for the simultaneous determination of both textile dyes at low levels, showing a notable increase in the anodic peaks. The repeatability of signals showed low relative standard deviation (RSD) values (<2.5%) and for the reproducibility of the sensor, manufactured independently, the RSD values were <4.0%. It was stored in the laboratory for two weeks, and the anodic peaks current variations were observed to be <3.0%. The authors reported that no interference affected the current of two dyes in the presence of several compounds on the surface of rGO–RuO_2–MWCNTs. The sample pretreatment consisted of centrifugation, filtration, and the determination was performed in the phosphate buffer solution (pH 6.0).

Hudari *et al* used commercial graphene-modified screen-printed carbon electrodes (GO/SPCE) for the voltammetric determination of Basic Blue 41 in drinking water from a treatment plant [54]. The authors validated the method by high-performance liquid chromatography with diode array detection. Water samples were analyzed directly in Britton–Robinson buffer solution (pH 4.5), without pretreatment steps. In addition, they highlight that the LOD and limit of quantitation values obtained using this sensor are compatible with that required for the evaluation of dye contamination in superficial water, for example, in which high dilution is expected.

3.4 Conclusion

This review shows that using graphene to modify carbon-based electrodes results in analytical determination methods with lower LOD values. This is of great importance because textile dyes are often found in low concentrations in environmental matrices. In addition, it is critical to evaluate concomitant compounds, as this type of environmental matrix may contain multiple textile dyes, inorganic ions, or other organic molecules that can potentially interfere with the redox potential or the measured current. Finally, it was verified that the number of papers published in the literature in the past decade reporting the use of graphene in the modification of carbon-based electrodes is quite small, and there are several possibilities for further studies and applications.

Acknowledgments

The authors gratefully acknowledge financial support and scholarships from the Brazilian funding agencies CNPq (Grant Nos. 408591/2018-8, 305320/2019-0), Fundação Araucária do Paraná, and CAPES (Grant Nos. 88887.674846/2022-00, 88887.830330/2023-00).

References

[1] Madhav S, Ahamad A, Singh A K, Kushawaha J, Chauhan J S and Sharma S *et al* 2020 Water pollutants: sources and impact on the environment and human health *Sensors in Water Pollutants Monitoring: Role of Material Advanced Functional Materials and Sensors* ed D Pooja, P Kumar, P Singh and S Patil (Springer) pp 43–62

[2] Jayaswal K, Sahu V and Gurjar B R 2018 Water pollution, human health and remediation *Water Remediation Energy, Environment, and Sustainability* ed S Bhattacharya, A Gupta, A Gupta and A Pandey (Springer) pp 11–27

[3] Haseena M, Malik M F, Javed A, Arshad S, Asif N, Zulfiqar S and Hanif J 2017 Water pollution and human health *Environ. Risk Assess. Rem.* **1** 16–9

[4] Islam T, Repon M R, Islam T, Sarwar Z and Rahman M M 2022 Impact of textile dyes on health and ecosystem: a review of structure, causes, and potential solutions *Environ. Sci. Pollut. Res.* **30** 9207–42

[5] Lellis B, Fávaro-Polonio C Z, Pamphile J A and Polonio J C 2019 Effects of textile dyes on health and the environment and bioremediation potential of living organisms *Biotechnol. Res. Innov.* **3** 275–90

[6] Adane T, Adugna A T and Alemayehu E 2021 Textile industry effluent treatment techniques *J. Chem.* 2021 5314404

[7] Madhav S, Ahamad A, Singh P and Mishra P K 2018 A review of textile industry: wet processing, environmental impacts, and effluent treatment methods *Environ. Qual. Manag.* **27** 31–41

[8] Al-Tohamy R, Ali S S, Li F, Okasha K M, Mahmoud Y A-G, Elsamahy T, Jiao H, Fu Y and Sun J 2022 A critical review on the treatment of dye-containing wastewater: ecotoxicological and health concerns of textile dyes and possible remediation approaches for environmental safety *Ecotoxicol. Environ. Saf.* **231** 113160

[9] Benkhaya S, M' rabet S and El Harfi A 2020 A review on classifications, recent synthesis and applications of textile dyes *Inorg. Chem. Commun.* **115** 107891

[10] Mahapatra N N 2016 *Textile Dyes* (Woodhead Publishing India PVT. Limited)

[11] Benkhaya S, Harfi S E and El Harfi A 2017 Classifications, properties and applications of textile dyes: a review *Appl. J. Environ. Eng. Sci.* **3** 311–20

[12] Abel A 2012 The history of dyes and pigments: from natural dyes to high performance pigments *Colour Design* ed J Best (Amsterdam: Elsevier) pp 557–87

[13] Hagan E and Poulin J 2021 Statistics of the early synthetic dye industry *Herit. Sci.* **9** 33

[14] Samsami S, Mohamadizaniani M, Sarrafzadeh M-H, Rene E R and Firoozbahr M 2020 Recent advances in the treatment of dye-containing wastewater from textile industries: overview and perspectives *Process Saf. Environ. Prot.* **143** 138–63

[15] Senthil Rathi B and Senthil Kumar P 2022 Sustainable approach on the biodegradation of azo dyes: a short review *Curr. Opin. Green Sustain. Chem.* **33** 100578

[16] Ben S H *et al* 2021 Diversity of synthetic dyes from textile industries, discharge impacts and treatment methods *Appl. Sci.* **11** 6255

[17] Zhao Y-G, Li X-P, Yao S-S, Zhan P-P, Liu J-C, Xu C-P, Lu Y-Y, Chen X-H and Jin M-C 2016 Fast throughput determination of 21 allergenic disperse dyes from river water using reusable three-dimensional interconnected magnetic chemically modified graphene oxide followed by liquid chromatography–tandem quadrupole mass spectrometry *J. Chromatogr.* A **1431** 36–46

[18] Simion Beldean-Galea M, Copaciu F-M and Coman M-V 2018 Chromatographic analysis of textile dyes *J. AOAC Int.* **101** 1353–70

[19] Lou C, Wu C, Zhang K, Guo D, Jiang L, Lu Y and Zhu Y 2018 Graphene-coated polystyrene-divinylbenzene dispersive solid-phase extraction coupled with supercritical fluid chromatography for the rapid determination of 10 allergenic disperse dyes in industrial wastewater samples *J. Chromatogr.* A **1550** 45–56

[20] Hakami A A H, Wabaidur S M, Ali Khan M, Abdullah Alothman Z, Rafatullah M and Siddiqui M 2020 Development of ultra-performance liquid chromatography–mass spectrometry method for simultaneous determination of three cationic dyes in environmental samples *Molecules* **25** 4564

[21] Ghasemi E and Kaykhaii M 2016 Application of a novel micro-cloud point extraction for preconcentration and spectrophotometric determination of azo dyes *J. Braz. Chem. Soc.* **27** 1521–6

[22] Ghasemi E and Kaykhaii M 2016 Application of micro-cloud point extraction for spectrophotometric determination of malachite green, crystal violet and Rhodamine B in aqueous samples *Spectrochim. Acta* A **164** 93–7

[23] Bişgin A T, Sürme Y, Uçan M and Narin İ 2016 Simultaneous spectrophotometric determination and column solid-phase extraction of two Lanaset textile dyes in environmental water samples *J. Ind. Eng. Chem.* **38** 186–92

[24] Hashemi S H, Kaykhaii M, Jamali Keikha A and Mirmoradzehi E 2019 Box-Behnken design optimization of pipette tip solid phase extraction for methyl orange and acid red determination by spectrophotometry in seawater samples using graphite based magnetic $NiFe_2O_4$ decorated exfoliated as sorbent *Spectrochim. Acta* A **213** 218–27

[25] Atsever N, Borahan T, Girgin A, Selali Chormey D and Bakırdere S 2021 A simple and effective determination of methyl red in wastewater samples by UV–Vis spectrophotometer with matrix matching calibration strategy after vortex assisted deep eutectic solvent based liquid phase extraction and evaluation of green profile *Microchem. J.* **162** 105850

[26] Hudari F F, Brugnera M F and Zanoni M V B 2017 Advances and trends in voltammetric analysis of dyes *Applications of the Voltammetry* ed M Stoytcheva and R Zlatev (InTech) pp 75–108

[27] Amrutha B M, Manjunatha J G, Bhatt A S, Raril C and Pushpanjali P A 2019 Electrochemical sensor for the determination of Alizarin Red-S at non-ionic surfactant modified carbon nanotube paste electrode *Phys. Chem. Res.* **7** 523–33

[28] Zainudin N S and Rahman N A A 2021 Determination of Reactive Black 5 (RB5) by differential pulse anodic stripping voltammetric technique *J Acad.* **9** 73–9

[29] Hareesha N, Manjunatha J G, Amrutha B M, Pushpanjali P A, Charithra M M and Prinith Subbaiah N 2021 Electrochemical analysis of indigo carmine in food and water samples using a poly(glutamic acid) layered multi-walled carbon nanotube paste electrode *J. Electron. Mater.* **50** 1230–8

[30] Honeychurch K C 2022 Voltammetric behaviour of Rhodamine B at a screen-printed carbon electrode and its trace determination in environmental water samples *Sensors* **22** 4631
[31] Kounaves S P 1997 Voltammetric techniques *Handbook of Instrumental Techniques for Analytical Chemistry* ed F A Settle (Prentice-Hall) pp 709–26
[32] Scholz F 2015 Voltammetric techniques of analysis: the essentials *ChemTexts* **1** 17
[33] Aleixo L M 2018 Voltametria: conceitos e técnicas *Rev. Chemkeys* **3** 1–21
[34] Moraes J T, Salamanca-Neto C A R, Švorc Ľ and Sartori E R 2017 Advanced sensing performance towards simultaneous determination of quaternary mixture of antihypertensives using boron-doped diamond electrode *Microchem. J.* **134** 173–80
[35] Mansano G R, Eisele A P P and Sartori E R 2015 Electrochemical evaluation of a boron-doped diamond electrode for simultaneous determination of an antihypertensive ternary mixture of amlodipine, hydrochlorothiazide and valsartan in pharmaceuticals *Anal. Methods* **7** 1053–60
[36] Coldibeli B, Campos N S M, Salamanca-Neto C A R, Scremin J, Mattos G J, Marcheafave G G and Sartori E R 2020 Feasibility of the use of boron-doped diamond electrode coupled to electroanalytical techniques for the individual determination of pravastatin and its association with acetylsalicylic acid *J. Electroanal. Chem.* **862** 113987
[37] Guaratini C C I, Fogg A G and Zanoni M V B 2001 Studies of the voltammetric behavior and determination of diazo reactive dyes at mercury electrode *Electroanalysis* **13** 1535–43
[38] Silva T A, Pereira G F, Fatibello-Filho O, Eguiluz K I B and Salazar-Banda G R 2016 Electroanalytical sensing of indigo carmine dye in water samples using a cathodically pretreated boron-doped diamond electrode *J. Electroanal. Chem.* **769** 28–34
[39] Radi A, Mostafa M R, Hegazy T A and Elshafey R M 2012 Electrochemical study of vinylsulphone azo dye Reactive Black 5 and its determination at a glassy carbon electrode *J. Anal. Chem.* **67** 890–4
[40] Radi A-E, Nassef H M and El-Basiony A 2013 Electrochemical behavior and analytical determination of Reactive Red 231 on glassy carbon electrode *Dye Pigment.* **99** 924–9
[41] Huang X, Zeng Z, Fan Z, Liu J and Zhang H 2012 Graphene-based electrodes *Adv. Mater.* **24** 5979–6004
[42] Wu S, He Q, Tan C, Wang Y and Zhang H 2013 Graphene-based electrochemical sensors *Small* **9** 1160–72
[43] He Q, Wu S, Yin Z and Zhang H 2012 Graphene-based electronic sensors *Chem. Sci.* **3** 1764
[44] Darabi R and Shabani-Nooshabadi M 2020 Determination of Acid Green 25 in the presence of Brilliant Blue in hair dye and wastewater using voltammetric sensor amplified with reduced graphene oxide/$CoFe_2$ *Environ. Technol. Innov.* **19** 101020
[45] Akbari S 2022 A new voltammetric sensor according to graphene quantum dots/ionic liquid modified carbon paste electrode for amaranth sensitive determination *Int. J. Environ. Anal. Chem.* **102** 789–803
[46] Moradpour H, Beitollahi H, Nejad F G and Di Bartolomeo A 2022 Glassy carbon electrode modified with N-doped reduced graphene oxide sheets as an effective electrochemical sensor for amaranth detection *Materials (Basel)* **15** 3011
[47] Gao Y, Li H, Tong J and Wang L 2021 A new voltammetric sensor based on poly (L-cysteine)/GR composite film modified electrode for the sensitive determination of amaranth in wastewater *Environ. Technol.* **42** 2385–90

[48] Shetti N P, Malode S J, Malladi R S, Nargund S L, Shukla S S and Aminabhavi T M 2019 Electrochemical detection and degradation of textile dye Congo red at graphene oxide modified electrode *Microchem. J.* **146** 387–92

[49] Revathi M and Kumar N S 2016 Voltammetric analysis of Reactive Blue 49 dye using graphene modified glassy carbon electrode as working electrode *J. Nanosci. Technol.* **2** 231–3

[50] Revathi M, Kavitha B, Vedhi C and Senthil Kumar N 2019 Electrochemical detection and quantification of Reactive Red 195 dyes on graphene modified glassy carbon electrode *J. Environ. Sci. Health Part* C **37** 42–54

[51] Jafari S, Dehghani M, Nasirizadeh N and Akrami H R 2017 Voltammetric determination of Basic Red 13 during its sonoelectrocatalytic degradation *Microchim. Acta* **184** 4459–68

[52] Jafari F, Nasirizadeh N and Mirjalili M 2020 Enhanced degradation of reactive dyes using a novel carbon ceramic electrode based on copper nanoparticles and multiwall carbon nanotubes *Chinese J. Chem. Eng.* **28** 318–27

[53] Khosropour H, Rezaei B and Ensafi A A 2019 A selective and sensitive detection of residual hazardous textile dyes in wastewaters using voltammetric sensor *Microchem. J.* **146** 548–56

[54] Hudari F F, Costa Ferreira S L and Zanoni M V B 2016 Multi-responses methodology applied in the electroanalytical determination of hair dye by using printed carbon electrode modified with graphene *Electroanalysis* **28** 1085–92

Chapter 4

Graphene electrochemical sensors for nucleotides

R Rajkumar, G Kavitha, S Parameswari, P Karpagavinayagam and C Vedhi

Nanomaterial-based sensors are being used more often in a wide range of industries, from straightforward wearable or medical sensors that can be used at home to monitor health to more complex sensors used by the aviation or border control industries. Because they are more portable, less expensive, and easier to use than traditional chromatography methods, nanoparticle-based sensors have just started to change drug detection methods. When compared to other nanomaterials, thin graphene films offer a much higher surface-to-weight ratio. This property has prompted the development of more sensitive and dependable sensors. With applications in environmental monitoring, medical diagnostics, and industries, graphene-based sensors have become one of the most well-known and extensively researched sensing materials of the past 20 years thanks to the exceptional properties of graphene and its potential to be tuned to target particular molecules. This chapter introduces many methods for producing graphene and making electrodes modified with graphene for electrochemical sensors. Also reviewed and contrasted are new research findings using various graphene-based materials as an electrochemical platform for the detection of various biomolecules and compounds. In the near future, further electrochemical research on this unique material should become available.

4.1 Introduction

It can be argued that the properties of the two-dimensional (2D) materials are of particular relevance to the detection of psychoactive drugs due to the common requirement for miniaturized systems that allow on-site, discrete drug monitoring. Over the years, 2D materials have been used to monitor a vast range of analytes, across a variety of disciplines. With potential drug analytes spanning legal, illicit,

doi:10.1088/978-0-7503-5578-0ch4

and medical drug compounds, the need for accurate, portable drug testing is typically assumed to be relevant to forensic work, but it is also highly relevant to clinical settings and workplace drug testing, especially for industries where safety is a top priority [1].

Traditional drug analysis techniques, such as chromatography [2], immunochromatography [3], mass spectrometry (MS) [4], nuclear magnetic resonance (NMR) [5], ultraviolet–visible spectroscopy (UV–vis) [6], Fourier-transform infrared spectroscopy (FTIR) and IR [7], surface-enhanced Raman spectroscopy (SERS) and Raman [8], circular dichroism spectroscopy (CD) (new techniques that can deliver quick, focused replies while being affordable and simple to use are still needed) [11, 12]. In some cases, next-generation 2D material-based sensors could take the place of these traditional techniques and do away with their shortcomings.

Here, we look at some of the 2D materials, including graphene, that have potential for use in the electrochemical detection of recreational drugs, as well as current developments in those sensors [9, 10].

4.2 Graphene-based electrochemical sensors

2D graphene has received the greatest attention among the many nanostructures [13]. This particular nanomaterial has sp^2-bonded carbon atoms with interactions, forming a thin layer with a densely conjugated carbon-based honeycomb structure [14–16]. Researchers have been very interested in this type of carbon-based material for the past 20 years due to its properties, which include quick electron transfer, electrical and thermal conductivity, mechanical strength, biocompatibility, low density, and high surface area [17–19]. Numerous types of electrochemical sensors have been developed using graphene [20, 21]. Functional groups can be added to graphene to improve its properties in any desired way, from increasing stability to making it interact solely with specific molecules. Over the years, several graphene-based materials have been studied in order to create sensors. Researchers have used a variety of techniques over the years to create improved sensors, including homogeneous mixing with polymers or metal organic frameworks, doping, bonding, or simply decorating graphene with metals or simple metallic nanoparticles, anchoring selective groups to the numerous functional groups (–OH, C–O–C, and –COOH), on graphene oxide (GO), and directly forming a bond between reduced graphene oxide (rGO) and another organic molecule through chemical reactions.

4.3 Graphene-based nanomaterials as a biosensor

In general, sensors are made up of a receptor and a transducer (figure 4.1). The chemical or inorganic component known as a receptor interacts specifically with the target molecule. The desired molecule could be an organic, inorganic, or even an entire cell. The transducer is the component of the sensor that transforms chemical data into a quantifiable signal. Biosensors that translate the interactions between the receptor and the target molecules into detectable measurements use graphene-based nanomaterials as transducers [22]. The bioreceptor (molecules like antibodies, ssDNA, and enzymes), which is required for this to happen, must be connected to

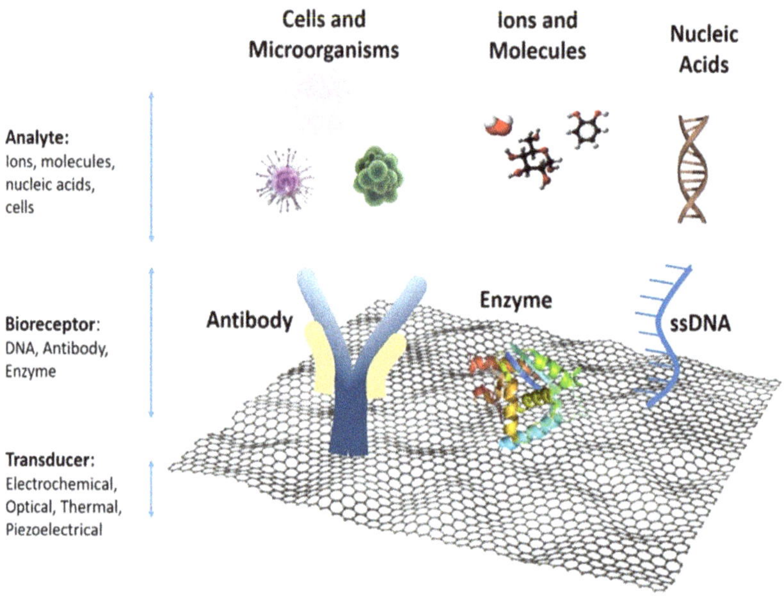

Figure 4.1. Examples of biosensors and components on a graphene platform. Reproduced from [51]. CC BY 4.0.

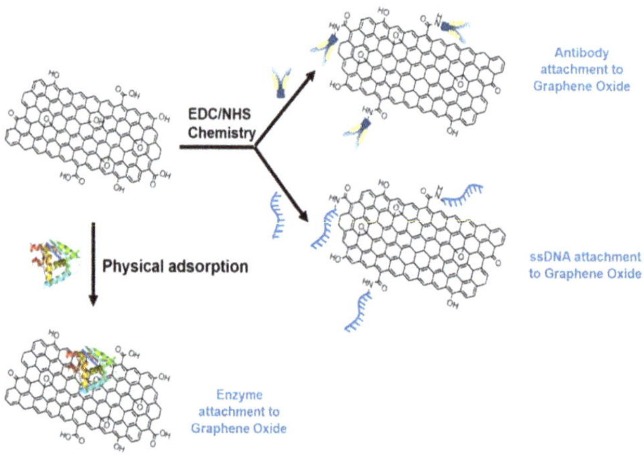

Figure 4.2. Schematic of the most common attachment methods of bioreceptors, such as antibody, DNA and enzymes onto graphene surfaces. Reproduced from [51]. CC BY 4.0.

the transducer surface. EDC/NHS chemistry is the most often used way for attaching antibodies and ssDNA to graphene and its derivatives (GO, rGO), whereas physisorption is the most frequently used method for immobilizing enzymes (figure 4.2).

Due to its extensive surface area, electrical conductivity, rapid electron transfer rate, and ability to immobilize a variety of compounds, graphene has been used in the creation of numerous biosensors for diverse transduction types [23]. As an illustration, the conjugated structure of graphene can enhance the flow of electrons between the bioreceptor and transducer, which can result in high signal sensitivity for electrochemical sensors [24, 25]. Additionally, graphene-based nanomaterials have the ability to function as a quencher in the transducer to produce fluorescent biosensors. According to research, graphene (G), GO, and rGO all exhibit extremely high fluorescence quenching efficiencies [26–28].

Some graphene features impacting the detection limit of the target molecules need to be taken into account when employing graphene nanoparticles to create sensors. For instance, different batches of graphene and its derivatives, as well as various synthetic techniques, can result in diverse graphene-based nanomaterials' characteristics and functions in the biosensors. The selectivity and sensitivity of the biosensors can also be directly impacted by the orientation of the G, GO, or rGO sheets in relation to the bioreceptor. Additionally, the number of layers, functional groups, and oxidation states of graphene and its derivatives will affect the bonding between the transducer and the bioreceptor as well as the sensing ability of various sensors. The interactions and detection limit of the target molecule can both be impacted by the amount of functional groups on the nanomaterials. To avoid the unintended attachment of biomolecules rather than the intended target molecules, it is frequently important in this context to block any nonspecific adsorption sites on the nanomaterial. This can be achieved by coating the sensor with a blocking agent, such as casein, superblock, or bovine serum albumin (BSA) [29], or by using tween surfactant. These restrictions must be taken into account in order for graphene-based nanomaterial biosensors to have high sensitivity, stability, and response times, which could lead to improvements in treatment and diagnosis [30].

4.4 Graphene-based nanomaterials and deoxyribonucleic acid (DNA)

Deoxyribonucleic acid (DNA) is a biomolecule that is ideally suited for biosensor technologies due to its wide variety of physical, chemical, and biological properties. DNA's flexibility, ease of synthesis, simple chemistry to connect to many platforms, ease of regeneration, and high specificity due to distinctive nucleotide sequences are some of its most important qualities as a biosensor [31, 32]. However, DNA biosensors have been found to have both benefits and drawbacks. High specificity, real-time analytical capabilities, tiny measurement systems, and the capacity to do multiplex measurements of many targets are all significant benefits of DNA biosensors [33, 34]. However, one of the main drawbacks of DNA biosensors is that DNA is easily destroyed, necessitating particular storage and analysis settings, such as certain media or a buffer, to keep the DNA stable and preserve its attachment to the transducer. Temperature and pH variations can also have an impact on the performance of DNA-based sensors [35]. For instance, the sensitivity of DNA biosensors is dependent on experimental temperatures since the hybridization event between the probe and the target molecules will only take place at the

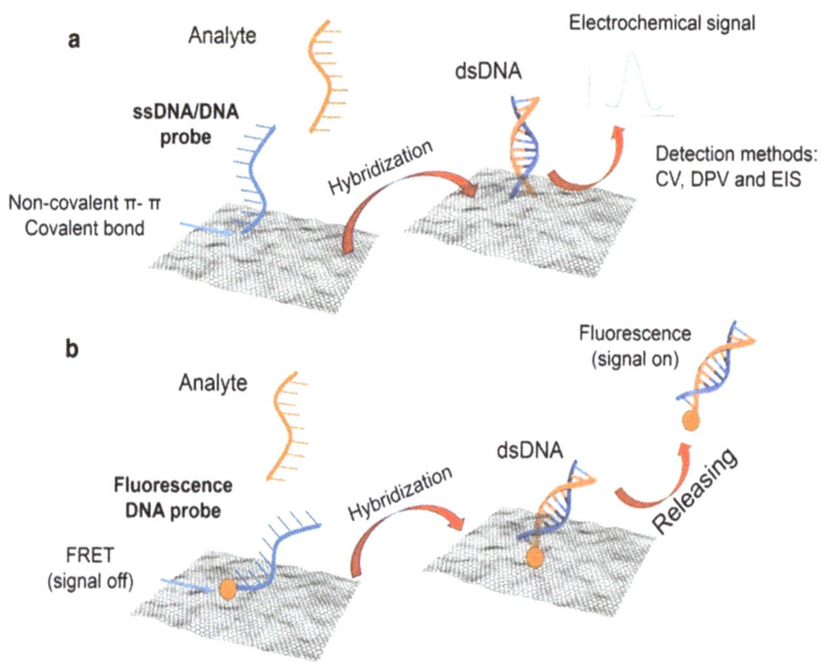

Figure 4.3. Scheme of graphene-based nanomaterials as a DNA biosensor. Electrochemical detection (a) and fluorescent detection (b). Reproduced from [51]. CC BY 4.0.

optimal temperatures that must be established before the sensor is put into use. In the case of pH, the current response exhibits the maximum signal at pH 7.0, and virtually no signal at pH below 7.0. As a result, a buffer containing potassium or sodium phosphate is required to improve the sensor's performance [36, 37]. Despite these drawbacks, nucleic acids have become more popular in the disciplines of biosensors and biological assays due to their uses in genetics, infectious diseases, and the clinical detection of pathogens [38]. Electrochemical sensors and fluorescence sensors are the two major types of sensors used in DNA biosensors that use graphene-based nanomaterials as transducers. The electrochemical sensor is based on measurements of changes in voltage, current, or impedance that can be brought on by electrochemical factors, such as electron loss, conductivity, or capacitance changes, which are brought on by DNA hybridization or the oxidation of adenine (A), thymine (T), cytosine (C), and guanine (G) in the DNA. Through the use of cyclic voltammetry (CV), differential pulse voltammetry (DPV), or electrochemical impedance spectroscopy (EIS), these biosensors can generate electrochemical signals that can be detected [39]. Utilizing interactions on the surface of nanomaterials based on graphene, the electrochemistry method immobilizes DNA (figure 4.3).

To connect covalently with the DNA, one can also utilize G edges, GO, or rGO with their functional groups (carboxylic, hydroxyl, and epoxide groups) [40]. EDC/NHS is the chemical most frequently employed to immobilize DNA on graphene-based nanomaterials; it is thoroughly explained in the antibody section. The majority of the research on enhancing the sensitivity and selectivity of electrochemical biosensors has been focused on transducer modification. For instance, utilizing the DPV approach at pH 7.0, the original glassy carbon electrode (GCE) can be changed with GO to allow for the direct detection of A, T, G, and C for dsDNA or ssDNA [41]. In a further investigation, DNA probes and rGO were added to the GCE to enable it to hybridize with a target DNA that could then be identified by EIS or CV. The vast surface area and excellent conductivity of rGO are utilized in this investigation. Another study looked at a DNA sensor that could detect dsDNA with sensitivity ranging from 0.1 fM to 10 mM by using the active and sharp edges of reduced graphene nanowalls (RGNW). According to this study's authors, the RGNW sheet's active edge locations may improve electron transport between DNA and the electrode in the DPV more uniformly [42]. The sensor's detection range and sensitivity can vary depending on the target and the detecting material.

The hybridization of two single-stranded DNA molecules is the foundation of the fluorescent DNA nanosensor. The complementary DNA for the target DNA is one of the ssDNAs, which is marked with a fluorescent dye. In order to improve the visualization and detection of the target ssDNA, this approach makes use of the optical quenching property of graphene-based materials. The direct adsorption of the DNA probe on the graphene-based surface allows for the immobilization of fluorescently-labeled DNA due to the interaction between the ring structure of the DNA bases and the graphene surface.

The GO-based sensor is one type of fluorescence biosensor that has been created. Multicolored DNA probes have been used in the production of this sensor to identify various DNA sequences. When hybridization takes place, our multiplex GO-based DNA sensor exhibits excellent emission signal from particular targets and negligible background fluorescence [43]. Fluorescence resonance energy transfer, often known as FRET or Förster, is a popular use of the fluorescence sensing strategy that can also make use of materials based on graphene. Using FRET, the fluorescently labeled DNA probe is first quenched to the surface of the graphene-based nanomaterials in this detection approach, turning off the fluorescent signal (figure 4.3).

The fluorescent molecule is released from the graphene surface together with the target DNA upon hybridization of the probe with the target DNA, and the fluorescent signal is activated for optical detection. For instance, a nanocomposite of gold nanoparticles (AuNPs) and GO was made and used as a quencher with the usage of fluorescent carbon dots (CDs) and a DNA probe, also known as nano quencher, in the endeavor to develop a reliable, biocompatible, and scalable biosensor for HIV-1 detection. The CDs/AuNPs/GO nanoprobe also employed the FRET technique. The fluorescent signal activates and hybridization takes place in the presence of target ssDNA. In the absence of the target DNA, AuNPs on the GO nanosheets act as a quencher for CD fluorescence. In DNA biosensors, AuNPs/GO demonstrates remarkable selectivity and sensitivity [44]. The detection limit of this sensor is 15 fM. Different graphene-based composites have been utilized to obtain the requisite sensitivity in the search for the ideal sensor. For instance,

Table 4.1. Graphene-based DNA biosensors with fluorescent detection. Reproduced from [51]. CC BY 4.0.

Detected element	Sensing material	Detection range	Refs.
ssDNA	GO	Detection limit 200 nM	[45]
ssDNA	GO and exonuclease III	Detection limit 20 pM	[46]
ssDNA	GO	200 nM	[47]
DNA and exonuclease activity	GO ethidium bromide (EB)	50–2500 nM Detection limit 32 nM	[48]
Staphylococcus aureus DNA	GO–DNA sensor	0.0125–3.125 nM Detection limit at 0.006 25 nM	[49]
ssDNA	Exonuclease III (ExoIII) and GO	Detection limit 0.5 pM	[50]
HIV-1 gene	AuNPs/GO nanocomposite	50.0 fM to 1.0 nM Detection limit at 15 fM	[44]
ssDNA	GO	0–25 nM Detection limit at 100 pM	[43]
T antigen gene of SV40 DNA	GO	40.0–260 nM Detection limit at nM	[43]

employing target recycling exonuclease III, ssDNA can be found using a fluorescent graphene sensor with a sensitivity as low as 0.5 pM (table 4.1). The summary of various works using graphene-based nanomaterials' quenching properties to boost or improve the fluorescence detection of DNA biosensors is shown in table 4.1.

4.5 DNA hybridization on graphene electrode

After the graphene electrodes were functionalized with a DNA probe and hybridized with a corresponding DNA target, changes were detected using electrochemical techniques. The peak current intensity and R_{ct} of the ferri/ferrocyanide redox-active species from the test electrolyte significantly changed for the samples that used higher graphene concentrations, such as the 0.2 and 0.4 mg ml^{-1} shown in figure 4.4, after immobilizing the DNA probe on the surface of the GCE modified with graphene. This phenomenon could be attributed to the negatively charged phosphate backbone of immobilized DNA, which provided $[Fe(CN)_6]^{3-/4-}$ with an electrostatic repulsive force. Additionally, a thicker graphene sheet promoted the buildup of immobilized DNA, which may potentially improve the biosensor's sensitivity. The samples that used larger graphene concentrations had enhanced signal alterations and a higher sensitivity as a result. The fact that the electrodes with graphene bases at 0.2 and 0.4 mg ml^{-1} showed a substantial response while those with other graphene concentrations showed little to no change in signal implied that more complementary DNA was adsorbed on those electrodes.

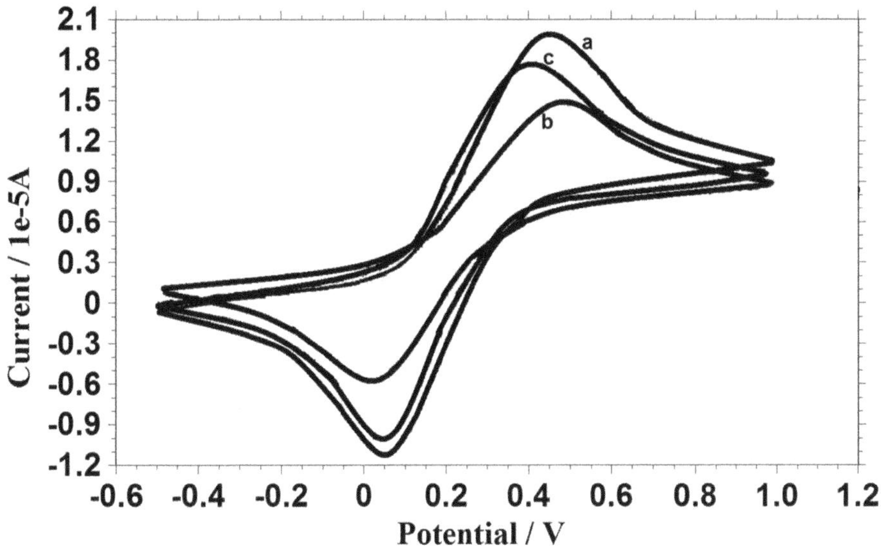

Figure 4.4. Cyclic voltammetric behavior of (a) 1 mM $[Fe(CN)_6]^{3-/4-}$, 0.1 M KCl, (b) 0.2 mg ml^{-1} graphene, after adsorption of 10 μM DNA, (c) 0.4 mg ml^{-1} graphene, after adsorption of 10 μM DNA.

4.6 Conclusion

In this brief overview, we summarize recent research on graphene and graphene-related biosensors that may have uses in medicine and the life sciences. We display the findings of the reported analytical performance of each sensor and highlight their application in the disciplines of the biological sciences and medicine. This study has discussed DNA, antibody, and enzyme-based biosensors because each has benefits and drawbacks. Overall, the kind of application will determine the sort of sensor that is chosen. For instance, using DNA in biosensing technologies can be an affordable way to quickly find cancer signs or microorganisms. The use of antibodies or enzymes in biosensors, however, can be more efficient in the detection or monitoring of some disorders due to the wide variety of molecules present in the body. Antibodies, for example, can be used to specifically detect viruses like the Zika virus, HIV, and influenza A virus, among others. On the other hand, enzymes have demonstrated a high potential for detecting glucose levels in only little volumes of material. Overall, due to its high surface area, electrical conductivity, electron transfer rate, and ability to encapsulate a range of different proteins, graphene and graphene-based nanomaterials have shown considerable promise in biosensor technologies. For usage in clinical settings, it is promising to design biosensors that are quick to process, sensitive, stable, and selective to their target molecule. However, much more research assessing the security and dependability of the sensors needs to be carried out in order to achieve uniform and dependable results and generate biosensors suitable for usage in the medical industry.

Despite the fact that graphene is a superior electrode material for use in medical sensing applications, unique techniques for the careful production and processing of graphene require more attention and should be researched in subsequent studies. Targeting particular analytes is successful with the present chemical methods for biomolecules to change the graphene surface. To prevent the adsorption of undesirable molecules on graphene and to enhance the orientation of biomolecules on graphene platforms, the sensing platform may still be improved. Therefore, improved knowledge of the physics and chemistry at the graphene surface as well as interactions with biomolecules at the interface will be crucial for developing graphene-based nanosensors. Additionally, the development of dependable, repeatable, and cost-effective sensors with high accuracy, sensitivity, and specificity is required for the miniaturization and manufacture of compact biosensors for diagnostic purposes. For use in remote locations during emergencies, some of these sensors must be made more affordable. The sensors' downsizing can also enable quick detection of bacterial and viral infections, as well as their application in self-monitoring biological implants to catch significant medical disorders. The aforementioned uses of the biological sciences will help safeguard lives and enhance human health. To confirm, guarantee, and corroborate the biocompatibility and non-toxicity of graphene-based nanomaterials, however, so that their prolonged use does not pose any health risks, much work still has to be done.

References

[1] Agius R and Kintz P 2010 Guidelines for European workplace drug and alcohol testing in hair *Drug Test. Anal.* **2** 367–76

[2] Plumb R S, Stumpf C L, Granger J H, Castro-Perez J, Haselden J N and Dear G J 2003 Use of liquid chromatography/time-of-flight mass spectrometry and multivariate statistical analysis shows promise for the detection of drug metabolites in biological fluids *Rapid Commun. Mass Spectrom.* **17** 2632–8

[3] Li W, Li X, Yang T, Guo X and Song Y 2020 Detection of saliva morphine using surface-enhanced Raman spectroscopy combined with immunochromatographic assay *J. Raman Spectrosc.* **51** 642–8

[4] Paul R, Tsanaclis L, Murray C, Boroujerdi R, Facer L and Corbin A 2019 Ethyl glucuronide as a long-term alcohol biomarker in fingernail and hair. Matrix comparison and evaluation of gender bias *Alcohol Alcohol.* **54** 402–7

[5] Peng S X 2000 Hyphenated HPLC-NMR and its applications in drug discovery *Biomed. Chromatogr.* **14** 430–41

[6] Ryu J, Manivannan R and Son Y-A 2021 Azo dye-based optical probe for the detection toward mimic molecule of date rape drug *Microchem. J.* **166** 106237

[7] McCrae K, Tobias S, Grant C, Lysyshyn M, Laing R, Wood E and Ti L 2020 Assessing the limit of detection of Fourier-transform infrared spectroscopy and immunoassay strips for fentanyl in a real-world setting *Drug Alcohol Rev.* **39** 98–102

[8] Wang H, Xue Z, Wu Y, Gilmore J, Wang L and Fabris L 2021 Rapid SERS quantification of trace fentanyl laced in recreational drugs with a portable Raman module *Anal. Chem.* **93** 9373–82

[9] Uchiyama N, Kikura-Hanajiri R, Shoda T, Fukuhara K and Goda Y 2011 Isomeric analysis of synthetic cannabinoids detected as designer drugs *J. Pharm. Soc. Jpn.* **131** 1141–7

[10] Ding Y, Li X, Guo Y, Yan J, Ling J, Li W, Lan L, Chang Y, Cai J and Zha L J A 2017 Rapid and sensitive detection of ketamine in blood using novel fluorescence genosensor *Anal. Bioanal. Chem.* **409** 7027–34

[11] Cinti S, Basso M, Moscone D and Arduini F 2017 A paper-based nanomodified electrochemical biosensor for ethanol detection in beers *Anal. Chim. Acta* **960** 123–30

[12] Zheng D, Hu H, Liu X and Hu S 2015 Application of graphene in electrochemical sensing *Curr. Opin. Colloid Interface Sci.* **20** 383–405

[13] Nodehi M, Baghayeri M and Veisi H J T 2021 Preparation of GO/Fe_3O_4@PMDA/AuNPs nanocomposite for simultaneous determination of As^{3+} and Cu^{2+} by stripping voltammetry *Talanta* **230** 122288

[14] Boroujerdi R, Abdelkader A and Paul R 2020 State of the art in alcohol sensing with 2D materials *Nano-Micro Lett.* **12** 33

[15] Akça A, Karaman O and Karaman C 2021 Mechanistic insights into catalytic reduction of N_2O by CO over Cu-embedded graphene: a density functional theory perspective *ECS J. Solid State Sci. Technol.* **10** 041003

[16] Karaman C, Karaman O, Atar N and Yola M 2021 Tailoring of cobalt phosphide anchored nitrogen and sulphur co-doped three-dimensional graphene hybrid: Boosted electrocatalytic performance towards hydrogen evolution reaction *Electrochim. Acta* **380** 138262

[17] Karaman C, Karaman O, Atar N and Yola M L 2021 Electrochemical immunosensor development based on core–shell high-crystalline graphitic carbon nitride@ carbon dots and $Cd_{0.5}Zn_{0.5}S$/d-$Ti_3C_2T_x$MXene composite for heart-type fatty acid–binding protein detection *Microchim. Acta* **188** 182

[18] Boroujerdi R and Paul R 2022 Introducing graphene–indium oxide electrochemical sensor for detecting ethanol in aqueous samples with CCD-RSM optimization *Chemosensors* **10** 42

[19] Boroujerdi R, Abdelkader A and Paul R 2022 Highly selective detection of ethanol in biological fluids and alcoholic drinks using indium ethylenediamine functionalized graphene *Sens. Diagn.* **1** 566–78

[20] Shao Y, Wang J, Wu H, Liu J, Aksay I A and Lin Y 2010 Graphene based electrochemical sensors and biosensors: a review *Electroanalysis* **22** 1027–36

[21] Sun D, Luo Y, Debliquy M and Zhang C 2018 Graphene-enhanced metal oxide gas sensors at room temperature: a review *Beilstein J. Nanotechnol.* **9** 2832–44

[22] Pumera M 2011 Graphene in biosensing *Mater. Today* **14** 308–15

[23] Rao C N R, Sood A K, Subrahmanyam K S and Govindaraj A 2009 Graphene: the new two-dimensional nanomaterial *Angew. Chem. Int. Ed.* **48** 7752–77

[24] Park C S, Yoon H and Kwon O S 2016 Graphene-based nanoelectronic biosensors *J. Ind. Eng. Chem.* **38** 13–22

[25] Kuila T, Bose S, Khanra P, Mishra A K, Kim N H and Lee J H 2011 Recent advances in graphene-based biosensors *Biosens. Bioelectron.* **26** 4637–48

[26] Kasry A, Ardakani A A, Tulevski G S, Menges B, Copel M and Vyklicky L 2012 Highly efficient fluorescence quenching with graphene *J. Phys. Chem.* C **116** 2858–62

[27] Batır G G, Arık M, Caldıran Z, Turut A and Aydogan S 2018 Synthesis and characterization of reduced graphene oxide/rhodamine 101 (rGO-Rh101) nanocomposites and their heterojunction performance in rGO-Rh101/p-Si device configuration *J. Electron. Mater.* **47** 329–36

[28] Wu X, Xing Y, Zeng K, Huber K and Zhao J X 2018 Study of fluorescence quenching ability of graphene oxide with a layer of rigid and tunable silica spacer *Langmuir* **34** 603–11

[29] Cheng S, Hideshima S, Kuroiwa S, Nakanishi T and Osaka T 2015 Label-free detection of tumour markers using field effect transistor (FET)- based biosensors for lung cancer diagnosis *Sensors Actuators* B **212** 329–34

[30] Wang F, Horikawa S, Hu J, Wikle H, Chen I H, Du S, Liu Y and Chin B A 2017 Detection of *Salmonella typhimurium* on spinach using phage-based magnetoelastic biosensors *Sensors* **17** 386

[31] Premkumar T and Geckeler K E 2012 Graphene–DNA hybrid materials: assembly, applications, and prospects *Prog. Polym. Sci.* **37** 515–29.

[32] Pedersen R *et al* 2014 Properties of DNA *Handbook of Nanomaterials Properties* (Berlin: Springer) pp 1125–57

[33] Recent K R 2017 Advances and applications of biosensors in novel technology *Biosens. J.* **06** 1–12

[34] Bora U, Sett A and Singh D 2013 Nucleic acid-based biosensors for clinical applications *Biosens. J.* **02** 1–8

[35] Koyun A, Ahlatcolu E and Koca Y 2012 Biosensors and their principles *A Roadmap of Biomedical Engineers and Milestones* (InTech)

[36] Kavita V 2017 DNA biosensors—a review *J. Bioeng. Biomed. Sci.* **7** 222

[37] Rahman M, Heng L Y, Futra D, Chiang C P, Rashid Z A and Ling T L 2017 A highly sensitive electrochemical DNA biosensor from acrylic-gold nano-composite for the determination of arowana fish gender *Nanoscale Res. Lett.* **12** 484

[38] Ray M, Ray A, Dash S, Mishra A, Achary K G, Nayak S and Singh S 2017 Fungal disease detection in plants: traditional assays, novel diagnostic techniques and biosensors *Biosens. Bioelectron.* **87** 708–23

[39] Liu A, Wang K, Weng S, Lei Y, Lin L, Chen W, Lin X and Chen Y 2012 Development of electrochemical DNA biosensors *Trends Anal. Chem.* **37** 101–11

[40] Song Y, Luo Y, Zhu C, Li H, Du D and Lin Y 2016 Recent advances in electrochemical biosensors based on graphene two-dimensional nanomaterials *Biosens. Bioelectron.* **76** 195–212

[41] Zhou M, Zhai Y and Dong S 2009 Electrochemical sensing and biosensing platform based on chemically reduced graphene oxide *Anal. Chem.* **81** 5603–13

[42] Akhavan O, Ghaderi E and Rahighi R 2012 Toward single-DNA electrochemical biosensing by graphene nanowalls *ACS Nano* **6** 2904–16

[43] He S *et al* 2010 A graphene nanoprobe for rapid, sensitive, and multicolor fluorescent DNA analysis *Adv. Funct. Mater.* **20** 453–9

[44] Qaddare S H and Salimi A 2017 Amplified fluorescent sensing of DNA using luminescent carbon dots and AuNPs/GO as a sensing platform: a novel coupling of FRET and DNA hybridization for homogeneous HIV-1 gene detection at femtomolar level *Biosens. Bioelectron.* **15** 773–80

[45] Liu F, Choi J Y and Seo T S 2010 Graphene oxide arrays for detecting specific DNA hybridization by fluorescence resonance energy transfer *Biosens. Bioelectron.* **25** 2361–5

[46] Zhao X H, Ma Q J, Wu X X and Zhu X 2012 Graphene oxide-based biosensor for sensitive fluorescence detection of DNA based on exonuclease III-aided signal amplification *Anal. Chim. Acta* **727** 67–70

[47] Liu C, Wang Z, Jia H and Li Z 2011 Efficient fluorescence resonance energy transfer between upconversion nanophosphors and graphene oxide: a highly sensitive biosensing platform *Chem. Commun.* **47** 4661–3

[48] Jiang Y, Tian J, Chen S, Zhao Y, Wang Y and Zhao S 2013 A graphene oxide- based sensing platform for the label-free assay of DNA sequence and exonuclease activity via long range resonance energy transfer *J. Fluoresc.* **23** 697–703
[49] Pang S, Gao Y, Li Y, Liu S and Su X 2013 A novel sensing strategy for the detection of *Staphylococcus aureus* DNA by using a graphene oxide-based fluorescent probe *Analyst* **138** 2749–54
[50] Peng L, Zhu Z, Chen Y, Han D and Tan W 2012 An exonuclease III and graphene oxide-aided assay for DNA detection *Biosens. Bioelectron.* **35** 475–8
[51] Peña-Bahamonde J, Nguyen H N, Fanourakis S K and Rodrigues D F 2018 Recent advances in graphene-based biosensor technology with applications in life sciences *J. Nanobiotechnol.* **16** 75

Chapter 5

Graphene-modified electrochemical sensors for estimation of food contaminants

B Chethan, V Prasad, H G Rajprakash, A Sunilkumar and S Thomas

In the present situation, it is essential to detect food pollutants to maintain the excellent health of humanity. In this regard, electrochemical sensors work effectively to sense food contaminants. The primary food contaminants are bio-molecules and food pollutants. Nowadays, two-dimensional (2D) materials are extensively used to fabricate the sensing device. Among these 2D materials, graphene, its derivatives and their composites are highly workable materials. The chapter extensively studies graphene derivatives and their composites-based electrochemical sensors to detect food pollutants. The present chapter focuses on the organic and inorganic pollutants contaminating food. The organic contaminants like food colouring agents, pesticides and drugs, and the inorganic pollutants like metal cations and anions influence contaminating food is deliberately discussed. The use of the nanocomposites and derivatives of graphene to develop sensing material is briefly outlined. In the last section, the challenges faced and the conclusion are sketched.

5.1 Introduction

Food is the primary source of energy for human beings and living organisms. Getting nutrient and pollutant-free food is the fundamental right of every individual [1–11]. Food pollution and adulteration occur naturally and artificially in every food production and processing step. The presence of pollutants damages the health condition of human beings. According to the World Health Organization, contaminated and adulterated food may lead to more than 200 diseases [12–18]. The main reason for food contamination is a polluted ecosystem. Hence, identifying, detecting and controlling food pollutants is challenging. Food security is essential to secure human beings' health and increase the standard of life [19].

doi:10.1088/978-0-7503-5578-0ch5

Today many analytical techniques, namely chromatography, spectrophotometry, and electrochemical techniques, are more popular to estimate and determine food pollution [20–22]. These techniques are precise, sensitive and selective for detecting and determining food pollutants. But these techniques find more disadvantages like high cost, sophisticated instrumentation, miniaturisation not being possible, and multiple pretreatments required [22]. So, the scientific community finds many challenges to fabricating low-cost, reliable, versatile, miniaturised, and accurate methods to detect pollution in food products. With great anticipation, the research community has proposed the electrochemical sensor. Electrochemical sensors have found significant applications in ecosystem monitoring, agriculture, and industrial applications in recent decades [23, 24]. The great interest in electrochemical sensors is due to their trustable sensitivity, timing characteristics, better selectivity and portability.

Electrochemical sensors work on the principle of the redox process to detect the electroactive analyte [25, 26]. The changes in the physical impulse are determined by measuring the changes in the applied potential. Usually, many researchers are working on metals and metal oxide-based sensors, but graphene, a carbon-based material, gives more promising applications than the rest of the other materials because of its astonishing applications, such as rapid timing behaviour, multiple detection ability and over potential. Compared to other 2D materials (carbon nanotubes (CNTs), graphite and fullerene), graphene-based materials have more potential [27]. Usually, in the electrochemical process, graphene nanosheets attach to the graphitic electrodes through a π–π interaction [28]. This interaction negatively affects the sensing ability of the graphene system. Because of these reasons, the research community has increased their studies on graphene oxide (GO) and reduced graphene oxide (RGO) materials. GO and RGO contain hydroxyl, carboxyl and epoxy functional groups, forming a good interaction with the analyte, facilitating the easier hopping of the electrons [29]. The polar functional groups also enhance the interlayer separation between the graphene nanosheets and convert it into a hydrophilic material. With all these efforts, graphene and its derivative-based material synthesise the electrochemical sensors for pollutant detection in food materials.

With these ideas, this chapter focuses on detecting particular types of pollutants. We discuss the graphene-based electrochemical sensor system for detecting organic contaminants like food colourants, drugs, and pesticides and inorganic pollutants like metal cations and anions in food items (figure 5.1). The last part of the chapter consists of the conclusion and the challenges faced in this area of research.

5.2 Graphene and its derivatives-based electrochemical sensors for organic food pollutants

5.2.1 Food colourants and preservatives

The colouring agents and preservatives are added to the food items to attract consumers. The colouring agents include two types, namely, natural and synthetic

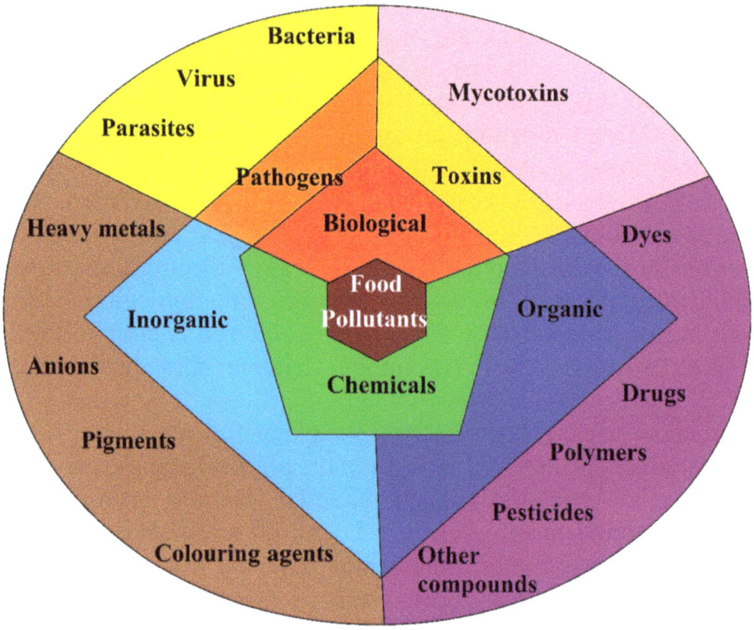

Figure 5.1. Food pollutants classification.

food colours. These artificial colouring agents are added to beverages, bakery items, ice cream and meat products to increase their flavour, colour and perishability [30, 31]. The natural colourants are extracted from plants, vegetables, fruit and animals. Usually, these natural colourants are non-toxic and safe. But the synthetic colouring agents are synthesised in artificial mode and are unrealistic.

5.2.1.1 Pure material for pollutants detection

Jampasa *et al* fabricated an RGO on screen-printed carbon electrode to detect sunset yellow and tartrazine in soft drinks, juice, and energy drink. At particular electrochemical conditions, the fabricated sensor has shown good sensitivity in detecting sunset yellow and tartrazine in food items. The fabricated sensor has seen a sunset yellow in the 0.01–20.0 mM concentration range and tartrazine concentration range from 0.02 to 20.0 mM [32].

5.2.1.2 Nanocomposites for pollutants detection

To enhance the sensitivity and timing behaviour and decrease the detection limit, the graphene-based composites are synthesised by compositing with polymers, metal nanoparticles and metal oxides. Many researchers have fabricated. The graphene-based sensor devices for the detection of pollutants in food products. Some of the reviews are provided in table 5.1.

Table 5.1. Graphene-based nanocomposites for various food pollutants detection.

Material	Dye detected	Detection limit	Food item	Ref.
Ag/graphene/screen-printed carbon electrodes (SPCEs)	Rhodamine B	1.9 μM	Crackers	[33]
Ag/RGO/paper electrode	Sudan I	41.3 nmol	Chilli powder	[34]
RGO/5-amino-1,3,4-thiadiazole-2-thiol/Pt/GCE	Orange II	3.4×10^{-10} M	Ketchup	[35]
NiO/Ag/RGO/GCE	Sunset Yellow	13 nM	Soft drinks	[36]
Poly(diallyldimethylammonium chloride) RGO/GCE	Quinoline Yellow	0.002 μM	Soft drink	[37]
Poly(diallyldimethylammonium chloride)–graphene/Ni/GCE	Allura Red	8.0 nmol l^{-1}	Strawberry juice	[38]

5.2.2 Pesticides

Chemicals like insecticides, fungicides, and herbicides are included in the category of pesticides [39]. The primary purposes of their addition to farming areas are to boost crop output and safeguard agricultural products from pests. Extensive use of pesticides in agriculture may result in health hazards such as neurological disorders, respiratory problems, bone marrow disorders, infertility, and carcinogenicity [40]. It has also been generally established that pesticides accumulate in food goods. Pesticides in food goods must be identified for safe food and healthy life.

5.2.2.1 Pure graphene material

Tasaltin *et al* published an article based on a fungicide called propamocarb that can be detected in cucumbers using an RGO/Au electrode. In contrast, this electrode does not respond to cypermethrin, deltamethrin, or malathion and only responds favourably to propamocarb. This electrode has good sensitivity and responds to propamocarb quickly within short intervals [41].

5.2.2.2 Doped graphene materials

Mousavi and Moradian and Hong *et al* have shown that the conductivity of host materials is typically improved by doping with impurity atoms at an ideal concentration [42, 43]. Chen *et al* reported that nitrogen (N)-doped holey graphene, a graphene derivative with nanopores on a basal plane, modified GCE to quantify methyl parathion. This study shows that the *N*-configurations (pyridinic N, pyrrolic N, and graphitic N) play an essential role in detecting methyl parathion (pH = 8). Doped holey graphene containing pyrrolic N exhibits a comprehensive dynamic response (1 ng mL to 150 g mL) and a meager detection limit (3.5 pg mL). This sensor's apple, grape, and cucumber samples exhibit respectable recoveries [44].

5.2.2.3 Graphene-based nanocomposites

Research on nanocomposites based on graphene derivatives as electrode modifying layers has also been found to be generally advantageous in pesticide detection. Nehru *et al* used Ce-doped TiO_2/GO/GCE as an example to detect methyl parathion [44, 45]. Nanocomposite electrode exhibits excellent sensitivity, a small detection limit (0.0016 M), and a broad dynamic range (0.002–48.327 M) for the measurement of methyl parathion. This composite electrode demonstrates excellent selectivity, stability, and real-world applications with positive recoveries. Notably, N-doped holey graphene/GCE has a picomolar level sensing capability towards methyl parathion [44].

Wang *et al* built an Ag/RGO/GCE-based pentachlorophenol sensor. This sensor responds to pentachlorophenol gradually in the 0.008–10.0 M range. By using this electrode, pentachlorophenol can be detected down to 0.001 M. With good recoveries (DPV, 94.0%–108.0%), this electrode successfully detects pentachlorophenol in vegetables. High-performance liquid chromatography (96.0%–103.3%) could also be used to verify this result. The appropriate surface area and electron-transfer capability of the Ag/RGO nanocomposite are credited with the improved sensing performance [46].

Ijaz *et al* showed that the insecticide paraquat (C12H14Cl2N2) is another significant one frequently used on crops like cotton and coconuts. However, research on the toxicity of paraquat has also been deliberately discussed [47].

Sant'Anna *et al* have created activated biochar/RGO/CPE for paraquat determination using the DPV approach. With a more excellent cathodic current responsiveness than bare CPE (pH = 6), the one-electron process reduces paraquat on this modified electrode. Oxygen and nitrogen chemical groups make paraquat interact more favourably, which improves detection. The detection limit, linearity, and sensitivity of activated biochar/RGO/CPE are 3.52 AL mol^{-1}, 0.74–9.82 mol^{-1}, and 23.1 nmol l^{-1}, respectively. This sensor could also be used successfully with honey, coconut water, lemon, and lettuce leaf samples [48].

5.2.3 Drugs

Antibiotics are examples of drugs that treat illnesses, soothe pain, and safeguard people's lives. However, their incorrect use has adverse effects on both people and animals. For instance, overuse of antibiotics will result in poor taste perception and inadequate nutrition in people [49, 50]. Sensors are being made to assess the presence of antibiotics in food samples [49, 50].

5.2.3.1 Pure graphene derivatives

Pan *et al*, by utilising differential pulse anodic stripping voltammetric methods, presented GO/SPCE for the detection of ciprofloxacin [51]. The complex that forms when ciprofloxacin and Mn^{2+} ions interact is the basis for the sensing technique. This technique has been performed on milk samples with recoveries ranging from 81.4% to 95.4% (1–8 M). In a study by de Faria *et al*, the detection threshold for the ciprofloxacin analysis was shown to be 0.3 M. An RGO/GCE sensor for chloramphenicol determination was explored in another study. The study's researchers utilised adsorption stripping differential pulse voltammetry as an analytical technique. This electrode's sensitivity is nearly ten times higher than plain GCE. Additionally, RGO/GCE has been used on milk samples (whole and skimmed) and has shown recovery rates of 93%–108% for the measurement of chloramphenicol [52].

5.2.3.2 Graphene derivatives-based nanocomposites

Many binary/ternary nanocomposites made from graphene and its derivatives have been employed as drug sensors. The following are the points derived from the earlier literature.

(a) Sensors with nano-molar detection limits could be easily identified by utilising graphene-based composites.
(b) The analytical performance of composite-modified electrodes is affected by drug structure.
(c) The sensing ability is also affected by the type of working electrodes.
(d) Most of the sensors have been tested in natural water and biological samples. They have also been found in food samples.
(e) Polymers or metal oxides have been discovered to be the best components to be added to graphene or its derivatives for the best sensing capability.

5.2.4 Other compounds

Willson *et al* showed that caffeine, or 1,3,7-trimethyl xanthine, is an active component in food. However, it causes various issues in people. Thus, its presence must be approximated as mandatory in food samples. As a result, a caffeine sensor was created [53]. Further, Shehata *et al* developed GO/reduced glutathione/CPE for the fast detection of caffeine in water. Caffeine was determined using DPV with a linear range of 8–800 M (detection limit = 0.153 M). This caffeine sensor's performance has been tested against black tea, coffee, Nescafe, and Cola with outstanding results [54].

Jose *et al* created a new caffeine sensor by altering the electrode material for GCE with nicotinic acid hydrazide/GO. This composite-modified GCE has a higher peak current for caffeine than bare GCE and a lower detection limit (8.7–10 9 M). This sensor also showed good recoveries for caffeine added to Imol plus and energy beverages. The detection ability of nicotinic acid hydrazine/GO is superior to that of GO/reduced glutathione/CPE [55].

Tian *et al* described a PtFe/graphene-based voltammetric sensor for detecting bisphenol A. De-alloying of PtFeAl ternary alloy results in nanoporous PtFe. Bisphenol A could be calculated linearly using this sensor in the 0.2–96 M range. The lowest concentration of bisphenol A detectable by PtFe/graphene is 0.17 M. Furthermore, a PtFe/graphene sensor was used to detect bisphenol A in natural water samples with high recoveries (96%–101%) [56].

Wang *et al* employed iodine-doped graphene/GCE for bisphenol A measurement. In a simulated solution (10 M bisphenol A in 0.1 M phosphate buffer solution, pH = 6), this electrode exhibits redox peaks with high peak currents and a lower detection limit (0.02 M). The electrode has been employed for bisphenol A sensing in milk samples with adequate recoveries (95%–110%) [57].

5.3 Electrochemical sensors based on graphene and its derivatives for inorganic food contaminants

5.3.1 Metal ions

Kharazi *et al* metal ions may reach the human body through the nutrient chain and deposit in tissues, producing severe health problems. As a result, graphene derivatives-based electrochemical detectors have been demonstrated to measure traces of metal ions precisely. In most cases, electrochemical build-up followed by a stripping technique will be used to quantify metal ions. The initial procedure is aggregation (or pre-enrichment), during which metal ions become embedded on the outermost layer of the working electrode by applying an appropriate negative voltage. The reduction process will deposit a specified concentration of metals at regular times [58].

The following procedure, stripping, includes oxidising (dissolving) the metal that has been applied as a layer to form equivalent metal ions. The peak current of anodic reactions will be commensurate to the amount of these ions containing metals. Highly dangerous metal cations such as mercury, lead, and cadmium ions were found in diverse food samples.

5.3.1.1 Pure graphene derivatives

Liendo *et al* built antimony (Sb^{3+}) sensors entirely from pristine eRGO. The surface area of the eRGO/GCE is 0.28 cm^2, which is greater than the surface area of the bare GCE (0.11 cm^2). The ideal pH, accumulation period, and voltage potential are 4.3, 150 s, and 1.0 V, respectively. This sensor performs well in determining Sb^{3+} ions in lettuce, celery, and canned drinks [59].

5.3.1.2 Graphene derivative-based nanocomposites

Ma *et al* demonstrated Ag/RGO-modified GCE for sensing Hg^{2+} ions. Hg^{2+} ions were placed for thirty seconds at 1.1 V (pH = 5.0). Exfoliating was done using DPV at possibilities ranging from 0 to +0.4 V. This electrode's wavelength detection limit and dynamic spectrum are 0.018 nM and 0.01–100 μg l^{-1}, respectively. The higher electrochemical surface space (0.13 cm^2) and the enhanced electron-transfer capabilities of the Ag/RGO composite contribute significantly to its sensing performance. With acceptable recoveries, this electrode also identifies Hg^{2+} ions in fish (head, body, and tail section) samples [60].

Wang *et al* discovered that sulphur-doped C_3N_4/graphene composites can detect trace quantities of Hg^{2+}, Cd^{2+}, and Pb^{2+} ions. This redesigned electrode has a detection threshold of 1.17, 0.38, and 0.61 nM for Cd2+, Pb^{2+}, and Hg^{2+} ions, respectively. This electrode's improved detection abilities are due to its multilayer permeable layer framework, distinctive tube bundle topology, and larger active centres [61].

Jeong *et al* recently developed graphene fibre using a laser-assisted technique for the simultaneous measurement of Cd^{2+} and Pb^{2+} ions. As the source and laser, an authorised polyimide film and a CO_2 laser (energy = 46 J cm^{-2}) were employed, respectively. Bismuth might additionally be electrochemically synthesised on graphene fibre. At pH = 4.5, the graphene fibre/Bi electrode exhibits an excellent linear response (1.0–140.0 g l^{-1}) and a meager detection limit (0.4 g l^{-1}) for both Cd^{2+} and Pb^{2+} ions [62].

Guo *et al* in another work investigated the RGO/MoS_2/chitosan/GCE as a Pb^{2+} ion sensor. This system's variable range and threshold for detection have been determined to be 0.005–0.05–2.0 and 0.0016 mM, correspondingly. This research found that this electrode may detect Pb^{2+} ions in tobacco leaves (sensitivity = 31.12 mA mM RGO and MoS_2 nanoflowers enhanced electrical conduction and a particular surface area, respectively [63].

5.3.2 Inorganic anions

Xu *et al* elevated quantities of inorganic anions in food, such as nitrite, sulphite, and others, are also hazardous to living creatures. The subsequent article covers electrochemical detectors for anions. Although a metal–graphene hybrid might improve electrochemical capabilities, current combinations cannot overcome the limitations of their intrinsic properties because metals and graphene are conductors. This paper describes the creation of a ruthenium–graphene quantum dot hybrid by reducing RuC13 with aspartic acid and arginine-functionalized graphene quantum dots (Asp–Arg–GQDs). Asp–Arg–GQDs are a synthesis reducing agent, stabiliser, linker, and semiconductor. The resulting Ru–Asp–Arg–GQDs have a three-dimensional network structure of tiny Ru nanoparticles and graphene sheets. The suggested technique was effectively employed in the electrochemical detection of carbendazim in strawberries. This research also provides the path for developing a metal–graphene hybrid with good catalytic performance in sensing, energy storage, and catalysis [64].

In work by Qu *et al* it is demonstrated that food contains several chemicals that impact daily life. Food additive safety guidelines currently focus on the toxicity of single additions; however, food additives are frequently combined and may have additive, synergistic, or antagonistic effects. The current study used high content analysis (HCA) to explore the toxicity of dietary additives and damage pathways in HepG2 cells. To assess cell viability, we used the CCK-8 test. They were providing an experimental basis for establishing the safety of food additives. All of the dietary additives tested were found to inhibit the development of HepG2 cells in a dose-dependent manner. Sunset yellow and sodium sulphite had IC50 values of 1.06 and 0.30 g l^{-1} after 24 h, respectively. According to HCA, sunset yellow and sodium sulphite synergistically affected cell number, membrane permeability, mitochondrial membrane potential, intracellular calcium level, oxidative stress, and high-dose group DNA damage [65].

5.3.2.1 Pure graphene

In a study by Zhang *et al* for the measurement of chloride ions, graphene nanosheets incorporating CPE have been developed. The researchers investigated various voltammetry techniques (cyclic voltammetry, DPV, linear sweep voltammetry, and square-wave voltammetry) and discovered that DPV is the most effective approach for determining chloride ions. Considering several chloride ion sources, the typical peak value for chloride ions was determined to be 0.05 V. Graphene/CPE attained a sensitivity of 0.156 A mM cm^2 in 30 s. The degree of linearity and threshold for detecting this sensor are 0.5–1000 and 0.2 mM, respectively. This technological innovation worked well as a chloride sensor in ordinary water but not when applied to food samples. This study additionally discovered that the reaction of CPE and graphene-modified CPE is superior to that of bare GCE [66].

5.3.2.2 Composite-based electrodes

Katabami *et al* and Berardi *et al* state that nitrite ions are a popular food additive; however, in large quantities, they have detrimental effects on people. Surplus nitrite ions can interact with the amines to generate unpleasant nitrosamines. Nitrite gauges have been developed utilising graphene derivatives such as distinctive nanomaterials. The electrochemical detection of nitrite is based on its oxidation, which requires two electrons and two proton reactions [67, 68].

Jian *et al*, for nitrite detection, for instance, described eRGO/Au composite-modified SPCE. The nanoparticles of Au act as nitrite electrocatalytic locations, while the eRGO provides a place for the Au nanoparticles to soak up nitrite ions. This nitrite sensor's sensitivity, linear response, and detection limit are 0.3048 A M^{-1} cm^{-2}, 1–6000 M, and 0.13 M, respectively. Nitrite detection entails the adsorption of nitrite, complex formation, NO_2 creation, and, finally, nitrate formation [69].

In recent years, Yang *et al* developed an amperometric sensor for nitrite measurement using an Au/MoS_2/RGO ternary nanocomposite. The nitrite detection performance of the Au/MoS_2/RGO/GCE sensor depended on the presence of Au nanoparticles. The optimised $Au_{4.5}$/MoS_2/RGO sample has good sensitivity (0.805 A M^{-1} cm^{-2}), linear range (0.2–2600 M), and low detection limit (0.038 M). This sensor could be used to detect nitrites in natural water. Nitrite sensors have been successfully tested in various food samples [70] (table 5.2).

5.4 Conclusions and current challenges

Analytical scientists have devised electrochemical ways for detecting food contaminants using nanotechnology to ensure food safety. Graphene and its derivatives, for illustration, have been extensively researched as electrochemical sensing probes for organic and inorganic food contaminants. Pure graphene, doped graphene, and graphene-based nanocomposites have been developed as electrode components. Graphene-based electrodes were used to determine artificial food colourants, medicines, insecticides, and other chemical substances. Metal ions of mercury, lead, and cadmium have also been precisely selected. Food specimens have also been found to include anions such as nitrite, nitrate, sulphite, and chloride.

Table 5.2. List of research articles published on graphene-based electrochemical drug sensors.

Graphene-based electrode	Drug	Method	Range	Detection limit	Real sample	Reference
Polyaniline/Au/GO/GCE	Azithromycin	DPV	0.3–920.0 nM	0.1 nM	Blood serum	[71]
Ag/polyindole/RGO	Benorilate	DPV	0.06–80 μM	6 nM	Tablets	[72]
Ce-doped WO_3/GO/GCE	Furazolidone	DPV	1–260 μM	0.054 μM	Urine	[73]
Zinc–graphene oxide/ZnO/SPCE	Promethazine	DPV	0.05–177.45 μM	0.3 nM	Human serum, urine, tablet, river water	[74]

The remarkable electrochemical sensing properties of graphene derivatives-based nanomaterials are related to their rapid transfer of electrons ability, various modes of contact, huge electroactive active sites, and stability. The concentration of modifying layers, the unique morphology and nature of the working electrode, the pH of the solution, the electrode surface area, and the structure of the organic pollutant all impact sensing performance. Despite great inspiration for graphene-based electrochemical sensors, obstacles still need to be overcome. The electrodes' operating stability and repeatability are critical concerns. To build a stable sensor, the distinctive morphology of the nanomaterials must be preserved during the redox process. Some graphene-based electrodes could be used in food samples, utilising routine addition procedures.

References

[1] Manjunatha J G 2020 A surfactant enhanced graphene paste electrode as an effective electrochemical sensor for the sensitive and simultaneous determination of catechol and resorcinol *Chem. Data Collect.* **25** 100331

[2] Gleiter H 2000 Nanostructured materials: basic concepts and microstructure *Acta Mater.* **48** 1–29

[3] Pushpanjali P A, Manjunatha J G, Nagarajappa H, D'Souza E S, Charithra M M and Prinith N S 2021 Voltammetric analysis of antihistamine drug cetirizine and paracetamol at poly (L-Leucine) layered carbon nanotube paste electrode *Surf. Interfaces* **24** 101154

[4] Hareesha N, Manjunatha J G, Amrutha B M, Sreeharsha N, Basheeruddin A, Md and Khalid A 2021 A fast and selective electrochemical detection of vanillin in food samples on the surface of poly (glutamic acid) functionalized multiwalled carbon nanotubes and graphite composite paste sensor *Colloids Surf.*, A **626** 127042

[5] Manjunatha J G, Deraman M, Nur Hamizah B and Talib I A 2014 Selective detection of dopamine in the presence of uric acid using polymerized Phthalo blue film modified carbon paste electrode *Adv. Mater. Res.* **895** 447–51

[6] Hareesha N and Manjunatha J G 2021 Electro-oxidation of formoterol fumarate on the surface of novel poly (thiazole yellow-G) layered multi-walled carbon nanotube paste electrode *Sci. Rep.* **11** 12797

[7] Chethan B, RajPrakash H G, Ravikiran Y T, Vijayakumari S C and Thomas S 2019 Polypyyrole based core-shell structured composite based humidity sensor operable at room temperature *Sensors Actuators* B **296** 126639

[8] Tigari G and Manjunatha J G 2019 Electrochemical preparation of poly(arginine)-modified carbon nanotube paste electrode and its application for the determination of pyridoxine in the presence of riboflavin: an electroanalytical approach *J. Anal. Test.* **3** 331

[9] Hareesha N, Manjunatha J G, Pushpanjali P A, Prinith Subbaiah N, Charithra M M, Sreeharsha N, Basheeruddin Asdaq S M and Md Khalid A 2022 Electrochemical sensing of antibiotic drug amoxicillin in the presence of dopamine at simple and selective carbon paste electrode activated with cetyltrimethylammonium bromide surfactant *Monatsh. Chem.* **153** 31–8

[10] Pal S L, Jana U, Manna P K, Mohanta G P and Manavalan R 2011 Nanoparticle: an overview of preparation and characterization *J. Appl. Pharm. Sci.* **06** 228–34

[11] Amrutha B M, Manjunatha J G, Aarti S B and Nagarajappa H 2021 Sensitive and selective electrochemical detection of vanillin at graphene-based poly (methyl orange) modified electrode *J. Sci.: Adv. Mater. Devices* **6** 415–24

[12] Li Z, Zhou J, Dong T, Xu Y and Shang Y 2021 Application of electrochemical methods for detecting abiotic stress biomarkers in plants *Biosens. Bioelectron.* **182** 113105

[13] Rather I A, Koh W Y, Paek W K and Lim J 2017 The sources of chemical contaminants in food and their health implications *Front. Pharmacol.* **8** 830

[14] González N, Marquès M, Nadal M and Domingo J L 2019 Occurrence of environmental pollutants in foodstuffs: a review of organic vs. conventional food *Food Chem. Toxicol.* **125** 370–5

[15] Volschenk C M, Gerber R, Mkhonto M T, Ikenaka Y, Yohannes Y B, Nakayama S, Ishizuka M, Van Vuren J H J, Wepener V and Smit N J 2019 Bioaccumulation of persistent organic pollutants and their trophic transfer through the food web: human health risks to the rural communities reliant on fish from South Africa's largest floodplain *Sci. Total Environ.* **685** 1116–26

[16] Wang Z, Li Y, Kong F, Li M, Xi M and Yu Z 2021 How do trophic magnification factors (TMFs) and biomagnification factors (BMFs) perform on toxic pollutant bioaccumulation estimation in coastal and marine food webs *Reg. Stud. Mar. Sci.* **44** 101797

[17] WHO 2020 Fact sheet on food safety (https://who.int/en/news-room/fact-sheets/detail/food-safety)

[18] Poma G, Fujii Y, Lievens S, Bombeke J, Gao B, Jeong Y, McGrath T J and Covaci A 2021 Occurrence, patterns, and sources of hazardous organic chemicals in edible insects and insect-based food from the Japanese market *Food Chem. Toxicol.* **154** 112311

[19] Medina D A V, Borsatto J V B, Maciel E V S and Lanças F M 2021 Current role of modern chromatography and mass spectrometry in the analysis of mycotoxins in food *TrAC Trends Anal. Chem.* **135** 116156

[20] Wang K, Li Z, Li J and Lin H 2021 Raman spectroscopic techniques for nondestructive analysis of agri-foods: a state-of-the-art review *Trends Food Sci. Technol.* **118** 490–504

[21] Zhang K, Gan N, Shen Z, Cao J, Hu F and Li T 2019 Microchip electrophoresis based aptasensor for multiplexed detection of antibiotics in foods via a stir-bar assisted multi-arm junctions recycling for signal amplification *Biosens. Bioelectron.* **130** 139–46
[22] Villamil C, Calderon M N, Arias M M and Leguizamon J E 2020 Validation of droplet digital polymerase chain reaction for *Salmonella* spp. quantification *Front. Microbiol.* **11** 1512
[23] Joshi P, Riley P, Goud K Y, Mishra R K and Narayan R 2022 Recent advances of boron-doped diamond electrochemical sensors toward environmental applications *Curr. Opin. Electrochem.* **32** 100920
[24] Park J, Kim J W, Kim H and Yoon W 2021 An electrochemical hydrogen peroxide sensor for applications in nuclear industry *Nucl. Eng. Technol.* **53** 142–7
[25] Amali R K A, Lim H N, Ibrahim I, Huang N M, Zainal Z and Ahmad S A A 2021 Significance of nanomaterials in electrochemical sensors for nitrate detection: a review *Trends Environ. Anal. Chem.* **31** e00135
[26] Karimi-Maleh H, Tahernejad-Javazmi F, Daryanavard M, Hadadzadeh H, Ensafi A A and Abbasghorbani M 2014 Electrocatalytic and simultaneous determination of ascorbic acid, nicotinamide adenine dinucleotide and folic acid at ruthenium(II) complex-ZnO/CNTs nanocomposite modified carbon paste electrode *Electroanalysis* **26** 962–70
[27] Wu L, Fu X, Liu H, Li J and Song Y 2014 Comparative study of graphene nanosheet- and multiwall carbon nanotube-based electrochemical sensor for the sensitive detection of cadmium *Anal. Chim. Acta* **851** 43–8
[28] Baig N, Waheed A, Sajid M, Khan I, Kawde A N and Sohail M 2021 Porous graphene-based electrodes: advances in electrochemical sensing of environmental contaminants *Trends Environ. Anal. Chem.* **30** e00120
[29] Monnappa A B, Manjunatha J G G, Bhatt A S and Nagarajappa H 2021 Sensitive and selective electrochemical detection of vanillin at graphene based poly (methyl orange) modified electrode *J. Sci.: Adv. Mater. Devices* **6** 415–24
[30] Dey S and Nagababu B H 2022 Applications of food colour and bio-preservatives in the food and its effect on the human health *Food Chem. Adv.* **1** 100019
[31] Spence C 2019 On the changing colour of food & drink *Int. J. Gastron. Food Sci.* **17** 100161
[32] Jampasa S, Siangproh W, Duangmal K and Chailapakul O 2016 Electrochemically reduced graphene oxide-modified screen-printed carbon electrodes for a simple and highly sensitive electrochemical detection of synthetic colourants in beverages *Talanta* **160** 113–24
[33] Kartika A E, Setiyanto H, Manurung R V, Jenie S N A and Saraswaty V 2021 Silver nanoparticles coupled with graphene nanoplatelets modified screen-printed carbon electrodes for rhodamine B detection in food products *ACS Omega* **6** 31477–84
[34] Meng F, Qin Y, Zhang W, Chen F, Zheng L, Xing J, Aihaiti A and Zhang M 2022 Amplified electrochemical sensor employing Ag NPs functionalized graphene paper electrode for high sensitive analysis of Sudan I *Food Chem.* **371** 131204
[35] Yun M, Choe J E, You J M, Ahmed M S, Lee K, Üstündag Z and Jeon S 2015 High catalytic activity of electrochemically reduced graphene composite toward electrochemical sensing of Orange II *Food Chem.* **169** 114–9
[36] Lv G, Shi B, Huang H, Chen H, Feng H, Zhou P P, Ye W, Zhang Z and Yang Z 2021 Ternary NiO/Ag/reduced graphene oxide nanocomposites as, a sensitive electrochemical sensor for nanomolarity detection of sunset yellow in soft drinks *J. Food Compos. Anal.* **104** 104136

[37] Fu L, Zheng Y, Wang A, Cai W and Lin H 2015 Sensitive determination of Quinoline yellow using poly(diallyldimethylammonium chloride) functionalized reduced graphene oxide modified grassy carbon electrode *Food Chem.* **181** 127–32

[38] Yu L, Shi M, Yue X and Qu L 2016 Detection of Allura red based on the composite of poly (diallyldimethylammonium chloride) functionalized graphene and nickel nanoparticles modified electrode *Sensors Actuators* B **225** 398–404

[39] Ajiboye T O, Oladoye P O, Olanrewaju C A and Akinsola G O 2022 Organophosphorus pesticides: impacts, detection and removal strategies *Environ. Nanotechnol. Monit. Manag.* **17** 100655

[40] Dhananjayan V and Ravichandran B 2018 Occupational health risk of farmers exposed to pesticides in agricultural activities *Curr. Opin. Environ. Sci. Health* **4** 31–7

[41] Tasaltın N, Karakus S, Tasaltın C and Baytemir G 2022 Highly sensitive and selective rGO based non-enzymatic electrochemical sensor for propamocarb fungicide pesticide detection *Food Chem.* **372** 131267

[42] Mousavi H and Moradian R 2011 Nitrogen and boron doping effects on the electrical conductivity of graphene and nanotube *Solid State Sci.* **13** 1459–64

[43] Hong S *et al* 2021 Carbon nanotube fibers with high specific electrical conductivity: synergistic effect of heteroatom doping and densification *Carbon* **184** 207–13

[44] Chen Z, Zhang Y, Yang Y, Shi X, Zhang L and Jia G 2021 Hierarchical nitrogen-doped holey graphene as sensitive electrochemical sensor for methyl parathion detection *Sens. Actuator. B Chem.* **336** 129721

[45] Nehru R, Hsu Y F, Wang S F, Dong C-D, Govindasamy M, Habila M A and AlMasoud N 2021 Graphene oxide@Ce-doped TiO_2 nanoparticles as electrocatalyst materials for voltammetric detection of hazardous methyl parathion *Microchim. Acta* **188** 216

[46] Wang L, Li X, Yang R, Li J and Qu L B 2020 A highly sensitive and selective electrochemical sensor for pentachlorophenol based *Food Anal. Methods* **13** 2050–8

[47] Ijaz M U, Batool M, Ashraf A, Al-Ghanim K A, Shahzadi S, Samad A, Al-Misned F, Mruthinti S S and Mahboob S 2022 A study on the curative effect of nobiletin on paraquat induced toxicity in rat *J. King Saud Univ. Sci.* **34** 101673

[48] Sant'Anna M V S, de Oliveira S. Silva J, Gevaerd A, Lima L S, Monteiro M D S, Carregosa I S C, Wisniewski A, Marcolino-Junior L H, Bergamini M F and Sussuchi E M 2022 Selective carbonaceous-based (nano)composite sensors for electrochemical determination of paraquat in food samples *Food Chem.* **373** 131521

[49] Kan Y, Nagai J and Uesawa Y 2021 Evaluation of antibiotic-induced taste and smell disorders using the FDA adverse event reporting system database *Sci. Rep.* **11** 9625

[50] Liu D, Xu Y Y, Junaid M, Zhu Y G and Wang J 2022 Distribution, transfer, ecological and human health risks of antibiotics in bay ecosystems *Environ. Int.* **158** 106949

[51] Pan M, Guo P, Liu H, Lu J and Xie Q 2021 Graphene oxide modified screen-printed electrode for highly sensitive and selective electrochemical detection of ciprofloxacin residues in milk *J. Anal. Sci. Technol.* **12** 55

[52] de Faria L V, Lisboa T P, Alves G F, Matos M A C, Muñoz R A A and Matos R C 2021 Adsorptive stripping voltammetric determination of chloramphenicol residues in milk samples using reduced graphene oxide sensor *Anal. Methods* **13** 5711

[53] Willson C 2018 The clinical toxicology of caffeine: a review and case study *Toxicol. Rep.* **5** 1140–52

[54] Shehata M, Azab S M and Fekry A M 2019 May glutathione and graphene oxide enhance the electrochemical detection of caffeine on carbon paste sensor in aqueous and surfactant media for beverages analysis *Synth. Met.* **256** 116122
[55] Jose J, Subramanian V, Shaji S and Sreeja P B 2021 An electrochemical sensor for nanomolar detection of caffeine based on nicotinic acid hydrazide anchored on graphene oxide (NAHGO) *Sci. Rep.* **11** 11662
[56] Tian C, Chen D, Lu N, Li Y, Cui R, Han Z and Zhang G 2018 Electrochemical bisphenol A sensor based on nanoporous PtFe alloy and graphene modified glassy carbon electrode *J. Electroanal. Chem.* **830–1** 27–33
[57] Wang K P, Hu J M and Zhang X 2022 Sensitive electrochemical detection of endocrine disruptor bisphenol A (BPA) in milk based on iodine-doped graphene *Microchem. J.* **173** 107047
[58] Kharazi A, Leili M, Khazaei M, Alikhani M Y and Shokoohi R 2021 Human health risk assessment of heavy metals in agricultural soil and food crops in Hamadan, Iran *J. Food Compos. Anal.* **100** 103890
[59] Liendo F, de la Vega A P, Aguirre M J, Godoy F, Martí A A, Flores E, Pizarro J and Segura R 2022 A simple graphene modified electrode for the determination of antimony (III) in edible plants and beverage *Food Chem.* **367** 130676
[60] Ma S, Zhang Q, Zhu J, Shi H, Zhang K and Shen Y 2021 Rational engineering of Ag-doped reduced graphene oxide as electrochemical sensor for trace mercury ions monitoring *Sensors Actuators* B **345** 130383
[61] Wang J, Yu P, Kan K, Lv H, Liu Z, Sun B, Bai X, Chen J, Zhang Y and Shi K 2021 Efficient ultra-trace electrochemical detection of Cd^{2+}, Pb^{2+} and Hg^{2+} based on hierarchical porous S-doped C_3N_4 tube bundles/graphene nanosheets composite *Chem. Eng. J.* **420** 130317
[62] Jeong S E, Kim S, Han J H and Pak J J 2022 Simple laser-induced graphene fibre electrode fabrication for high-performance heavy-metal sensing *Microchem. J.* **172** 106950
[63] Guo C, Wang C, Sun H, Dai D and Gao H 2021 A simple electrochemical sensor based on rGO/MoS_2/CS modified GCE for highly sensitive detection of Pb(II) in tobacco leaves *RSC Adv.* **11** 29590–7
[64] Xu J, Wang M, Liu Q, Lin X, Pu K and He Z 2022 Gut microbiota mediated the toxicity of high concentration of dietary nitrite in C57BL/6 mice *Ecotoxicol. Environ. Saf.* **231** 113224
[65] Qu D, Gu Y, Feng L and Han J 2017 High content analysis technology for evaluating the joint toxicity of sunset yellow and sodium sulfite *in vitro Food Chem.* **233** 135–43
[66] Zhang M, Wang C, Zhang Z, Ye J and Fang P 2019 A novel carbon paste electrode for sensitive, selective and rapid electrochemical determination of chloride ion based on three-dimensional graphene *Sensors Actuators* B **299** 126951
[67] Katabami K, Hayakawa M and Gando S 2016 Severe methemoglobinemia due to sodium nitrite poisoning *Case Rep. Emerg. Med.* **2016** 9013816
[68] Berardi G, Albenzio M, Marino R, D'Amore T, Taranto A D, Vita V and Iammarino M 2021 Different use of nitrite and nitrate in meats: a survey on typical and commercial Italian products as a contribution to risk assessment *LWT – Food Sci. Technol. (Lebensm. Wiss. Technol.)* **150** 112004
[69] Jian J M, Fu L, Ji J, Lin L, Guo X and Ren T L 2018 Electrochemically reduced graphene oxide/gold nanoparticles composite modified screen-printed carbon electrode for effective electrocatalytic analysis of nitrite in foods *Sensors Actuators* B **262** 125–36

[70] Yang Y, Lei Q, Li J, Hong C, Zhao Z, Xu H and Hu J 2022 Synthesis and enhanced electrochemical properties of AuNPs@MoS_2/rGO hybrid structures for highly sensitive nitrite detection *Microchem. J.* **172** 106904
[71] Jafari S, Dehghani M, Nasirizadeh N and Azimzadeh M 2018 An azithromycin electrochemical sensor based on an aniline MIP film electropolymerized on a gold nano urchins/graphene oxide modified glassy carbon electrode *J. Electroanal. Chem.* **829** 27–34
[72] Li R F, Zhang S F, Kong F Y, Wang Z X, Fang H L and Wang W 2021 Highly sensitive determination of benorilate using silver nanoparticles/polyindole/reduced graphene oxide nanocomposite modified electrode *J. Electroanal. Chem.* **903** 115771
[73] Ramadhass K D, Ganesan M, Chen T W, Chen S M, Hao Q, Lei W and Gopalakrishnan G 2021 Porous-coral-like cerium doped tungsten oxide/graphene oxide micro balls: a robust electrochemical sensing platform for the detection of antibiotic residue *Colloids Surf.* A **628** 127275
[74] Sebastian N, Yu W C, Hu Y C, Balram D and Yu Y H 2021 Morphological evolution of, nanosheets-stacked spherical ZnO for GO-Zn/ZnO ternary nanocomposite preparation a novel electrochemical platform for nanomolar detection of antihistamine promethazine hydrochloride *J. Alloys Compd.* **890** 161768

Chapter 6

Electro-analysis of hormones in a graphene-modified electrochemical sensor

Vinayak Adimule, Nidhi Manhas and Kalpana Sharma

In recent years there the need for the efficient determination of hormones and their quantitative estimation have arisen. In order to develop new low cost, efficient and reliable electrode system for the identification of hormones, especially estriol present in the water, development in electrochemical sensors arises with the fabrication of copper nanoparticles (CuNPs) over reduced graphene oxide (rGO). As-synthesized Cu decorated over rGO serves as a new electrochemical sensor that is an effective tool for quantification of hormones. This electrochemical sensor has shown a linear regime and detection, with quantification limits to 0.17–0.56 ml^{-1}. The hormone estriol is hydrophobic and its determination and probable mechanism are discussed in detail.

6.1 Introduction

Currently, one big problem faced by chemists all over the world is to effectively determine hormones. The quantification of hormones is required, especially estriol, a steroid estrogenic hormone that causes malfunctioning of sexual and reproductive systems, bone structure and affects some organs [1]. The hormones (estriol) are also capable of developing degenerations in the tissues, cancer, etc [2]. Estriol is commonly found in mammals during the pregnancy period and it is helpful in menstruation and the reproductive cycle [3]. In the human body estriol is conjugated with other molecules and on excretion will be released into water. Therefore, it is essential to monitor hormones and their quantitative determination [4]. Different methodologies are used to estimate hormones such as electrophoresis, immunoassays, electroanalytical methods, chromatography, etc [5–9]. Among these methods electrochemical sensor determination and estimation become easier due to their easy operation, fast response, low cost, high and enhanced sensitivity, lower detection limits, quantification limits, automation, development of hybrid electrodes, etc

doi:10.1088/978-0-7503-5578-0ch6

[1, 10]. The electrode systems are incorporated with hybrid structures in order to improve their sensitivity and selectivity [4]. The electrode modifiers include some of the nanoparticles deposited over the substrate such as palladium, silver, copper, antimony, gold, etc, as they possess high conductivity, selectivity, fast response in the electrochemical system and chemical stability. Recent studies have made use of reduced graphene oxide (rGO) modified with Sb and AgNPs as modifiers for the effective determination of estriol [1, 3]. As per the literature, there are some other nanomaterials reported that can be employed for the estimation and detection of estriol [11–21]. The present chapter discusses the synthesis of rGO modified with CuNPs, their cyclic voltammetric determination of estriol hormone, quantification and brief mechanism.

6.2 Experimentation

6.2.1 Materials and characterization

Differential pulse voltammetry (DPV) and cyclic voltammetry (CV) comprise a Potentiostat (PGSTAT-128N) Autolab electrochemical system from The Netherlands with installed NOVA-2.1 software. A three-electrode system was used for the analysis, glassy carbon (GC) loaded with CuNPs@rGO as the first electrode, Pt used as the auxiliary electrode and as reference electrode Ag/AgCl/KCl (3 mol l^{-1}) was used. Morphology and microstructure were performed using a scanning electron microscope (SEM) instrument from JEOL (JSM-7500F). Energy dispersive x-ray (EDS) was performed using an e-beam current of 3.5 nA, with time constant of 6400, accelerating voltage of 10 kV, and using an Ultradry detector from Thermo Fischer Scientific Ltd.

6.2.2 Preparation of solutions

All the reagents, chemicals, solvents were bought from S.D. Fine Chem Ltd, Spectrochem Ltd and Sigma Aldrich, India and used directly without performing any further purifications. Millipore (Mili-Q-system) water was used for all the experiments and the reagents were not previously purified. Graphene oxide (GO) (99.8% purity, Sigma Aldrich India), $CuCl_2$ (99.6% purity, Sigma Aldrich India), Estriol (99% purity, Sigma Aldrich, India), Ethanol (99.9% purity, Chengashu, China), Acetone (99.5% purity, Spectrochem Ltd, India), sodium hydroxide (NaOH, 98.6% purity, S.D. Fine Chem Ltd, India), potassium hydroxide (KOH, 99.1% purity, S.D. Fine Chem Ltd, India), sodium dodecyl sulphate (SDS, 99.0% purity, Spectrochem Ltd, India), sodium borohydride ($NaBH_4$, 98.4% purity, S.D. Fine Chem Ltd, India).

6.2.3 Synthesis of rGO–CuNPs

Synthesis of rGO–CuNPs was carried out as per the previously established procedures reported in the literature [22]. In the first step, GO (4 mg ml^{-1}) was suspended in the solution of SDS in the proportion of 10:4. Absolute ethanol was used as a solvent medium. The entire contents were kept in ultra sonication for a time period of 20 min. To the contents, sodium borohydride ($NaBH_4$) was added in excess portions (15 g) and kept in ultra sonication for a time period of 1 h. $CuCl_2$ (cuprous chloride, 10 ml of ethanol) solution in ethanol was added and sonicated for a period of 20–30 min. The material was centrifuged and washed with ethanol (50 ml × 2 times), water (10 ml × 2 times), dried at 100 °C–200 °C, calcinated at 600 °C–700 °C. During the electrode preparation the nanostructure was sonicated for 5 min to render the homogeneity of the dispersed CuNPs over the reduced graphene sheet.

6.2.4 Preparation of the electrodes

The surfaces of the GC electrode surfaces were polished with polystyrene silicon paper. The sonication of electrodes was carried out in ethanol and in water for a time period of 5–10 min. To the polished, cleaned electrode about 10 l of CuNPs@rGO has been loaded and dried at room temperature. The electrode is then taken to the measurement of electrochemical reactions.

6.3 Result and discussion

6.3.1 Morphology and electrochemical characterization of estriol

Morphological change in the nanomaterial is observed by SEM. Figure 6.1 refers to (A) GO and (B) rGO. The SEM image shows wrinkled and twisted structure. In the rGO molecule, after the reduction process active sites are exposed [23]. The extent of disorder, defective levels and the doping levels in the rGO are determined from Raman spectroscopic investigations [24]. Insets of figures 6.1(A) and (B) show the Raman spectra of GO and rGO, respectively, with unblocking of the active sites after the reduction of GO. The edge of the GO after the reduction and the higher sp^2 hybridization can be determined from the Raman spectral analysis. Figure 6.1(C) shows the CuNPs distributed over the graphene surface non-uniformly. The inset of figure 6.1(D) shows EDS elemental mapping of the as-synthesized CuNPs@rGO. The size of the NPs varied from 5 to 30 nm. Electrochemical characterization of the CuNPs@rGO was carried out using GC/CuNPs@rGO using 0.1 mol l^{-1} PBS in $p^H = 7.0$ solution. The potential is varied from −0.5 to +0.8 V. On the other hand, CV peaks show oxidation, reduction peaks (due to the oxidation of Cu^0 to Cu^{1+}) and this confirms its presence over the surface of the rGO. The observed results are in good agreement with the previously published works [22, 24].

6.3.2 Electrochemical behaviour of rGO–CuNPs

A CV experiment is designed in order to evaluate the oxidation and reduction behaviour of the hormone estriol using 0.1 mol l^{-1} in PBS solution at pH = 7.0. A sweep speed of 50 mV S^{-1} was employed during the experimentation. The dotted line in figure 6.2 shows the electrochemical CV spectrum in the absence of the hormone, and the solid line indicates the oxidation of the hormone estriol (10 mol l^{-1} addition) +0.54 V. The obtained results are in good agreement with the previously reported works [1, 26].

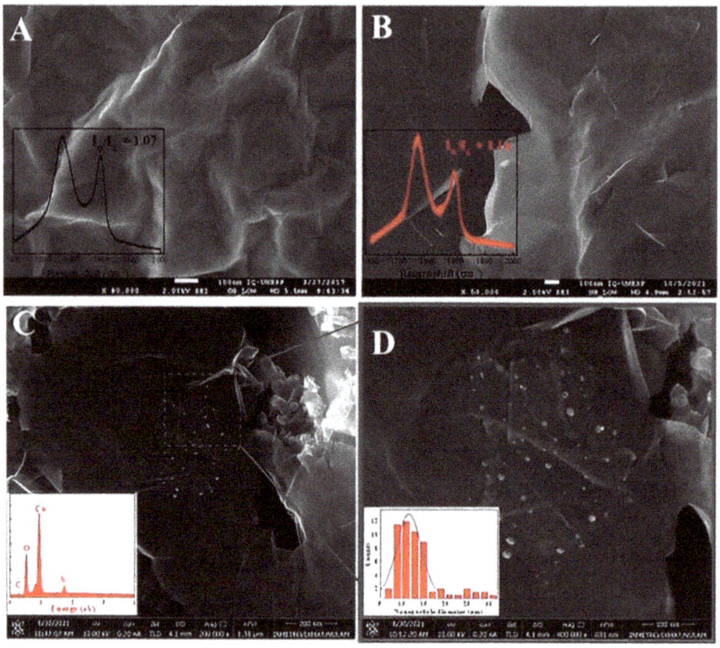

Figure 6.1. SEM images of (A) GO and (B) rGO materials (inset: Raman spectra). (C) rGO–CuNPs nanocomposite (inset: EDS spectrum). (D) rGO–CuNPs at higher magnitude (inset: particle size distribution histogram). Reproduced from [26]. CC BY 4.0.

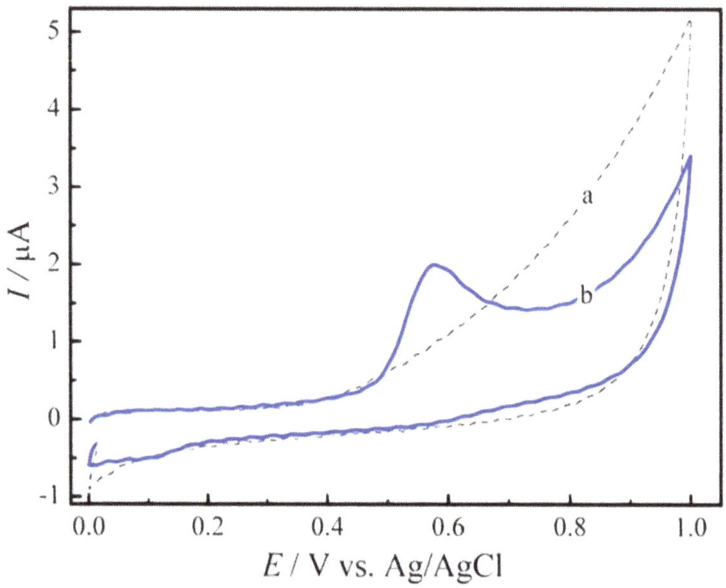

Figure 6.2. CV in the absence (curve a) and presence (curve b) of 10 μmol l^{-1} estriol hormone (0.1 mol l^{-1} PBS pH 7.0). Reproduced from [26]. CC BY 4.0.

6.3.3 Optimization parameters and analytical characteristics

Table 6.1 depicts optimization parameters for the measurement of sensitivity of the hormone estriol. Previous reports stated that different concentrations of the nanomaterials render different selectivity and sensitivity in CV, and CuNPs dispersed over the rGO show different selectivity with higher proportions of nanomaterials loadings. Therefore, concentrations of CuNPs–rGO evaluated by DPV, in 0.1 mol l^{-1} PBS solution, at neutral p^H, were sweep speed of 10 mV s^{-1}, amplitude of 25 mV and potential window of +0.4 to +0.8 V. The estriol used was 10 μmol l^{-1}. From the table the best concentration of the electrode was found to be 0.02 mg mol^{-1}.

The hormone estriol may have the tendency to get adsorbed on the surface of the electrodes. The study of pre-concentration was carried out to understand the best potential window to be set across electrochemical investigation [26, 27]. The studied potentials for the selectivity of 30 s and from the experiments the best potential window selected was −1.1 V. In order to understand the oxidation products of the estriol hormone and to understand the oxidation mechanism, a different p^H range of the solution 5.0–9.0 in DPV in PBS solution was carried out. With an amplitude of 25 mV, sweep speed of 10 mV s^{-1}, and at p^H = 7.0 it shows greater sensitivity and selectivity. A plot of E_{pa} V s p^H is shown in figure 6.3, which shows a linear relationship with a slope of −61 V and also the same number of protons and electrons are involved in the oxidation mechanism of estriol [28]. The oxidation

Table 6.1. Voltammetry analysis with GC/rGO–CuNPs. Reproduced from [26]. CC BY 4.0.

Parameters	Optimization range	Optimized range
rGO–CuNPs concentration (mg ml^{-1})	0.01–0.04	0.02 mg ml^{-1}
pH	5–9	7
Pre-concentration potential (V)	−1.2 to −0.5	−1.1 V

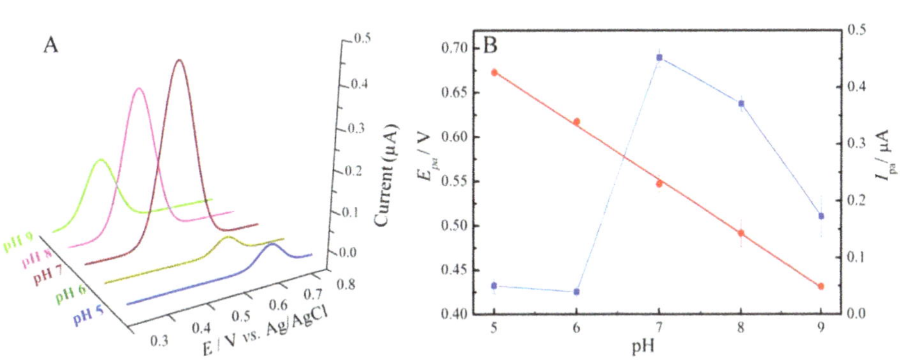

Figure 6.3. (A) DPV voltammograms of 10.0 μmol l^{-1} of estriol (0.1 mol l^{-1} PBS at pH ranging from 5 to 9). (B) Effect of pH at the anodic peak potential (● Epa) and current (■ Ipa) of estriol. Reproduced from [26]. CC BY 4.0.

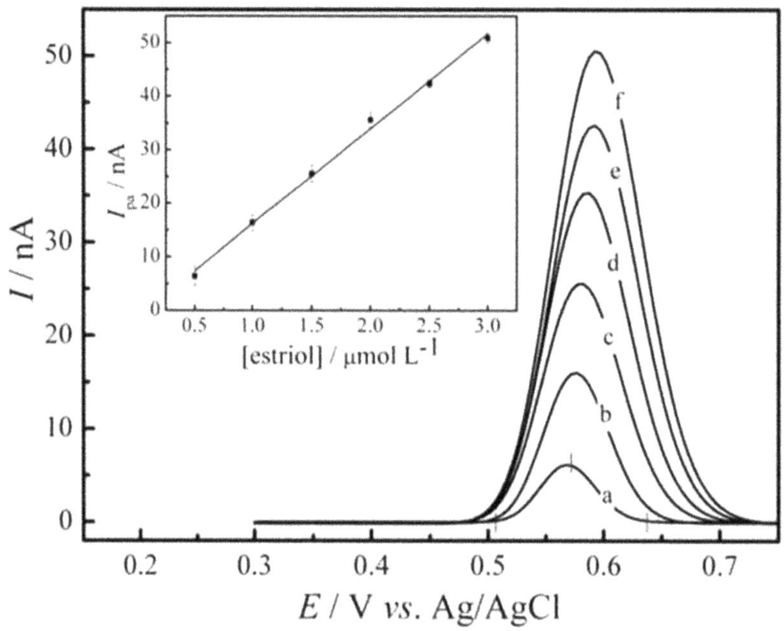

Figure 6.4. DPV responses for the detection of estriol in the remediation samples, as follows: (s) sample, (a), (b), (c) represent three successive additions of 0.5 μmol l^{-1} of estriol (inset: linear dependence of anodic peak current with estriol concentration). Reproduced from [26]. CC BY 4.0.

Table 6.2. Comparison table of previously reported literature. Reproduced from [26]. CC BY 4.0.

Electrode	Linear range	LOD (μmol l^{-1})	LOD (μmol l^{-1})	Matrix sample	Reference
CPE/Fe_3O_4NPs[a]	3.0–111.0	2.75	8.35	Pharmaceutical samples	[29]
GC/CNB/AgNPs[b]	0.2–3.0	0.16	0.50	Creek water	[30]
GC/rGO–SbNPs[c]	0.2–1.4	0.005	Not reported	Natural waters	[1]
GC/rGO–AgNPs[d]	0.1–3.0	0.02	Not reported	Synthetic urine and tap water	[3]
GC/rGO–CuNPs	0.5–3.0	0.17	0.56	Bioremediation system water	[25]

[a] Carbon paste electrode modified with ferrimagnetic nanoparticles.
[b] Carbon black nanoballs decorated with silver nanoparticles.
[c] Antimony nanoparticles.
[d] Silver nanoparticles.

mechanism of estriol is evaluated by molecular dynamics and computer simulation studies as reported in the previous literature [1, 3].

The concentration of the hormone estriol was varied from 0.5 to 3.0 μmol l^{-1} in order obtain the analytical curve displayed in figure 6.4. DPV was set with the following parameters, amplitude of 25 mV, scanning speed of 10 mV s^{-1}, potential window of +0.3 to +0.8 V, 0.1 mol l^{-1} PBS solution at pH = 7.0 were used. The conditions were optimized as per table 6.1. Estriol concentration and anodic peak current are plotted and linear relationship for the 0.5–3.0 μmol l^{-1} concentration of the estriol is obtained. Table 6.2 depicts the literature reports of the various nanomaterials, detection and quantification limits, and target matrix for estriol detection. It is worth mentioning that several research works make use of graphene and nanomaterials-decorated graphene for the detection of estriol in natural water and clinical samples. The present work describes sensor-based CuNPs–graphene-decorated hybrid structures for the detection of estriol hormone at low concentration. A great challenge arises in the detection limit, of several ions, for molecules present in the solution. Some of the literature works report the usage of rGO–AgNPs [3], rGO–SbNPs [1] as electrode modifiers for the detection of limits of detection (LOD) in the analyte. 2D-carbon-based materials such as graphene possess higher sensitivity and selectivity for the detection of the hormone estriol in low concentration and are widely used as electrode modifiers [31]. This tool provides the possibility of applications in environmental monitoring that are quite versatile, low cost, and easy to fabricate and apply in the field of electrochemical sensing applications [32–37].

6.4 Conclusion

In summary, in the present chapter, CuNPs decorated with rGO hybrid nanomaterials have been synthesized by simple chemical precipitation method and incorporated in an electrochemical sensor for the effective detection of the hormone estriol. As demonstrated, the sensor showed better response towards detection of estriol in low concentration and at neutral p^H condition in aqueous solution. The electrochemical sensor is used for the quantitative determination of estriol in low concentration and its utilization in the effective decontamination of estriol hormones present in water resources for bioremediation have been explored. In this chapter effective usage of electrochemical sensors in the evaluation of efficiency in the bioremediation process, morphology of the synthesized samples, real sample analysis, and their analytical characteristics has been discussed in detail.

Authors contributions

Dr Vinayak Adimule and Dr Kalpana Sharma wrote and carried out corrections in the manuscript. Miss Nidhi Manhas was involved in editing and correcting grammatical and typo errors in the manuscript.

Conflict of interest

All authors declare that they do not have any conflict of interest.

Data availability

All data incorporated in the present chapter are extracted from open access publications.

Funding

All authors declare that they have not received any funding from any source or institutions or organizations.

Acknowledgements

All authors are thankful to School of Sciences (SOS), IGNOU, New Delhi, India, KLE Dr MSRIT, Department of Physics, Bangalore, Karnataka, India for constant encouragement and motivation for this work.

References

[1] Cesarino I, Cincotto F H and Machado S A S 2015 A synergistic combination of reduced graphene oxide and antimony nanoparticles for estriol hormone detection *Sensors Actuators* B **210** 453–9

[2] Manjunatha J G 2017 Electroanalysis of estriol hormone using electrochemical sensor *Sens. Bio-Sens. Res.* **16** 79–84

[3] Donini C A, da Silva M K L, Simoes R P and Cesarino I 2018 Reduced graphene oxide modified with silver nanoparticles for the electrochemical detection of estriol *J. Electroanal. Chem.* **809** 67–73

[4] Gomes E S, Leite F R F, Ferraz B R L, Mourao H A J L and Malagutti A R 2019 Voltammetric sensor based on cobalt-poly(methionine)-modified glassy carbon electrode for determination of estriol hormone in pharmaceuticals and urine *J. Pharm. Anal.* **9** 347–57

[5] Dong S, Chi L, Yang Z, He P, Wang Q and Fang Y 2009 Simultaneous determination of dihydroxybenzene and phenylenediamine positional isomers using capillary zone electrophoresis coupled with amperometric detection *J. Sep. Sci.* **32** 3232–8

[6] Pistonesi M F, Di Nezio M S, Centurion M E, Palomeque M E, Lista A G and Fernandez B S 2006 Determination of phenol, resorcinol and hydroquinone in air samples by synchronous fluorescence using partial least-squares (PLS) *Talanta* **69** 1265–8

[7] Li S F, Li X Z, Xu J and Wei X W 2008 Flow-injection chemiluminescence determination of polyphenols using luminol–$NaIO_4$–gold nanoparticles system *Talanta* **75** 32–7

[8] Gaudry D, Carlsen S, Khemani L, Miotto J and Alkalay D 1989 Determination of the dopamine agonist CGS 15855A in human plasma by gas chromatography/mass spectrometry *Anal. Lett.* **22** 2307–21

[9] Li N, Guo J, Liu B, Yu Y, Cui H, Mao L and Lin Y 2009 Determination of monoamine neurotransmitters and their metabolites in a mouse brain microdialysate by coupling high-performance liquid chromatography with gold nanoparticle-initiated chemiluminescence *Anal. Chim. Acta* **645** 48–55

[10] Moreira F, Santana E R and Spinelli A 2020 Ionic liquid-supported magnetite nanoparticles as electrode modifier materials for estrogens sensing *Sci. Rep.* **10** 1955

[11] Adimule V, Yallur B C, Bhowmik D and Gowda A H 2022 Dielectric properties of P3BT doped ZrY_2O_3/$CoZrY_2O_3$ nanostructures for low cost optoelectronics applications *Trans. Electr. Electron. Mater.* **23** 288–303

[12] Manjunatha J G, Deraman M, Basri N H, Nor N S, Talib I A and Ataollahi N 2014 Sodium dodecyl sulfate modified carbon nanotubes paste electrode as a novel sensor for the simultaneous determination of dopamine, ascorbic acid, and uric acid *C.R. Chim.* **17** 465–76

[13] Bagihalli G, Unki S N and Adimule V 2022 Synthesis, structural and optical properties of Co doped α-Sb_2O_4 nanocomposites *Adv. Mater. Res.* **1173** 3–11

[14] Adimule V, Banakar P and Naik V H 2018 Preparation, characterization and optical properties of chromium oxide and yttrium nanocomposites *AIP Conf. Proc.* **1989** 020001 (AIP Publishing LLC) p 020001

[15] Adimule V, Vageesha P, Bagihalli G, Bowmik D and Adarsha H J 2019 Synthesis, characterization of hybrid nanomaterials of strontium, yttrium, copper doped with indole Schiff base derivatives possessing dielectric and semiconductor properties *Emerging Research in Electronics, Computer Science and Technology* ed V Sridhar, M Padma and K Rao (Springer) 1131–40

[16] Adimule V 2018 Synthesis characterization of Sr-Gd nanocomposites doped with zirconium possessing electrical and optical properties *AIP Conf. Proc.* **1989** 030001

[17] Tigari G and Manjunatha J G 2020 A surfactant enhanced novel pencil graphite and carbon nanotube composite paste material as an effective electrochemical sensor for determination of riboflavin *J. Sci.: Adv. Mater. Devices* **5** 56–64

[18] Hareesha N and Manjunatha J G 2020 A simple and low-cost poly (dl-phenylalanine) modified carbon sensor for the improved electrochemical analysis of Riboflavin *J. Sci.: Adv. Mater. Devices* **5** 502–11

[19] Manjunatha J G 2017 A new electrochemical sensor based on modified carbon nanotube-graphite mixture paste electrode for voltammetric determination of resorcinol *Asian J. Pharm. Clin. Res.* **10** 295

[20] Charithra M M, Manjunatha J G and Raril C 2020 Surfactant modified graphite paste electrode as an electrochemical sensor for the enhanced voltammetric detection of estriol with dopamine and uric acid *Adv. Pharm. Bull.* **10** 247

[21] Hareesha N and Manjunatha J G 2020 Surfactant and polymer layered carbon composite electrochemical sensor for the analysis of estriol with ciprofloxacin *Mater. Res. Innovations* **24** 349–62

[22] Fernandes V Q, Silva M K L and Cesarino I 2020 Determination of isotretinoin (13-cis-retinoic acid) using a sensor based on reduced graphene oxide modified with copper nanoparticles *J. Electroanal. Chem.* **856** 113692

[23] Pei S and Cheng H M 2012 The reduction of graphene oxide *Carbon N.Y.* **50** 3210–28

[24] Wu J B, Lin M L, Con X, Liu H N and Tan P H 2018 Raman spectroscopy of graphene-based materials and its applications in related devices *Chem. Soc. Rev.* **47** 1822–73

[25] Setznagl S and Cesarino I 2022 Copper nanoparticles and reduced graphene oxide modified a glassy carbon electrode for the determination of glyphosate in water samples *Int. J. Environ. Anal. Chem.* **102** 293–305

[26] Barreto F C, Silva M K L and Cesarino I 2022 An electrochemical sensor based on reduced graphene oxide and copper nanoparticles for monitoring estriol levels in water samples after bioremediation *Chemosensors* **10** 395

[27] Adimule V, Yallur B C, Keri R, Chinna B and Sheetal B 2023 Microstructure, photoluminescence and electrical properties of SmxGd(1−x): SrO hybrid nanomaterials synthesized via facile coprecipitation method *Electron. Mater. Lett.* **19** 278–97

[28] Manjunatha J G 2020 A surfactant enhanced graphene paste electrode as an effective electrochemical sensor for the sensitive and simultaneous determination of catechol and resorcinol *Chem. Data Collect.* **25** 100331

[29] Adimule V, Nandi S S and Gowda A H J 2021 Enhanced power conversion efficiency of the P3BT (poly-3-butyl thiophene) doped nanocomposites of Gd-TiO_3 as working electrode *Techno-Societal 2020* (Berlin: Springer) pp 55–68

[30] Nandi S S, Suryavanshi A, Adimule V and Maradur S R 2020 Semiconductor current–voltage characteristics of some novel perovskite ionic nanocomposites of Sr 0.5 Cu 0.4 Y 0.1 and Sr 0.5 Mn 0.5 and their electronic sensor applications *AIP Conf. Proc.* **2274** 020006

[31] da Silveira J P, Piovesan J V V and Spinelli A 2017 Carbon paste electrode modified with ferrimagnetic nanoparticles for voltammetric detection of the hormone estriol *Microchem. J.* **133** 22–30

[32] Raymundo-Pereira P A *et al* 2017 Sensitive detection of estriol hormone in creek water using a sensor platform based on carbon black and silver nanoparticles *Talanta* **174** 652–9

[33] Adimule V, Yallur B C, Batakurki S R and Gowda A H J 2021 Microwave-assisted synthesis of Cr-doped Gd_2O_3 nanostructures and investigation on morphology, optical, and photoluminescence properties *Nanosci. Technol.* **13** 45–59

[34] Adimule V, Nandi S S, Yallur B C, Bhowmik D and Jagadeesha A H 2021 Enhanced photoluminescence properties of $Gd_{(x-1)}$ Sr_xO: CdO nanocores and their study of optical structural and morphological characteristics *Mater. Today Chem.* **20** 00438

[35] Adimule V, Nandi S S, Yallur B C, Bhowmik D and Jagadeesha A H 2021 Optical, structural and photoluminescence properties of Gd_xSrO: CdO nanostructures synthesized by co-precipitation method *J. Fluoresc.* **31** 487–99

[36] Manjunatha J G 2018 A novel voltammetric method for the enhanced detection of the food additive tartrazine using an electrochemical sensor *Heliyon* **4** e00986

[37] Adimule Vianyak R M G and Adarsha H J 2020 Synthesis and fabrication of Y-doped ZnO nanoparticles and their application as a gas sensor for the detection of ammonia *J. Mater. Eng. Perform.* **29** 4–5

Chapter 7

Graphene composite electrodes for electrochemical determination of drugs

Serdar Sanli and Sevinc Kurbanoglu

Graphene is one of the most famous materials and research topics in fundamental science and applications. Graphene is mainly used in sensor applications. Sensors are devices that can measure chemical or physical effects such as light, sound, heat, humidity, distance, force, pressure, and acceleration and convert them into electrical signals. Since sensors can be designed in very small sizes and offer high sensitivity and selectivity, they are used in almost every aspect of our daily lives. Chemical sensors can be broadly defined as devices that can qualitatively or quantitatively convert the chemical interaction or process between the medium (analyte) and the sensor device into valuable signals. Among all sensors, electrochemical sensors are commonly applied to pharmaceutical analytes, as the drug-active compounds are generally electroactive. Moreover, electrochemical methods are much simpler and cheaper than other analysis methods and have the advantages of miniaturization and comparative instrumental sensitivity. By modifying the surface of the working electrode, different sensors can be obtained with unique properties that are the main concern in electrochemical nanosensors. From this aspect, this chapter provides an overview of the fundamental aspects of graphene, setting the stage for the subsequent chapters that delve deeper into specific applications and advancements in drug analysis.

7.1 Graphene

Graphene is one of the most famous materials and research topics in fundamental science and applications. Before examining graphene's superior properties, it is necessary to look at its structure. Graphene can be defined as the single carbon plane of graphite. Naturally occurring graphite is formed by stacking plates of carbon atoms. Each sp^2 hybridized carbon atom forms strong covalent σ bonds with three neighboring carbon atoms and creates a hexagonal (honeycomb), one-atom-thin,

doi:10.1088/978-0-7503-5578-0ch7

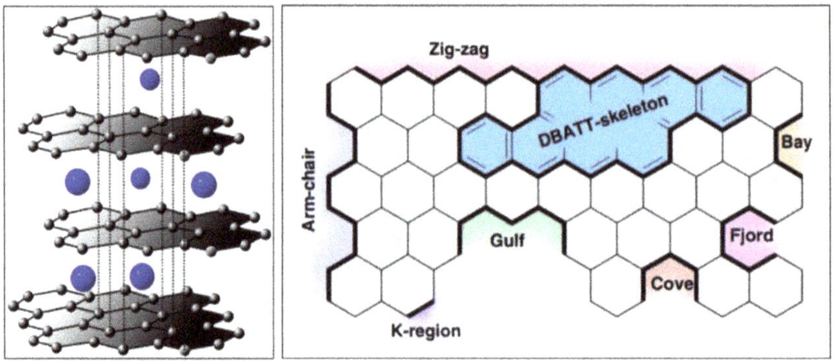

A B

Figure 7.1. Simplified structure of graphite layers (A) and schematic depiction of relevant edge structures (B). Reprinted from [12], copyright (2020), with permission from Elsevier.

two-dimensional structure. The first graphene isolation was accomplished by breaking the weak van der Waals interactions between the layers of graphite just with a piece of Scotch tape. Isolated graphene's popularity stems from being the thinnest and strongest material ever measured. Besides that, the low dimensional carbon layer comes with unique optical, electrical, and physical properties, positioning it as a promising material in various fields, including electronics, energy storage, composites, and biomedicine [1–5].

Graphene is a single-layer, single-atom-thick nano-sized particle in which regular carbon atoms in a hexagonal structure are sp^2 hybridized and are the basic building block of all existing carbon allotropes (figure 7.1). Graphene has attracted great interest due to its superior properties such as flexibility, relatively high surface area, high electron mobility, unparalleled strength, and toughness [6–11].

Various methods have been evolved for graphene synthesis, including mechanical exfoliation, chemical vapor deposition (CVD), epitaxial growth on silicon carbide, and reduction of graphene oxide (GO). Mechanical exfoliation involves repeatedly cleaving graphite using adhesive tapes to obtain thin graphene flakes. While this method yields high-quality graphene, it is not scalable for large-scale production. CVD allows graphene growth on metal substrates by decomposing hydrocarbons at high temperatures. This method enables the production of large-area graphene films, but the transfer process and impurities may affect its properties. Epitaxial growth on silicon carbide offers a route to obtain graphene of high quality, but it is limited to small-scale production. Reduction of GO involves the chemical reduction of GO, which is easier to produce in large quantities but often leads to reduced electrical conductivity and structural defects [6, 7, 10, 13].

7.1.1 Properties of graphene

In the most basic sense, graphene can be defined as a material with a hexagonal honeycomb lattice and sp^2 hybridization. The so-called sp^2 hybridization can be

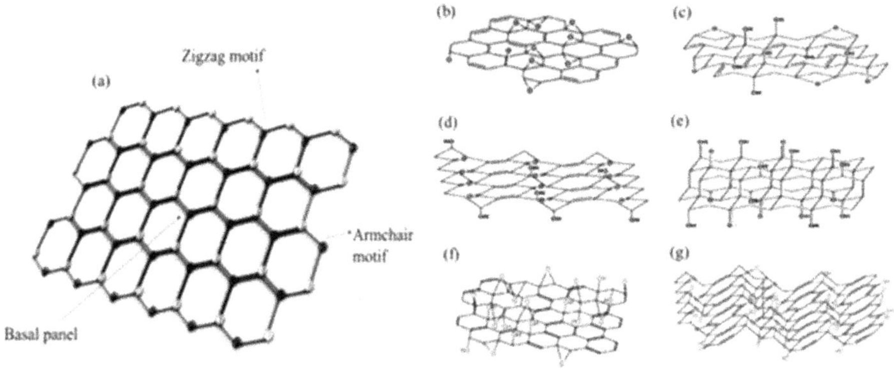

Figure 7.2. Structures of pristine graphene (a) and its GO derivatives based on Hofmann (b), Ruess (c), Scholz–Boehm (d), Nakajima–Matsuo (e), Lerf–Klinowski (f), and Szabo (g) models. Reprinted with permission from [16] CC BY 4.0.

summarized as a mixture of valence electronic states. Each carbon atom has three neighboring atoms and the hexagonal lattice is thus formed (figure 7.2). Graphene exhibits exceptional properties, making it a unique material with great potential for various applications. Graphene is known for its exceptional mechanical properties, with a tensile strength of about 130 GPa and Young's modulus of approximately 1 TPa. It is the strongest material ever measured, surpassing even the most robust metals. Additionally, graphene is highly flexible and can be bent and stretched without breaking, making it suitable for flexible electronics and composite materials [8, 14–16].

Moreover, graphene possesses remarkable electrical properties, including high electrical conductivity and exceptional electron mobility. It can conduct electricity more efficiently than copper and exhibits a high electron mobility of up to 200 000 $cm^2 (V s)^{-1}$ at room temperature. These properties make graphene a promising candidate for next-generation electronic devices, such as transistors, sensors, and transparent conductive films [9, 16–18].

Graphene also exhibits excellent thermal conductivity, with values exceeding 5000 $W (m K)^{-1}$, making it an exceptional heat conductor. Its two-dimensional nature and low atomic mass contribute to efficient phonon transport. These properties enable the development of graphene-based thermal management systems and heat dissipation materials. Furthermore, graphene possesses intriguing optical properties, such as high transparency over a broad wavelength range and light solid absorption. It absorbs around 2.3% of incident light across the visible spectrum, making it a suitable material for photodetectors, solar cells, and optical modulators [17, 19].

Besides chemical and optical properties, graphene exhibits excellent chemical stability due to its strong carbon–carbon bonds. It resists various chemicals, including acids, bases, and organic solvents. Moreover, graphene's two-dimensional structure makes it an effective barrier against gases and liquids, offering potential applications in protective coatings and membranes [16, 20, 21].

7.1.2 Potential applications of graphene

Graphene's extraordinary properties have led to a wide range of potential applications in various fields: such as electronics, energy storage, composite material design, biomedicine, etc. Graphene-based transistors have the potential to revolutionize the semiconductor industry by enabling faster and more energy-efficient devices. Graphene's high electron mobility and transparency make it suitable for flexible displays, wearable electronics, and high-frequency transistors. Moreover, graphene-based materials hold promise for advanced energy storage devices, such as supercapacitors and batteries. Graphene's large surface area, high electrical conductivity, and fast charge transport make it an ideal material for enhancing energy storage capacity and improving charging rates. Graphene can be incorporated into polymer matrices to enhance mechanical strength, electrical conductivity, and thermal properties. Graphene-based composites have potential applications in the aerospace, automotive, and construction industries [11, 13, 19]. Furthermore, graphene and its derivatives show potential in various biomedical applications, including drug delivery systems, bioimaging, biosensors, and tissue engineering. Graphene's biocompatibility, large surface area, and ability to interact with biomolecules make it a versatile material in this field [7, 17, 22].

7.2 Electrochemical sensors

Sensors are devices that can measure chemical or physical effects such as light, sound, heat, humidity, distance, force, pressure, and acceleration and convert them into electrical signals. Since sensors can be designed in very small sizes and offer high sensitivity and selectivity, they are used in almost every aspect of our daily lives [23–26]. Today, technological developments in microelectronics technology have enabled the development of fast and highly sensitive micro-sized sensors. Technological developments have enabled sensors to be used in a wide range of fields, such as biotechnology, automotive, mining, health, robotics, and machinery. Chemical sensors can be broadly defined as devices that can qualitatively or quantitatively convert the chemical interaction or process between the medium (analyte) and the sensor device into valuable signals [25]. Electrochemical sensors, on the other hand, are devices that convert the electrochemical interaction between chemical compounds or ions with an identifying interface into electrical signals [27–29] (figure 7.3). Electrochemistry is a branch that deals with ion distribution, ion transport, and electron transfer reactions between a solid conductor (electrode) and a solution interface. The sensors are also called 'biosensors' when they are used to utilize biological-based substances such as tissues, DNA, enzymes, antibodies, and cells [28, 30–32]. The transducer detects the chemical interaction between the analyte and the receptor and converts it into a meaningful electrical signal. Finally, the amplifier sends this electrical signal as a current to a display. Electrochemical sensors are commonly applied to pharmaceutical analytes, as the drug-active compounds are generally electroactive [33–37]. Electrochemical techniques such as differential pulse voltammetry (DPV), square-wave voltammetry, potentiometry, cyclic voltammetry (CV), and chronoamperometry are the most used techniques in electrochemical sensors [38, 39]. Potentiometry is based

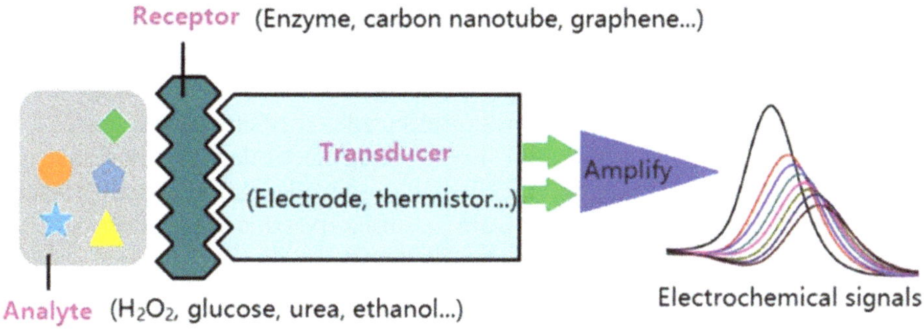

Figure 7.3. Schematic diagram of the electrochemical sensor. Reprinted with permission from [59] CC BY 4.0.

on the measurement of a potential difference between reference and working electrodes. The ion-sensing element in potentiometric sensors is a membrane containing ion-selective molecular acceptors or acceptor sites on a solid material. This membrane is placed between two solutions, one containing a sample and the other containing a fixed concentration of analyte ions. Ion exchanges on both membrane surfaces cause a potential difference between the edges of the membrane. This potential difference defines the concentration of analyte ions in the sample [40–43]. In general terms, amperometry is based on measuring current strength at the potential. In amperometric sensors, the current strength resulting from oxidation–reduction reactions occurring on the working electrode is a function of the concentration of electroactive structures in the sample medium. The concentrations of the structures in the medium are determined by analyzing the current strength with the help of a second electrode (reference electrode) in the medium [44–50]. Voltammetry is a set of electroanalysis methods used in various industrial processes and analytical chemistry. In voltammetry, information about the analyte is obtained by measuring currents at different potentials. Analytical data are obtained from a voltammogram, which plots the current produced by the electrochemical reactions of the analyte against the potential of the working electrode. Since biological reactions generally cause a decrease in impedance and an increase in capacitance and conductivity, the analysis of impedance, conductivity, or capacitance changes is performed with voltammetric sensors. Various voltammetric methods include CV, square-wave voltammetry, DPV, linear sweep voltammetry (LSV), and hydrodynamic voltammetry [51–55]. Electroanalytical systems allow the use of very small volumes (nL) of samples and have very low detection limits (10^{-12}). In addition, the electrode system regularly receives signals from the analyte medium, allowing tracking. Therefore, these systems are frequently used in clinical studies, control of industrial microbial processes, and specific pathogen diagnostics. Electrochemical methods are much simpler and cheaper than other analysis methods and have the advantages of miniaturization and comparative instrumental sensitivity [38, 56–58].

The specific chemical structures of electrochemical sensors have led to an extraordinary interest in electrochemistry-based devices, as they offer the possibility of highly sensitive and selective detection and monitoring. Electrochemical sensors generally consist of a sensor (or catalyst) and a transducer. The transducer, known as the electrode, converts chemical energy into electrical energy. The sensor displays this electrical energy on the screen and continues to generate electrical signals of the detected analyte for further analysis. Electrochemical sensors consist of a three-electrode system consisting of a working electrode, a reference electrode, and a counter electrode (auxiliary electrode). The working electrode is the most important one. By modifying the surface of the working electrode, different sensors can be obtained with unique properties that are the primary concern in electrochemical nanosensors [26, 60–62].

7.3 Recent graphene composite electrodes for electrochemical determination of drugs

In the literature, many types of research exist based on graphene-modified sensors for detecting several drugs. In this biosensor study, graphene was utilized as a quantum dot for the purpose of developing a morphine biosensor [63]. The fluorescence properties of the quantum dot served as a marker for the biosensor. Graphene quantum dots (GQDs) were obtained through the exposure of citric acid to high heat at 200 °C for a duration of 20 min The resulting material, exhibiting an orange color, was defined as GQD. Subsequently, the GQD solution was neutralized with NaOH, and particles were functionalized with antimorphine antibodies using EDC and sulfo-NHS crosslinkers called antimorphine-GQDs. The characterization of the functionalization stages involved x-ray photoelectron spectroscopy (XPS), Fourier transform infrared (FTIR) spectroscopy, AFM, and biosensor trials. Masteri-Farahani and Mosleh observed that antimorphine-GQDs displayed a lower fluorescence intensity compared to GQDs. This decrease in fluorescence intensity was attributed to the occurrence of Forster resonance energy transfer (FRET) from excited state GQDs to antimorphine antibodies. The developed sensor relied on this FRET phenomenon. In the presence of morphine, binding to the antimorphine receptors resulted in the inhibition of FRET and an increase in the fluorescence efficiency of GQDs. Consequently, an enhanced fluorescence intensity could be observed corresponding to the concentration of morphine added to the solution. The authors established a linear correlation between the concentration of morphine and the F/F0 ratio, representing the fluorescence change. The sensor exhibited a detection limit of 0.06 μM within the linear range of 0–35 μM ($r > 0.9977$) when comparing the fluorescence intensity of GQDs in the presence (F) and absence (F0) of morphine. Additionally, GQD assays were conducted to assess the binding of codeine and methamphetamine, revealing that the antimorphine antibody did not fully detect the binding sites of these substances. As a result, the FRET mechanism was not inhibited since codeine and methamphetamine could not effectively bind. This specificity of the sensor demonstrated its efficacy in selectively detecting morphine (figure 7.4). [63].

Researchers investigated the drug piroxicam in another study involving a graphene-based sensor [64]. They noted that while this drug is widely used as an analgesic and anti-inflammatory agent for various diseases, it has also been prescribed in conjunction with azithromycin to control fever in patients during the

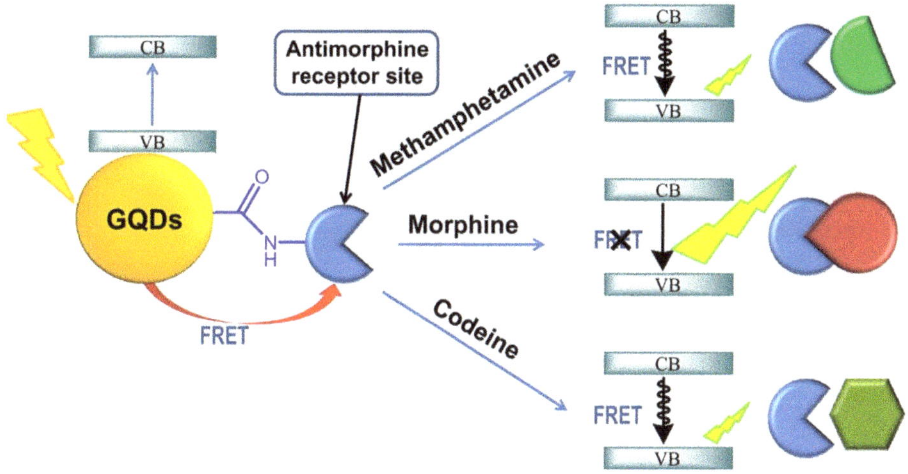

Figure 7.4. Functionalization of graphene quantum dots with antimorphine for detection of morphine. Reprinted from [63], copyright (2019), with permission from Elsevier.

recent COVID pandemic. The researchers emphasized the importance of detecting piroxicam due to its potential misuse and interaction with the environment. To this end, they developed a modified sensor using nitrogen-doped graphene (NGr) for the determination of piroxicam. GO was synthesized using the Hummers' method from graphite powder. Urea was used as a reducing agent, and nitrogen-doped graphene (NGr) was produced by subjecting GO to hydrothermal treatment at 160 °C for 8 h. The obtained NGr was drop-casted and dried onto a GC electrode, resulting in the NGr/GC sensor. Additionally, all sensor studies were repeated using thermally reduced graphene oxide (TRGO). The results were compared for GC, TRGO/GC, and NGr. Morphological and structural characterizations were performed using scanning transmission electron microscopy (STEM), energy-dispersive spectroscopy (EDS), x-ray powder diffraction, UV–visible (UV–vis) spectroscopy, and FTIR. For the measurement of piroxicam, the LSV electrochemical technique was preferred. The calibration graphs of the NGr/GC sensor exhibited two linear regions between concentrations of 10^{-6} to 5×10^{-5} M and 10^{-4} to 10^{-3} M for piroxicam. The limit of detection (LOD) was calculated as 3.03×10^{-7} M [64].

Ansari *et al* used a glassy carbon electrode (GCE) modified with a nanocomposite of WO_3-decorated graphene for the detection of ranolazine (RZ) which is an anti-anginal agent [65] (figure 7.5). WO_3 nanoparticles were synthesized using a conventional hydrothermal synthesis method. The resulting WO_3 nanoparticles were combined with graphene in a ratio of 1:2 and dispersed in *N*,*N*-dimethylformamide (DMF). The mixture was then subjected to ultrasonic treatment for a duration of 2 h. Subsequently, the resulting mixture was drop-casted onto a GCE and dried, resulting in the preparation of the WO_3/graphene/GCE. The structural and morphological characteristics of the modified electrode were thoroughly examined using techniques such as x-ray diffraction (XRD), FE-SEM, EDAX, HR-TEM, XPS, Raman spectroscopy, and FTIR spectroscopy, which confirmed the successful

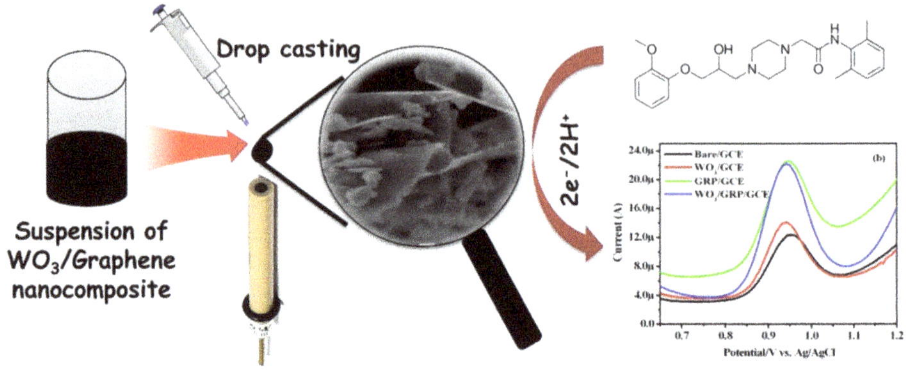

Figure 7.5. WO_3 decorated graphene nanocomposite based electrochemical sensor: a prospect for the detection of anti-anginal drug. Reprinted from [65], copyright (2019), with permission from Elsevier.

formation of the nanocomposite. The voltammetric response of RZ in Britton–Robinson buffer at pH 4.5 has been found well-defined, allowing for its quantification within the concentration range of 0.2–1.4 and 1.4–14 mM using square-wave voltammetry. The developed sensor has demonstrated a low detection limit of 0.13 mM. The WO_3/graphene/GCE sensor has exhibited excellent electron interceding behavior, attributed to the high electrical conductivity and large effective surface area provided by WO_3 nanoparticles and graphene nanosheets. The effectiveness of the proposed methodology has been further validated by successfully quantifying RZ in commercially available pharmaceutical tablets, achieving satisfactory recovery rates ranging from 99.8% to 100.2% [65].

In a recent study, a graphene-based biosensor for the detection of oxytocin was developed by Thomas and Balachandran [66]. Oxytocin is a hormone produced in the human body and is also used in the treatment of conditions such as autism, anxiety, depression, and other related health issues. Moreover, studies have shown its applications in livestock farming to increase milk production and in agriculture to promote fruit development. However, the misuse of oxytocin in various fields poses a threat to human health. The authors emphasize the need for a reliable diagnostic test and, accordingly, developed an electrochemical oxytocin sensor based on nitrogen and phosphorus co-doped graphene (NPG). In the study, graphite oxide was extracted using the Hummers' method. The obtained graphite oxide was mixed with phosphoric acid and ethylenediamine in an ultrasonic bath for one hour. Subsequently, the solution was subjected to a hydrothermal reaction at 180 °C in an autoclave to obtain NPG. The same procedure was applied with variations in components to obtain NG, PG, and reduced graphene oxide (rGO). For oxytocin measurements, NPG was immobilized on a platinum electrode using a Nafion binder. Surface characterizations were performed using scanning electron microscope (SEM), TEM, XPS, Raman spectroscopy, and CV. Following electrochemical optimization, oxytocin detection was carried out. Square-wave stripping voltammetry (SWSV) was employed, and a calibration curve was plotted, demonstrating linearity in two ranges: 0.1–10 and 15–95 nM. The LOD of the NPG electrode was calculated as 40 pM. Oxytocin was successfully determined in real samples at different concentrations, resulting in recoveries ranging from 90.01% to 105.23% [66].

In another study by Anupriya *et al*, a sensor was developed for the determination of dimetridazole (DMZ), which is used as an anti-protozoal, antifungal, and antibiotic drug, using nano-sized praseodymium tungstate and rGO (PW NPs/rGO) composite [67]. In the synthesis process, Na_2WO_4 and $Pr(NO_3)_3$ were mixed, acidified by adding HCl, and praseodymium peroxo tungstic acid was obtained by adding H_2O_2. Finally, $Pr_6W_2O_{15}$ (PW NPs) was synthesized using the solvent evaporation technique. rGO was synthesized using a modified Hummers' method. PW NPs and rGO were mixed in a 1:2 ratio and sonicated for 1 h. The resulting suspension was named PW NPs/rGO and drop-casted and dried on a GCE for use (PW NPs/rGO/GCE). Surface characterization of the PW NPs/rGO/GCE sensor was performed using FE-SEM, XRD, HR-TEM, Raman spectroscopy, and XPS techniques. Electrochemical characterization was conducted using CV and electrochemical impedance spectroscopy (EIS) techniques. DPV was used as the method for DMZ determination. Peak potential graphs against DMZ concentration showed a linear increase in two different regions. As a result, a DMZ sensor based on PW NPs/rGO/GCE was developed by researchers with a wide linear range of 0.001–965 μM DMZ concentrations and a low LOD of 1.2 nM [67].

Tseng *et al* developed a sensor for flutamide (FD), a crucial anticancer drug and non-steroidal anti-androgen chemical [68]. They proposed the synthesis of sphere-like, perovskite-type, bimetallic $CaTiO_3$ nanoparticles decorated with GO sheets through the following procedure. In the typical preparation, titanium(IV) butoxide had been dissolved in ethylene glycol, resulting in a homogeneous solution. Subsequently, $Ca(OH)_2{\cdot}8H_2O$ (Ca^{2+}) powders were added to the homogeneous solution, followed by the addition of deionized water. Finally, GO powder, prepared from graphite with a sonication bath, was added. The reaction solution was subjected to ultrasonic irradiation using a high-intensity ultrasonic probe at room temperature in air for 1 h. The resulting GOS/$CaTiO_3$-based perovskite nanomaterial had been collected through centrifugation and had undergone washing with an ethanol solution and deionized water. The GOS/$CaTiO_3$ nanocomposites were characterized morphologically and chemically using SEM, XRD, and EDS analytical methods. Subsequently, the authors modified the as-prepared nanocomposites onto a GCE to enable the electrocatalytic and electrochemical sensing of the chemotherapeutic drug flutamide (FD). For the preparation of the GOS/$CaTiO_3$ modified electrode, GOS/$CaTiO_3$ nanomaterials were drop-casted onto a cleaned carbon surface and allowed to dry at room temperature. The resulting electrodes had been denoted as GOS/$CaTiO_3$/GCE. Following the completion of optimization steps, the detection of flutamide (FD) was carried out utilizing an amperometric technique by measuring electrochemical reduction peaks of FD at a potential of −0.42 V. The designed sensor exhibited a wide linear range of 0.015–1184 μM. A LOD had been calculated at 5.7 nM. Furthermore, the GOS/$CaTiO_3$ modified electrodes were tested in human blood and urine samples for the detection of the anticancer drug [68] (figure 7.6).

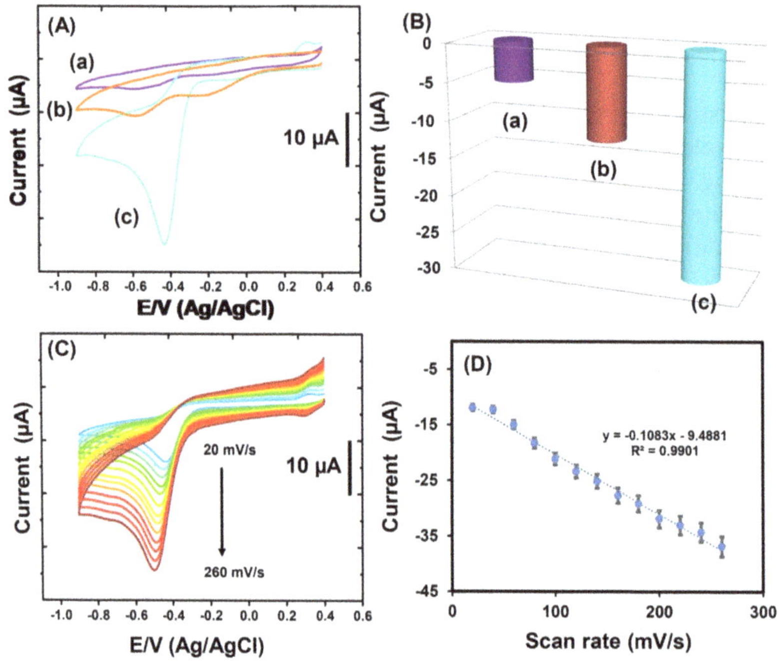

Figure 7.6. CVs of bare GCE (a), GOS/GCE (b), and GOS/$CaTiO_3$ NC/GCE (c) containing 100 μM flutamide in 0.05 M PB at the scan rate at 50 mV s^{-1} (A) and corresponding calibration plot (B). Different scan rate with flutamide (20–260 mV s^{-1}) in 0.05 M PB at GOS/$CaTiO_3$ NC/GCE (C). A plot between scan rate versus cathodic peak current (D). Reprinted from [68], copyright (2020), with permission from Elsevier.

In their study, the authors focused on the determination of acyclovir (ACV), an antiviral medication. While this antiviral drug is commonly used in the treatment of various skin diseases, the authors emphasized the potential health risks associated with its misuse and incorrect dosages. The main objective of the study was to develop a sensor for ACV monitoring by modifying a GCE with ferrous molybdate nanorods and GO composite. To obtain $FeMoO_4$ nanorods, a mixture of $FeCl_2$ and Na_2MoO_4 was stirred in a triethylene glycol solution, followed by hydrothermal treatment at 120 °C for 24 h. Subsequently, the $FeMoO_4$ nanorods were sonicated with GO in an *N,N* dimethylformamide solution to form the $FeMoO_4$–GO composite material. The modified $FeMoO_4$–GO/GCE sensor was prepared by drop-casting the $FeMoO_4$–GO onto a pre-cleaned GCE and allowing it to dry. The morphological characterization of the sensor was conducted using SEM, XRD, and EDS. The development steps of the sensor were electrochemically characterized using EIS and CV techniques. For the determination of ACV, LSV was employed as the electrochemical method. LSV was used to optimize parameters such as loading and deposition time, pH, and scan rate. The optimum conditions were determined as pH = 4.0, deposition time of 90 s, and scanning rate of 180 mV·s^{-1}, which were subsequently utilized in the subsequent stages of the experiment. $FeMoO_4$–GO/GCE sensor showed good linearity within the ranges of 0.1–10 μM and 10–100 mM ACV concentrations. The LOD was determined to be 20 nM. The sensor was successfully employed for the analysis of ACV drug samples, yielding high recoveries ranging from 95.0% to 100.0% [69].

Another study by Jesu Amalraj *et al*, focuses on the electrochemical development of a furazolidone (FZD) sensor utilizing an SnS_2–SnO_2/graphene composite [70]. FZD belongs to the nitrofuran group, exhibiting antibiotic properties with anti-bacterial and anti-protozoal activities. A core–shell-like structure was successfully synthesized via a facile sonication-assisted hydrothermal method by researchers. The synthesis involved the reaction of $SnCl_2$ and thioacetamide to obtain SnS_2–SnO_2. Subsequently, the SnS_2–SnO_2 was combined with graphene, and the resulting mixture underwent ultrasonication for a duration of 1 h to yield the SnS_2–SnO_2/graphene composite. Authors have examined the formation of the SnS_2–SnO_2/graphene composite through various physical characterizations, including x-ray crystallography (XRD), SEM, and FTIR spectroscopy. Subsequently, the synthesized SnS_2–SnO_2/graphene composite was drop-casted onto a pre-cleaned GCE surface to investigate the electrochemical oxidation of FZD. CV and DPV methods were used for optimization studies (figure 7.7). The optimum working potential has

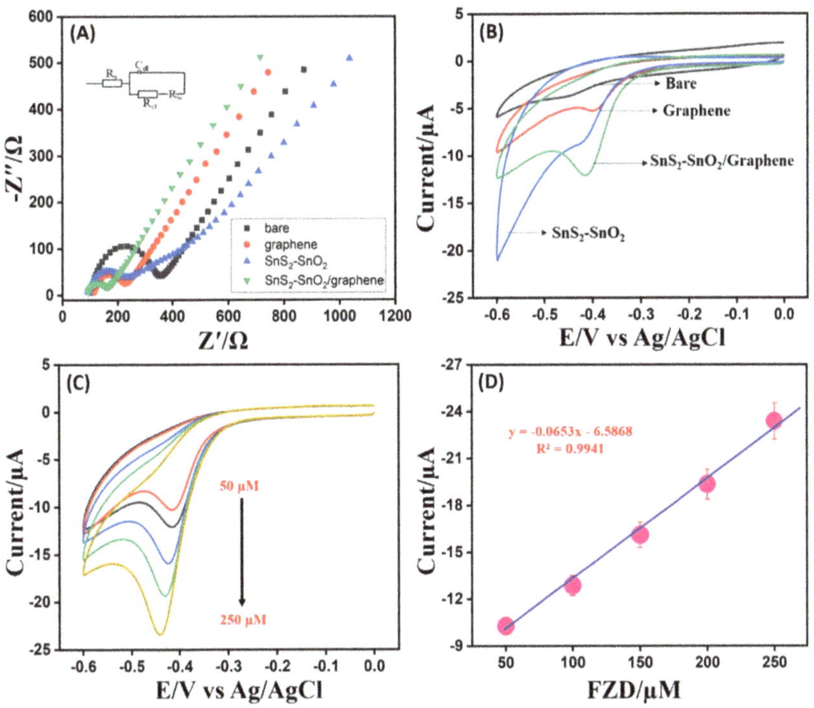

Figure 7.7. (A) Electrochemical impedance spectroscopy (EIS) of different modified electrodes at 0.1 M KCl comprising $[Fe(CN)_6]^{3-/4-}$ and inset image represents Randel's equivalent circuit model. (B) CVs at bare GCE, graphene, SnS_2–SnO_2, SnS_2–SnO_2/graphene-modified GCEs in 0.1 M PB (pH 7.0) containing 100 μM FZD (scan rate = 50 mV s^{-1}). (C) CVs observed at SnS_2–SnO_2/graphene/GCE in the presence of 50–250 μM FZD (electrolyte = 0.1 M PB solution (pH 7.0)) at the scan rate of 50 mV s^{-1}. (D) Calibration plot of concentration of FZD versus current obtained. Reprinted from [70], copyright (2020), with permission from Elsevier.

been found to be −0.516 V (versus Ag/AgCl) in a 0.1 M phosphate buffer (PB) solution at pH 7.0. The DPV results obtained with the SnS_2–SnO_2/graphene composite demonstrated a wide linear range of 0.015–190.5 μM with a lower LOD of 1.42 nM. The authors noted that SnS_2–SnO_2/graphene/GCE sensor had high selectivity, rapid response time, excellent repeatability, reproducibility, and long-term stability. Furthermore, the fabricated SnS_2–SnO_2/graphene/GCE sensor was employed for the determination of FZD in a commercially available furoxone tablet, and recovery results were obtained between 97.6% and 98.95% [70].

In another study, the authors focused on antibiotic determination, and ciprofloxacin (CIP) was selected as the analyte [71]. Graphene, in conjunction with sodium dodecyl sulfate (SDS), was utilized to fabricate the sensor. Initially, carbon paste electrodes (CPEs) were produced by blending graphite and parafilm at a ratio of 70:30 and filling the mixture into a Teflon tube. The CPE was then modified by mixing graphene and parafilm as a binder, resulting in the formation of the Gr/CPE electrode. The final form of the sensor, denoted as SDS·Gr/CPE, was achieved by introducing SDS into the electrolytic cell. Following the calculation of the active surface area, SEM and AFM techniques were employed to investigate the topography and morphology. The modification steps were further examined using CV and EIS techniques. Optimization of the sensor's accumulation time, pH, and electrolytic solution was conducted, yielding the most favorable results at pH 4.2 in a 0.2 M phosphate-buffered saline (PBS) medium. The researchers successfully determined CIP using the DPV technique with the developed sensor. The experimental findings demonstrated a linear relationship within the concentration range of 0.3–100 μM, with a LOD of 0.03 μM. Moreover, real case studies were performed following preliminary tests, reporting a recovery rate of 93.8% for pharmaceutical samples employing SDS·Gr/CPE and recovery rates of 98.7%, 95.0%, and 93.3% for samples containing 1, 2, and 3 μM CIP, respectively, in urine samples [71].

Another antibiotic sensor developed by Velusamy *et al* is about Metronidazole (MT, a renowned antibiotic drug extensively employed in the treatment of various medical conditions such as pelvic inflammatory disease, endocarditis, and bacterial vaginosis [72]. However, it is worth noting that excessive dosages and prolonged administration of MTZ can lead to adverse effects on human health. The interest is in accurately monitoring MTZ concentrations in real samples to ensure proper therapeutic management and minimize the occurrence of drug-related complications. For this purpose, the authors prepared CuO nanoparticle (CuO NPs) decorated graphene/b-cyclodextrin (GR-b-CD) composites using a simple electrochemical methodology for the detection of MTZ in this study (figure 7.8). The electrodeposition of CuO NPs onto modified electrodes composed of GR-b-CD composites was carried out. A stable GR-b-CD composite was briefly prepared by sonication of GR nanoflakes in a b-CD solution. The resulting GR-b-CD composite was drop-casted onto pre-cleaned GCEs and dried at room temperature. For the production of CuO NPs, an electrochemical cell containing $CuCl_2$ and KCl was prepared, and the pH of the solution was adjusted to 11 using NaOH. Subsequently, the GR-b-CD composite-modified GCE was immersed in the electrochemical cell, and CV was performed for electrodeposition. The synthesized GR-b-CD/CuO NPs

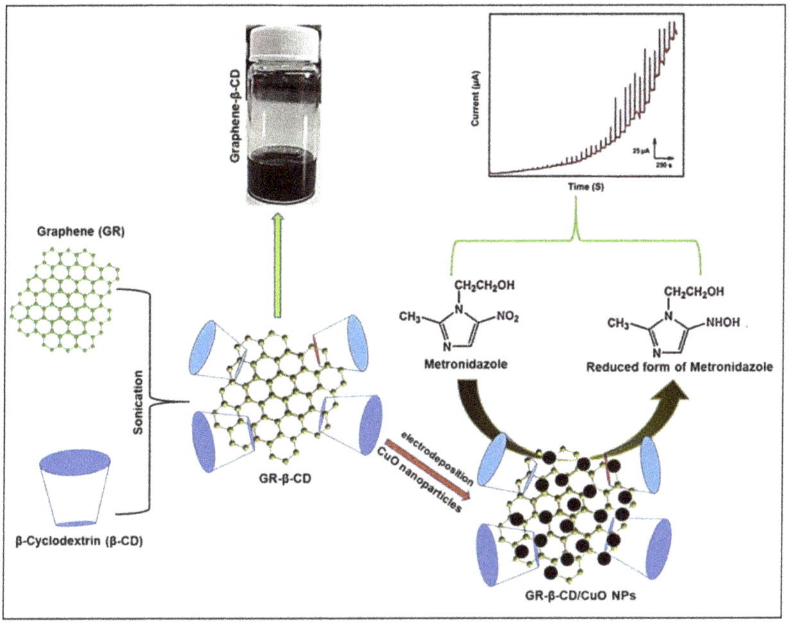

Figure 7.8. Schematic representation of novel electrochemical synthesis of copper oxide nanoparticles decorated graphene-β-cyclodextrin composite for trace-level detection of antibiotic drug metronidazole. Reprinted from [72], copyright (2018), with permission from Elsevier.

composites were characterized using high-resolution SEM, XRD, and Raman spectroscopy techniques. Electrochemical characterizations like catalytic activity and reduction potential, pH, and scan rate effects of the GR-b-CD/CuO NPs composite-modified electrodes for the electrochemical detection of metronidazole (MTZ) were investigated using CV measurements. Under optimized conditions, amperometry was employed for the determination of MTZ using the GR-b-CD/CuO NPs composite-modified electrodes. The composite electrodes exhibited a linear response in the concentration range of 0.002–210.0 mM for MTZ. The sensor demonstrated a remarkably low LOD of 0.6 nM. Additionally, the sensor exhibited high sensitivity, selectivity, and durability in the presence of a range of potentially interfering electroactive compounds [72].

In an alternative study by Sebastian *et al*, the researchers focused on the determination of the antibiotic chloramphenicol (CAP) using a nanocomposite of GO and 3D hierarchical zinc oxide (ZnO) [73]. CAP is a broad-spectrum antibiotic that was previously used in animal food production but is now prohibited in many European countries due to its adverse effects on human health. However, its use still persists in certain livestock areas and underdeveloped countries, necessitating the need for effective control measures in food products. During the development of the CAP sensor, GO was synthesized using a modified Hummers' method starting from graphite. Subsequently, 3D hierarchical ZnO nanostructures were synthesized through the reaction of a mixture of zinc nitrate and sodium citrate with sodium hydroxide. The synthesized GO and ZnO nanostructures were then combined and subjected to ultrasonic treatment for 2 h, forming GO/3D ZnO nanostructures. The composite material obtained was carefully deposited onto a cleaned GCE and dried to prepare the sensor for further analysis. The biosensor was characterized using SEM, EDX, FTIR, Raman spectroscopy, and UV–vis spectroscopy techniques. The electrochemical performance of the biosensor was evaluated using CV. Notably, the CV experiments for 100 μM CAP demonstrated that the GO/ZnO/GCE sensor exhibited superior anodic and cathodic peak currents compared to bare GCE or electrodes containing only ZnO or GO. The electroactive surface area was determined by measuring the CV scanning rate within the range of 20–200 mV s^{-1} in the presence of a 5 mM $[Fe(CN)_6]^{3-/4-}$ solution. The GO/ZnO/GCE modified electrode exhibited approximately seven times higher electroactive surface area compared to GCE alone, indicating the positive influence of the high surface area on the electrocatalytic performance. Following the optimization of pH and scan rate, a pH of 7.0 was identified as the optimum condition. Finally, CAP determination was carried out using the GO/ZnO/GCE electrode in a 0.05 M PBS solution with pH 7.0, employing the DPV technique. A calibration curve was constructed using the cathodic peak currents obtained at −0.58 V in DPV measurements conducted between −1 and +1 V using various concentrations of CAP. A linear relationship was observed in the concentration range of 0.212–7.212 μM CAP, described by the equation I (μA) = 0.117CCAP + 7.14 (R^2 = 0.99). The LOD was calculated as 0.01 μM using the formula LOD = 3s/b. Furthermore, the sensor demonstrated resilience against potential interfering substances in the environment, as evidenced by entrepreneurial trials [73].

In another study, researchers have developed a sensor for the determination of clonazepam (CNZ), a medication widely used in the treatment of mental disorders [74]. This drug is commonly prescribed for sleep problems, anxiety, and similar mental conditions. The authors emphasize the importance of accurately determining CNZ, particularly in beverages, due to its association with addiction, sexual abuse, and other criminal cases. To address this, they have developed an electrochemical sensor based on a hybrid nanocomposite of cobalt oxyhydroxide nanoflakes and rGO. In this work, GO was obtained from graphite powder using a modified Hummers' method and then transformed into rGO through the use of hydrazine hydrate. The cobalt oxyhydroxide nanoflakes were synthesized by reacting a $CoCl_2$ solution prepared in an alkaline environment with NaClO in an ultrasonic bath. The pH of the solution was subsequently neutralized with HCl, and the resulting nanoflakes were separated from the solution using centrifugation. The CoOOH nanoflakes were mixed with rGO in deionized water for approximately one hour to obtain the CoOOH–rGO nanocomposite. This mixture was then drop-casted onto screen-printed carbon electrodes (SPCEs) and dried at room temperature to develop the modified CoOOH–rGO/SPCE sensor (figure 7.9). The modified CoOOH–rGO/SPCE electrode was characterized using CV, EIS, as well as SEM, XRD, Raman spectroscopy, XPS, FTIR spectroscopy, and UV–vis spectroscopy for morphological and spectroscopic characterization. Following optimization of scan rate and pH, the determination of CNZ was initiated using the developed CoOOH–rGO/SPCE

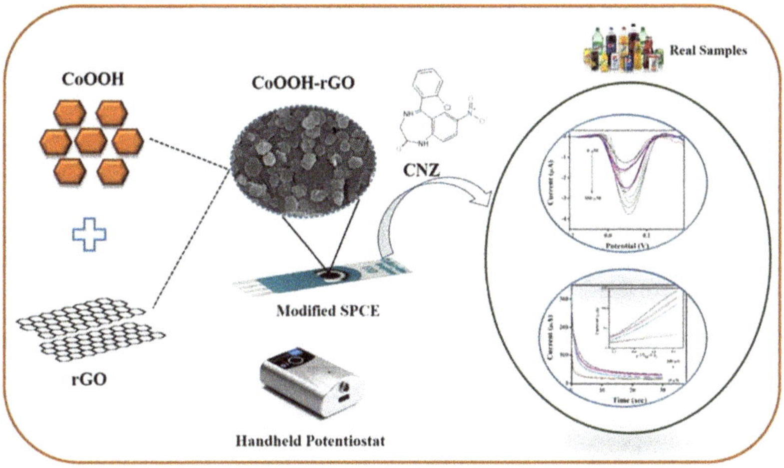

Figure 7.9. An electrochemical sensor based on cobalt oxyhydroxide nanoflakes/reduced graphene oxide nanocomposite for detection of illicit drug clonazepam. Reprinted from [74], copyright (2022), with permission from Elsevier.

sensor. The sensor that was developed demonstrated an impressive detection limit of 38 nM with excellent performance over a linear range of 0.1–350 μM. Furthermore, the sensor was successfully utilized for the determination of CNZ in spiked beverage samples, yielding satisfactory recovery rates ranging from 87% to 115%.

In a recent study, Saisahas *et al* developed a portable device for the determination of Xylazine [75]. This device is designed to work in conjunction with paper-based electrodes. The paper electrodes are produced in the form of nanocoral-like polyaniline-modified graphene-based electrodes. The authors applied wax barriers to chromatography papers using a wax printer. A cutting printer was used to create LTT (low-tack transfer tape) masks by cutting the LTT masks into the desired shape, which corresponds to the circuit pathways of the electrodes to be formed on the surface. Graphene ink was applied using screen printing with the LTT masks to create the circuit pathways. By removing the LTT masks, the desired design of graphene-based paper electrodes was obtained. To develop the modified electrodes, a coral-like composite of polyaniline (PANI) was synthesized through the polymerization of the aniline monomer in the presence of a 25% sodium chloride (NaCl) solution. The obtained PANI was drop-casted onto the active centers of the electrodes and dried at 70 °C. Morphological characterization of the developed sensors was performed using SEM, FE-SEM, and FTIR techniques. Xylazine determination experiments were carried out using these modified electrodes. Electrochemical characterizations were conducted using CV and EIS techniques. Xylazine measurements were performed using a sensor device developed by the authors, which was connected to a mobile phone. The DPV technique was employed. The researchers obtained linear calibration graphs in two different concentration ranges for Xylazine measurements: 0.2–5 and 5–100 μg ml^{-1}. The LOD was calculated as 0.06 μg ml^{-1}. In experiments where Xylazine was added to certain beverages, the produced sensor achieved a recovery rate ranging from 84 ± 4% to 105 ± 2% [75].

7.4 Conclusion

Graphene's unique structure, exceptional properties, and potential applications have captured the interest of scientists and engineers worldwide. Through various synthesis methods, researchers continue to explore the vast possibilities of graphene in electronics, energy storage, composites, and biomedicine. This chapter provides an overview of the fundamental aspects of graphene, setting the stage for the subsequent chapters that delve deeper into specific applications and advancements in drug analysis.

Acknowledgments

We express our deep gratitude to our great leader M Kemal Atatürk, as we commemorate the 100th anniversary of the Republic of Türkiye in 2023.

References

[1] Pareek A and Mohan S V 2019 Graphene and its applications in microbial electrochemical technology *Microbial Electrochemical Technology Sustainable Platform for Fuels, Chemicals and Remediation Biomass, Biofuels and Biochemicals* (Amsterdam: Elsevier) pp 75–97

[2] Manzetti S, Vasilache D and Francesco E 2015 Emerging carbon-based nanosensor devices: structures, functions and applications *Adv. Manuf.* **3** 63–72

[3] Siqueira J R and Oliveira O N 2016 Carbon-based nanomaterials *Nanostructures* (Hoboken, NJ: Wiley) pp 233–49

[4] Geim A K and Novoselov K S 2007 The rise of graphene *Nat. Mater.* **6** 183–91

[5] Boukhvalov D W and Katsnelson M I 2008 Modeling of graphite oxide *J. Am. Chem. Soc.* **130** 10697–701

[6] Geim A K 2009 Graphene: status and prospects *Science* **324** 1530–4

[7] Tahriri M, Del Monico M, Moghanian A, Tavakkoli Yaraki M, Torres R, Yadegari A and Tayebi L 2019 Graphene and its derivatives: opportunities and challenges in dentistry *Mater. Sci. Eng. C.* **102** 171–85

[8] Kuila T, Bose S, Mishra A K, Khanra P, Kim N H and Lee J H 2012 Chemical functionalization of graphene and its applications *Prog. Mater Sci.* **57** 1061–105

[9] Peres N M R 2009 The electronic properties of graphene and its bilayer *Vacuum* **83** 1248–52

[10] Allen M J, Tung V C and Kaner R B 2010 Honeycomb carbon: a review of graphene *Chem. Rev.* **110** 132–45

[11] Choi W, Lahiri I, Seelaboyina R and Kang Y S 2010 Synthesis of graphene and its applications: a review *Crit. Rev. Solid State Mater. Sci.* **35** 52–71

[12] Nebol'sin V A, Galstyan V and Silina Y E 2020 Graphene oxide and its chemical nature: multi-stage interactions between the oxygen and graphene *Surf. Interfaces* **21** 100763

[13] Dacheng Wei B, Liu Y, Wei D C and Liu Y Q 2010 Controllable synthesis of graphene and its applications *Adv. Mater.* **22** 3225–41

[14] Yan J A and Chou M Y 2010 Oxidation functional groups on graphene: structural and electronic properties *Phys. Rev.* B **82** 125403

[15] Dreyer D, Park S, Christopher B and Ruaff R 2010 The chemistry of graphene oxide *Chem. Soc. Rev.* **39** 228–40
[16] Catania F, Marras E, Giorcelli M, Jagdale P, Lavagna L, Tagliaferro A and Bartoli M 2021 A review on recent advancements of graphene and graphene-related materials in biological applications *Appl. Sci.* **11** 614
[17] Huang X, Yin Z, Wu S, Qi X, He Q, Zhang Q, Yan Q, Boey F and Zhang H 2011 Graphene-based materials: synthesis, characterization, properties, and applications *Small* **7** 1876–902
[18] Ţucureanu V, Obreja C A, Crăciun G, Romaniţan C, Mihăilescu C M, Stan D and Matei A 2022 Preparation and evaluation of nanocomposites based on transitional oxides and carbon materials for electrochemical applications *Ceram. Int.* **48** 27201–12
[19] Kumar A, Sharma K and Dixit A R 2019 A review of the mechanical and thermal properties of graphene and its hybrid polymer nanocomposites for structural applications *J. Mater. Sci.* **54** 5992–6026
[20] Liu J, Liu Z, Barrow C J and Yang W 2015 Molecularly engineered graphene surfaces for sensing applications: a review *Anal. Chim. Acta* **859** 1–19
[21] Ananda Murthy H C, Gebremedhn Kelele K, Ravikumar C R, Nagaswarupa H P, Tadesse A and Desalegn T 2021 Graphene-supported nanomaterials as electrochemical sensors: a mini review *Results Chem.* **3** 100131
[22] Jastrzębska A M, Kurtycz P and Olszyna A R 2012 Recent advances in graphene family materials toxicity investigations *J. Nanopart. Res* **14** 1320
[23] Tran L T, Tran H V, Dang H T M, Huynh C D and Mai T A 2020 Silver nanoparticles decorated polyaniline nanowires-based electrochemical DNA sensor: two-step electrochemical synthesis *J. Electrochem. Soc.* **167** 087508
[24] Singh E, Meyyappan M and Nalwa H S 2017 Flexible graphene-based wearable gas and chemical sensors *ACS Appl. Mater. Interfaces* **9** 34544–86
[25] Hulanicki A, Glab S and Ingman F 1991 Chemical sensors: definitions and classification *Pure Appl. Chem.* **63** 1247–50
[26] Kurbanoglu S, Bakirhan N K, Shah A and Ozkan S A 2019 Chemical nanosensors in pharmaceutical analysis *New Developments in Nanosensors for Pharmaceutical Analysis* (Amsterdam: Elsevier) pp 141–70
[27] Wang Y, Xu H, Zhang J and Li G 2008 Electrochemical sensors for clinic analysis *Sensors* **8** 2043–81
[28] Kimmel D W, Leblanc G, Meschievitz M E and Cliffel D E 2012 Electrochemical sensors and biosensors *Anal. Chem.* **84** 685–707
[29] Hanrahan G, Patil D G and Wang J 2004 Electrochemical sensors for environmental monitoring: design, development and applications *J. Environ. Monit.* **6** 657–64
[30] Zhu Y C, Mei L P, Ruan Y F, Zhang N, Zhao W W, Xu J J and Chen H Y 2019 Enzyme-based biosensors and their applications *Advances in Enzyme Technology* 1st edn (Amsterdam: Elsevier) pp 201–23
[31] Kurbanoglu S, Erkmen C and Uslu B 2020 Frontiers in electrochemical enzyme based biosensors for food and drug analysis *TrAC – Trends Anal. Chem.* **124** 1–24
[32] Anik U 2017 Electrochemical medical biosensors for POC applications *Medical Biosensors for Point of Care (POC) Applications* (Amsterdam: Elsevier) pp 275–92
[33] Stradiotto N R, Yamanaka H and Zanoni M V B 2003 Electrochemical sensors: a powerful tool in analytical chemistry *J. Braz. Chem. Soc.* **14** 159–73

[34] Zhao Q, Gan Z and Zhuang Q 2002 Electrochemical sensors based on carbon nanotubes *Electroanalysis* **14** 1609–13

[35] Ahammad A J S, Lee J-J and Rahman M A 2009 Electrochemical sensors based on carbon nanotubes *Sensors* **9** 2289–319

[36] Ozkan S A and Uslu B 2016 From mercury to nanosensors: past, present and the future perspective of electrochemistry in pharmaceutical and biomedical analysis *J. Pharm. Biomed. Anal.* **130** 126–40

[37] Agrawal S and Prajapati R 2012 Nanosensors and their pharmaceutical applications: a review *Int. J. Pharm. Sci. Nanotechnol.* **4** 1528–36

[38] Erkmen C, Unal D N, Kurbanoglu S and Uslu B 2022 Basics of electrochemical sensors *Organic Electrodes, Fundamental to Advanced Emerging Applications* (Springer) pp 81–99

[39] Simões F R and Xavier M G 2017 Electrochemical sensors *Nanoscience and Its Applications* (Amsterdam: Elsevier) pp 155–78

[40] Issa Y M, Khalil M M, Zayed S I M and Hussein A 2009 Flow injection potentiometric sensor for determination of phenylpropanolamine hydrochloride *Arab. J. Chem.* **2** 41–6

[41] Badr I H A, Saleh G A, Sayed S M and Nour El-Deen D A M 2014 A novel membrane sensor for batch and flow injection potentiometric determination of cefazolin sodium in pharmaceutical preparations *Int. J. Electrochem. Sci.* **9** 1621–36

[42] Stradiotto N R, Yamanaka H and Zanoni M V B 2003 Review electrochemical sensors: a powerful tool in analytical chemistry *J. Braz. Chem. Soc.* **14** 159–73

[43] Koncki R 2007 Recent developments in potentiometric biosensors for biomedical analysis *Anal. Chim. Acta* **599** 7–15

[44] Rocchitta G *et al* 2016 Enzyme biosensors for biomedical applications: strategies for safeguarding analytical performances in biological fluids *Sensors* **16** 780

[45] Bóka B, Korózs M, Nánási M and Adányi N 2015 Novel amperometric tri-enzyme biosensor for lysine determination in pharmaceutical products and food samples *Electroanalysis* **27** 817–24

[46] Amatatongchai M, Sroysee W, Chairam S and Nacapricha D 2017 Amperometric flow injection analysis of glucose using immobilized glucose oxidase on nano-composite carbon nanotubes-platinum nanoparticles carbon paste electrode *Talanta* **166** 420–7

[47] Stefano J S, Tormin T F, da Silva J P, Richter E M and Munoz R A A 2017 Amperometric determination of omeprazole on screen-printed electrodes using batch injection analysis *Microchem. J.* **133** 398–403

[48] Salimi A, Roushani M, Haghighi B and Soltanian S 2006 Amperometric detection of insulin at renewable sol-gel derived carbon ceramic electrode modified with nickel powder and potassium octacyanomolybdate(IV) *Biosens. Bioelectron.* **22** 220–6

[49] Amine A and Mohammadi H 2019 Amperometry *Encyclopedia of Analytical Science* (Amsterdam: Elsevier) pp 85–98

[50] Patel B A 2020 Amperometry and potential step techniques *Electrochemistry for Bioanalysis* (Amsterdam: Elsevier) ch 2 pp 9–26

[51] Islam M N and Channon R B 2019 Electrochemical sensors *Bioengineering Innovative Solutions for Cancer* (Amsterdam: Elsevier) pp 47–71

[52] Uslu B and Ozkan S A 2011 Electroanalytical methods for the determination of pharmaceuticals: a review of recent trends and developments *Anal. Lett.* **44** 2644–702

[53] Mirceski V, Gulaboski R, Lovric M, Bogeski I, Kappl R and Hoth M 2013 Square-wave voltammetry: a review on the recent progress *Electroanalysis* **25** 2411–22

[54] Compton R G and Banks C E 2010 *Understanding Voltammetry* (Singapore: World Scientific)

[55] Gosser D K 1994 *Cyclic Voltammetry* (New York: VCH)

[56] Scholz F *et al* 2010 *Electroanalytical Methods: Guide to Experiments and Applications* 2nd edn (Berlin: Springer)

[57] Brajter-toth A and Chambers J 2002 *Electroanalytical Methods of Biological Materials* (Boca Raton, FL: CRC Press)

[58] Ozkan S A 2012 *Electroanalytical Methods in Pharmaceutical Analysis and Their Validation* 1st edn (HNB Publishing)

[59] Chen K, Chou W, Liu L, Cui Y, Xue P and Jia M 2019 Electrochemical sensors fabricated by electrospinning technology: an overview *Sensors (Switzerland)* **19** 3676

[60] Aftab S, Kurbanoglu S, Ozcelikay G, Shah A and Ozkan S A S A 2019 NH_2-functionalized multi walled carbon nanotubes decorated with ZnO nanoparticles and graphene quantum dots for sensitive assay of pimozide *Electroanalysis* **31** 1083–94

[61] Santos A M, Silva T A, Vicentini F C and Fatibello-Filho O 2020 Flow injection analysis system with electrochemical detection for the simultaneous determination of nanomolar levels of acetaminophen and codeine *Arab. J. Chem.* **13** 335–45

[62] Soman G, Vandana M and Hegde G 2023 Molecularly imprinted graphene based biosensor as effective tool for electrochemical sensing of uric acid *Sens. Int.* **4** 100243

[63] Masteri-Farahani M and Mosleh N 2019 Functionalization of graphene quantum dots with antimorphine: design of selective nanosensor for detection of morphine *Mater. Lett.* **241** 206–9

[64] Varodi C, Coros M, Pogăcean F, Ciorîţă A, Turza A and Pruneanu S 2022 Nitrogen-doped graphene-based sensor for electrochemical detection of piroxicam, a NSAID drug for COVID-19 patients *Chemosensors* **10** 47

[65] Ansari S, Ansari M S, Satsangee S P and Jain R 2019 WO_3 decorated graphene nanocomposite based electrochemical sensor: a prospect for the detection of anti-anginal drug *Anal. Chim. Acta* **1046** 99–109

[66] Thomas R and Balachandran M 2023 Heteroatom engineered graphene-based electrochemical assay for the quantification of high-risk abused drug oxytocin in edibles and biological samples *Food Chem.* **400** 134106

[67] Anupriya J, Karuppusamy N, Chen S M and Lin K Y 2022 Synergistically improved electrochemical performance by the assembly of nanosized praseodymium tungstate on reduced graphene oxide for the detection of dimetridazole in biological and aquatic samples *J. Environ. Chem. Eng.* **10** 108800

[68] Tseng T W, Rajaji U, Chen T W, Chen S M, Huang Y C, Mani V and Irudaya Jothi A 2020 Sonochemical synthesis and fabrication of perovskite type calcium titanate interfacial nanostructure supported on graphene oxide sheets as a highly efficient electrocatalyst for electrochemical detection of chemotherapeutic drug *Ultrason. Sonochem.* **69** 105242

[69] Wei Y, Yao L, Wu Y, Liu X, Feng J, Ding J, Li K and He Q 2022 Ultrasensitive electrochemical detection for nanomolarity Acyclovir at ferrous molybdate nanorods and graphene oxide composited glassy carbon electrode *Colloids Surf.* A **641** 128601

[70] Jesu Amalraj A J, Umesh N M and Wang S F 2020 Synthesis of core–shell-like structure SnS_2-SnO_2 integrated with graphene nanosheets for the electrochemical detection of furazolidone drug in furoxone tablet *J. Mol. Liq.* **313** 113554

[71] Sawkar R R, Shanbhag M M, Tuwar S M, Mondal K and Shetti N P 2022 Sodium dodecyl sulfate–mediated graphene sensor for electrochemical detection of the antibiotic drug: ciprofloxacin *Materials (Basel)* **15** 7872
[72] Velusamy V, Palanisamy S, Kokulnathan T, Chen S W, Yang T C K, Banks C E and Pramanik S K 2018 Novel electrochemical synthesis of copper oxide nanoparticles decorated graphene-β-cyclodextrin composite for trace-level detection of antibiotic drug metronidazole *J. Colloid Interface Sci.* **530** 37–45
[73] Sebastian N, Yu W C and Balram D 2019 Electrochemical detection of an antibiotic drug chloramphenicol based on a graphene oxide/hierarchical zinc oxide nanocomposite *Inorg. Chem. Front.* **6** 82–93
[74] Sachdev G A and Matai I 2022 An electrochemical sensor based on cobalt oxyhydroxide nanoflakes/reduced graphene oxide nanocomposite for detection of illicit drug-clonazepam *J. Electroanal. Chem.* **919** 116537
[75] Saisahas K, Soleh A, Promsuwan K, Saichanapan J, Phonchai A, Sadiq N S M, Teoh W K, Chang K H, Abdullah A F L and Limbut W 2022 Nanocoral-like polyaniline-modified graphene-based electrochemical paper-based analytical device for a portable electrochemical sensor for xylazine detection *ACS Omega* **7** 13913–24

IOP Publishing

Graphene-Based Sensors

Jamballi G Manjunatha

Chapter 8

Graphene-based sensing platform for analysis of food flavours and additives

Raunak Katiyar, Yashi Dwivedi, Titiksha Kumar Sagar and Sandeep Chandrashekharappa

The exceptional characteristics of graphene with the connected ordered framework, including its huge specific surface area, ample gaps, and free binder, suggest that it might provide an effective electrode material for an electrochemical sensor. The 3D graphene-based electrochemical sensor does, however, have some drawbacks, including poor selectivity, extended dynamic responses, and insufficient sensitivity. In addition, numerous nanomaterials have been gradually added to the graphene-based electrode, including metals and metal compounds, metal–organic frameworks, polymers, carbon-based materials, etc. The electrochemical sensor was given the distinct benefits of cheap cost, convenience, high sensitivity, great selectivity, and high efficiency for the detection of analytes thanks to the thoughtful design and manufacture of these components. As a result, there are now more applications for food analysis, with a focus on quality and safety. These include heavy metal ions, pesticide residues, food additives, pathogenic bacteria, etc.

8.1 Introduction

Food safety is still a primary issue affecting individuals everywhere. Unsafe concentrations of chemicals and food containing biological toxins pose a major risk to the public and food supplies and health. The economies of developing countries are most impacted by foodborne infections, according to the World Health Organization (WHO). The problems with food safety in underdeveloped nations are widely known; estimates show that there are approximately 1500 cases of diarrhea each year worldwide, 75% of which are caused by biological food poisoning and cause 3 million deaths. Food safety is one of the WHO's top 11 priorities. According

doi:10.1088/978-0-7503-5578-0ch8

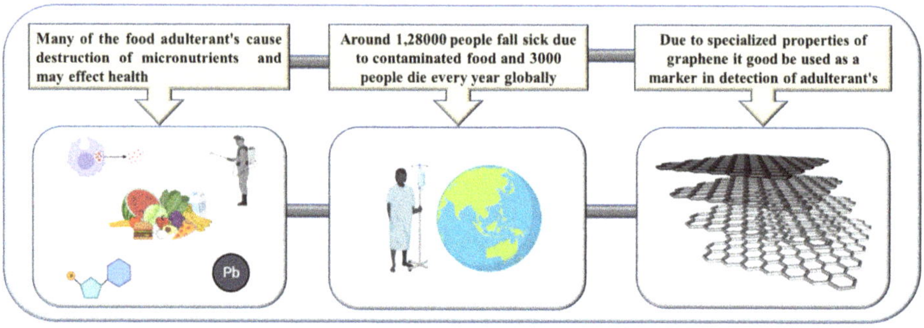

Figure 8.1. Current scenario of food safety.

to the US Centres for Disease Control and Prevention, 128 000 people are hospitalized, 3000 people pass away from foodborne illnesses, and 48 million Americans get sick because of contaminated food [1].

The development of quick and simple tests that can accurately determine the presence of hazardous pollutants in food is crucial for managing and resolving the issues associated with foodborne illnesses. This allows for the implementation of corrective measures. Only a small number of samples can be analysed, and most analysis of chemical and biological pollutants is done in centralised laboratories [2]. To ease the remote screening of possibly contaminated samples at food production and handling facilities where resources or specialized equipment are not available, more efficient procedures are required. While there are a limited number of applications for food safety, a variety of enzyme-based and bioaffinity assays have been published as alternatives to traditional analytical instruments. Despite advancements, these tests are still time-consuming, expensive, need many analytical processes, inclusion of delicate chemicals, expensive equipment, and are not portable. The monitoring and control of food quality and safety might be greatly enhanced by inexpensive, simple techniques that could be implemented remotely to site locations [3] (figure 8.1 explains about the current global situation of food safety).

The wonder substance known as graphene has numerous accolades to its credit. It is the strongest substance ever measured and the thinnest material known to exist. At normal temperature, its charge carriers demonstrate enormous intrinsic mobility, have no effective mass, and can move for micrometres without scattering. In addition to having a record thermal conductivity, stiffness, and gas permeability, graphene can support current densities six orders of magnitude higher than those of copper while resolving the tension between its brittleness and ductility [4]. A Dirac-like equation that describes electron transport in graphene enables benchtop experiments to study relativistic quantum events. A single graphite atomic layer called graphene is sufficiently isolated from its surroundings to be regarded as freestanding, and this is crucial. Of course, everyone is aware with atomic planes as the building blocks of bulk crystals, but one-atom-thick materials like graphene were unknown. The main cause of this is that low-dimensional (low-D) crystal formation is severely prohibited by Nature. The stability of macroscopic 1D and two-dimensional (2D) objects is negatively impacted by thermal fluctuations, which are implied by crystal formation at high temperatures (T). Nanometer-sized crystals and flat molecules can both be grown [5].

8.2 Chemistry behind graphene

The perfect surface is graphene, which has two faces with no bulk in between. Although the physics of this surface is currently in the spotlight, its chemistry has largely gone undiscovered. According to what we currently know about the chemistry of graphene, it can adsorb and desorb a variety of atoms and molecules, including NO_2, NH_3, K, and OH, just like the surface of graphite [6]. Weakly connected adsorbates frequently serve as acceptors or donors and the face of graphene is not flat but typically exhibits nanometer-scale corrugations. Local reactivity might be significantly influenced by the related strain and curvature. Third, chemical bonds that would be unstable if just one graphene surface were exposed can be formed because reagents can connect to both graphene faces, which changes the energetics [7]. Consider graphene as a massive flat molecule as an alternative to the surface chemistry viewpoint. Graphene can take part in chemical processes just like any other molecule. The key distinction between the two points of view is that in the latter scenario, it is implicitly believed that adsorbates will connect to the carbon scaffold in a stoichiometric fashion, or periodically as opposed to randomly. New 2D crystals with distinctive electronic structures and unique electrical, optical, and chemical capabilities could come from this. Graphene, a 2D hydrocarbon with one hydrogen atom connected to each site of the honeycomb lattice, is the first example that is currently understood [7, 8]. Because adsorbates are anticipated to self-organize into periodic patterns, like the case of graphite, which is widely known for its surface superstructures, many different graphene-based crystals should be feasible. In the search for new graphene-based 2D crystals, doping with atomic hydrogen (as in graphane), F–, OH–, and numerous functional groups seem to be promising candidates [9] (figure 8.2 rules out the properties of graphene, which differentiates it from other different fabricating platforms).

To regulate the electrical structure, for instance, stoichiometric derivatives are useful in many applications, including electronics. Probably even locally, chemical changes can be induced. Imagine a circuit built entirely of graphene where the interconnects are composed of unaltered graphene and other regions are changed to make them semiconducting and allow transistors [10]. Despite an abundance of potential discoveries and considering its potential applications, scientists have not shown much interest in graphene chemistry. It should be noted that graphene not being standard is one factor, not a typical molecule or a surface. The absence of samples appropriate for conventional chemistry has likely been the biggest challenge, though. But there have been recent developments in creating graphene suspensions [9, 11]. There is now access to liquid-phase chemistry, and preferably this will lead to the expert guidance that graphene researchers have anxiously awaited.

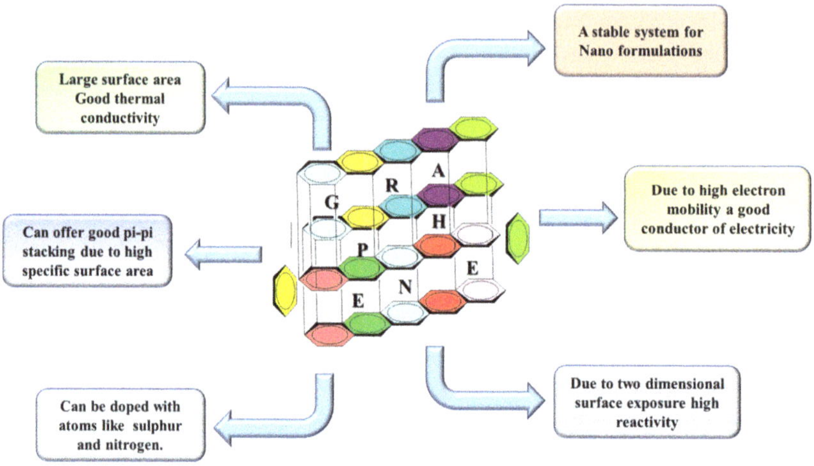

Figure 8.2. Properties of graphene.

8.3 Quantum stability

The special characteristics of graphene result from the behaviour of electrons. That, in and of itself, is nothing new because we already know that unexpected collective motions can arise when many particles interact strongly with one another, as summed up in Philip Anderson's famous adage 'more is different.' However, in the case of graphene, the electrons behave as if they have zero mass due to their interaction with the honeycomb structure [12]. As a result, the electrons in graphene are known as Dirac fermions and are governed by the Dirac equation, which is a quantum-mechanical description of electrons travelling relativistically [13]. The Canadian theorist Philip Russell Wallace made the first prediction on the relativistic behaviour of electrons in graphene in 1947. Wallace, however, selected the graphene model as his starting point to explore graphite because at the time no one thought that a solid could be one atom thick. On the foundation of his work, extensive research into the thermodynamic and transport properties of graphite was conducted in the 1960s [14]. The remarkable concordance between theoretical predictions and experimental data for properties like the heat capacity is regarded as one of the pinnacles of condensed-matter physics [15]. In high-energy particle physics, neutrinos are examples of massless Dirac fermions that we are already accustomed to. Since neutrinos have no electric charge, they interact weakly with all types of matter. In contrast, the Dirac fermions in graphene have a single electric charge and may be controlled by electromagnetic fields. Since the manipulation of electrons within materials is at the core of contemporary electronics, graphene's significantly unusual electron behaviour may enable us to develop semiconductor technology that is more advanced than silicon-based alternatives [14].

For instance, the unique Hall effect is the distinguishing characteristic that sets a graphene sheet apart from a typical metal. In the original Hall effect, which was discovered in 1879, a current flowing over a metal's surface in the presence of a transverse magnetic field results in a drop in potential that is perpendicular to the current and magnetic field [16]. The Hall effect is used to detect magnetic fields because the ratio of the potential drop to the current flowing (known as the Hall resistivity) is directly proportional to the applied magnetic field. In graphene, electrons behave as though they have no mass due to interactions with the honeycomb-like arrangement of carbon atoms. A periodic potential is produced by the ordered atomic arrangement in any crystal, which causes electrons to flow in waves known as Bloch waves [17].

For many years, it was predicted that carbon nanotubes, with their miniscule dimensions and extremely low electrical resistance, would result in a revolution in nanoelectronics. These hopes, however, have not yet been realised due to several challenges. These include scaling up the manufacture of well-defined nanotube diameters, overcoming the high resistance at the interfaces between nanotubes and the metal contacts connecting them to circuits, and assembling nanotube-based electrical devices on a large scale [12].

8.4 Production of graphene

Progress in producing graphene with the right properties for each individual application is essentially what drives the market for graphene applications, and this situation is likely to persist for the foreseeable future or at least until each of the numerous potential applications for graphene can meet its own requirements [18]. Any procedure for producing or extracting graphene, based on the required size, purity, and efflorescence of the particular result, is referred to as synthesising graphene. In the earlier stage, numerous methods for creating thin graphitic films were discovered. On the surfaces of transition metals in the late 1970s, carbon precipitated as thin graphitic coatings [19]. Few-layer graphite was created in 1975 on a single crystal of platinum using chemical breakdown procedures, but it wasn't recognised as graphene due to a lack of characterization methods or perhaps because of its constrained potential uses.

Because they were difficult to isolate and transfer to insulating substrates, their electronic properties were never studied during those times [20].

Here, we solely focus on those that can be scaled. According to the quality of the resulting graphene (and consequently the potential applications), these can be divided into three categories: (1) graphene or reduced graphene oxide flakes for composite materials, conductive paints, and other products; (2) planar graphene for active and passive devices with lower performance; and (3) planar graphene for high-performance electronic devices. The qualities of a given grade of graphene (and consequently the range of applications that can make use of it) depend greatly on the calibre of the material, type of flaws, substrate, etc, all of which are significantly influenced by the manufacturing process [21]. For the synthesis of graphene, numerous methods have been developed recently, though mechanical chemical cleaving, thermal chemical vapour, chemical synthesis, and exfoliation (cleaving). The most widely utilised techniques now are chemical vapour deposition (CVD) synthesis. Microwave synthesis and unzipping nanotubes are two more approaches that have been documented. Although it was discovered that mechanical exfoliation using an AFM cantilever could produce few-layer graphene, the technique constraint was that graphene thickness only varied to *10 nm. Here we briefly describe all the methods for graphene synthesis [22].

8.5 Mechanical exfoliation

The most uncommon and important method for obtaining single-layer graphene flakes on selected substrates may be mechanical exfoliation. It is the first known technique for producing graphene. This top-down nanotechnology method stresses the materials that make up the layered structure's surface longitudinally or transversely [23].

Weak van der Waals forces cause monoatomic graphene layers to stack together to create graphite. The *in situ* Joule heating of suspended few-layer graphene within a transmission electron microscope causes sublimation. The graphene sublimation fronts primarily include 1100 zigzags [24]. A fractal-like 'coastline' morphology was seen under the right circumstances. At the graphene edges, extensive multiple-layer reconstructions created novel carbon nanostructures such sp^2-bonded bilayer edges (BLEs) and nanotubes attached to BLEs. Nanotubes tunnelling multiple layers of graphene sheets and flat fullerenes/nanopods were also seen. Surprisingly, BLEs rather than monolayer edges (MLEs) account for >99% of the graphene edges seen during sublimation, demonstrating that BLEs are the stable graphene edges at high temperatures. By using kinetic Monte Carlo (kMC) simulations, we replicated the 'coastline' sublimation morphologies. The simulation demonstrated geometrical and topological characteristics specific to reconstructions and sublimations of 2D graphene. These reconstructions were made possible via bending, which is not possible. These findings suggest that the multiple-layer graphene substrate can provide special potential for designing innovative nanodevices with complicated topologies made of carbon-based nanostructures [24].

8.6 Chemical exfoliation

One of the finest methods for creating graphene is chemical synthesis. Producing colloidal suspension through a chemical process that modifies graphite and a graphite intercalation component into graphene. A chemical process has already been utilised to produce graphene in several types of paper, including material, polymer composites energy storage materials, and transparent conductive electrodes [26, 27]. Exfoliation is caused by a highly exothermic reaction between potassium intercalated graphite and aqueous solvents, The separation of the graphitic layers is aided by the potassium ion solution and the evolution of hydrogen gas. The production of potassium ethoxide makes the resulting solution basic, which is eliminated by washing with ethanol until a neutral pH is attained. Over six times as much space as the initial graphite material is taken up by the carbon nanoscrolls. According to thermal studies, the nanoscrolls oxidise to carbon dioxide in the air at 450 °C. This is 200 °C cooler than pure graphite, which can withstand air temperatures of up to 650 °C. Lowering of thermal stability is anticipated because the carbon nanoscrolls significantly have fewer van der Waals contacts than the beginning graphite powder [27].

8.7 Chemical method by reduction of graphene oxide

One of the common methods for producing significant amounts of graphene is chemical reduction of graphite oxide. According to the Brodie, Staudenmaier, and Hummers' methods, concentrated sulfuric acid, nitric acid, and potassium permanganate are used to oxidise graphite to produce graphite oxide (GO) [11]. Sonication and reduction of graphene oxide (GO) is a different method of producing graphene. Along with the extrusion of nitrogen gas and the addition of H_2 across the alkenes, a significant amount of $NaBH_4$ has been utilised as a reducing agent. Phenyl hydrazine, hydroxylamine, glucose, ascorbic acid, hydroquinone, alkaline solutions, and pyrrole are further reducing agents that have been utilised [28]. Then, graphene films are produced by thermally or chemically reducing the graphene oxide using a straightforward, one-step solvothermal reduction process to create reduced graphene oxide dispersion in organic solvent [25]. Another method for producing graphene on a big scale is electrochemical reduction. Reduced graphene oxide monolayer flakes were originally discovered in 1962 [29].

To create GO nanoplatelets, the graphite oxide solution can next be sonicated. The oxygen groups can then be eliminated by using hydrazine as reducing agent, it was found that the reduction procedure was ineffective, and some oxygen was still present. GO is advantageous because, unlike graphite, each of its layers is hydrophilic [30].

8.8 Chemical vapour deposition

CVD involves chemical reactions in which precursor molecules are heated and transformed into a gaseous state. A substrate is dispersed on thermally disintegrating precursors at high temperature in this CVD technique. On the substrate's surface, it deposits thin films made of crystalline, solid, liquid, or gaseous precursors.

High-quality graphene is often deposited using the CVD technique onto a variety of transition-metal substrates, including Ni, Pd, Ru, Ir, and Cu. There are many different types of CVD procedures, including thermal, plasma enhanced (PECVD), cold wall, hot wall, reactive, and others, depending on the material quality, precursors, width, and structure needed [32].

The temperature is generally constant throughout CVD process reactors like hot wall reactors, and in cold wall systems, these walls never become heated. On a Cu thin film, graphene is primarily created using a cold wall system [21].

8.9 Growth on copper and nickel

Growth of graphene on copper indicates that it might develop more monolayer coverage as an alternative path to scalable growth of graphene. 2009 saw the first CVD whereby utilising the heat catalytic breakdown of methane, polycrystalline Cu foils were used to generate uniform and large area (*cm^2) graphene on a metal surface [31].

The growth of single-layer graphene sheets on copper foil was even more successful. Although copper is a less expensive alternative to other metals, it can also be easily extracted without harming graphene chemically [33, 34]. The carbon deposition mechanism was discovered to be mainly self-limiting for a relatively modest solubility of carbon in copper. Even at 1000 °C, the solubility of carbon in copper is minimal from the perspective of ppm, therefore the carbon precursor grows directly into graphene on the copper's surface [34]. The researcher had to look for new substrates that were Ni-based and could replace Cu due to a few undesirable characteristics of Cu, such as surface roughening and sublimation [35]. Ni foil, polycrystalline nickel thin film, and patterned Ni thin film were used to create graphene. The foils were exposed naked to a CH_4–Ar–H_2 atmosphere for 20 min at a temperature of 1000 °C after being first annealed in hydrogen. With few-layers graphene, it was discovered that the cooling rate was dependent on the thickness of the graphene layers [36]. Slower cooling prevents carbon from separating to the surface of the Ni foil, but faster cooling results in larger graphite layers; essentially, any graphene that is formed on the surface may not be uniformly distributed throughout. Time-consuming exposure to the carbon precursor, non-self-limiting catalysed development, and a lot of creases and folds were the issues with Ni synthesis [37].

8.10 Graphene at a glance

Due to its 2D structure and lack of bulk, graphene possesses unique characteristics that are particularly sensitive to its surroundings. As a result, it makes sense to think of employing graphene for sensor applications, such as strain gauges, DNA sequencing, and magnetic field measurements as well as magnetic field measurements. The latter application—with either electrical or optical readouts—is most likely the most competitive. The only crystal that can be extended by 20% is graphene, thus expanding the operating range of such sensors [38]. Graphene gas detectors currently only have a slight competitive advantage over existing devices while being incredibly sensitive. Their range of application is limited by low selectivity and water toxicity, but such detectors can be made so cheaply that they could be used in a few specialised applications. Although functionalization may increase the selectivity of graphene sensors, it is a somewhat expensive process that is probably best suited for biosensing [39]. The versatility of graphene sensors is by far their greatest benefit. Multidimensional measurements (such as strain, gas environment, pressure, and magnetic field) can be made with a single instrument. In this regard, graphene has special possibilities. As consumer electronics become more interactive, such sensors will undoubtedly be used in a wide range of items. The development of the universal resistance standard based on the quantum Hall effect was made possible by the special band structure of graphene [40]. Numerous characteristics of graphene make it potentially interesting for use in biological applications. Opportunities are presented by its expansive surface area, chemical purity, and potential for simple functionalization to distribute drugs. Its distinct mechanical characteristics point to possibilities in tissue engineering and

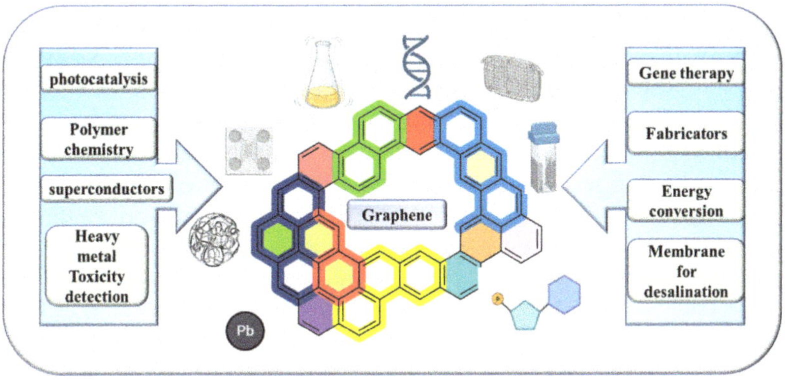

Figure 8.3. Applications of graphene.

regenerative medicine. It is the perfect support for imaging biomolecules in transmission electron microscopy because of its combination of extreme thinness, conductivity, and strength [41]. Additionally, chemically functionalized graphene may result in quick and sensitive detection tools that can detect a variety of biological components, including DNA, glucose, cholesterol, haemoglobin, and cholesterol. Graphene derivatives can solubilize and bind drug molecules due to their large surface area and delocalized p electrons, and if suitable *in vivo* drug distribution and release profiles can be achieved, they have the potential to be drug delivery vehicles (figure 8.3).

Because graphene is also lipophilic, it may be able to overcome membrane barrier penetration, a problem in medication administration. Most of the work that has been done thus far has been devoted to examining the loading and *in vitro* behaviour for anticancer medications like doxorubicin [42]. Although just a limited number of prospective products have reached clinical trials so far, tissue engineering is an emerging field of science with the potential to have a big influence on patient treatment across a variety of illness areas. The elastic and mechanical (strength and elasticity) and selective barrier properties of the scaffold materials used for tissue engineering could be enhanced by the incorporation of graphene, and their biological performance in areas like cell adhesion, proliferation, and differentiation could possibly be modulated. In mouse xenograft models, passive tumour targeting has been demonstrated by intravenously administering GO that has been treated with polyethylene glycol and marked with a near-infrared fluorescent dye but does not contain any drugs. A low-power near-infrared laser was used to treat the cancers, and the tumours were destroyed, demonstrating the promise of graphene derivatives for photothermal cancer treatment [42].

Delivering target genes safely and effectively to a particular site is essential for successful gene therapy. The main goal of gene therapy is often to treat diseases like cancer by replacing a defective gene or restoring a mutated one, and it has gained significant attention in the past 20 years due to numerous successful cases [43].

However, non-viral gene vectors have been thoroughly investigated for gene delivery due of the biosafety risks associated with viral vectors. Graphene-based carriers have become strong contenders for latent vectors of gene transfer due to their structural plasticity [44]. The method of packing of a gene into a non-viral vector is defined by the electrostatic contact between the negatively charged DNA and the positively charged vector. Cationic polymer produced chemically as a non-viral gene vector, polyethyleneimine (PEI) is a distinctive byproduct that has come to be regarded as the industry benchmark for other synthetic vectors. In addition, when combined with GO, PEI is the most frequently utilised polymer as a non-viral gene vector [45].

Human health is at danger due to the rise of food pollutants. Every stage of food production, including plant cultivation, processing, packing, transit, and storage, is susceptible to contamination [46]. A quick, sensitive, and multi-analyte analytical technique is required to evaluate the food sample with complex matrices because of the detrimental effect. Several components in food can be detected using 3D graphene composites-based electrochemical sensors. An overview of the 3D graphene composites-based sensing material for the electrochemical detection of heavy metal ions is described in brief [47].

8.11 Heavy metal detection

In contrast to organic substances, heavy-metal ions are easily accumulated in the human body through the food chain and are nonbiodegradable, which would result in serious chronic illness and disorders [48]. Many international and national organisations have restricted the maximum presence levels due to the rising concern regarding the pollution of heavy-metal ions, linear sweep voltammetry, square wave voltammetry, and differential pulse voltammetry are the most common types of electrochemical voltammetry utilised for the detection of heavy metals [49].

Reviewing the graphene-derived nanomaterial-based electrochemical sensors to identify the heavy metal ions based on the merits and benefits of 3D graphene, it was found that 3D graphene was also used for the electrochemical sensing of heavy metal ions. Due to the strong catalytic activity of noble metal nanoparticles via the size effect, 3D graphene modified with these particles is frequently employed for the sensing and detection of heavy metal ions. For detecting Hg^{2+}, a 3D graphene/Au electrode with porous frameworks was developed. To speed up the catalytic reaction during the sensing process, 3D graphene can transfer the electrons it has obtained from the Au electrode's catalytic process [50].

However, because of their very hydrophobic surfaces, 3D graphene materials have a low affinity for heavy metal ions. Consequently, several active species have been added to the 3D graphene framework. As an illustration, consider a nanocomposite of 3D graphene foam and sodium dodecyl sulfonate as a Pb^{2+} electrochemical sensor. The developed electrode displayed remarkable sensitivity for Pb^{2+} detection by square wave combining the exceptional electrical and specific area features of 3D graphene and the good accumulating ability of hemimicelles [51].

To detect the accumulated Cd^{2+} within bacterial and rice tissues, as well as to further examine the impact of Cd^{2+} uptake in rice plants, a 3D graphene-based electrochemical sensor has recently been adopted. Reduced Cd^{2+} uptake by rice seedlings would lessen the toxicity of Cd^{2+} in agricultural fields, improving productivity—a necessity for food security [52].

8.12 Pesticides detection

Pesticide use is growing in agriculture in order to protect ecosystems from dangerous bacteria and improve yield. At present, there are many reports on the use of electrochemical sensors to find pesticides. The 3D graphene-based electrochemical sensor could also identify several commonly used pesticides [53].

Due to their high level of pest toxicity, carbamates were frequently used in the agricultural sector. The residue and breakdown products, though, exhibited unanticipated toughness, toxicity, and bio accumulative traits. A novel sensing layer for the direct electrochemical measurement of carbaryl pesticides was produced from Au-modified sulfur and nitrogen atoms-doped 3D graphene. With good relative recoveries, this sensor was effectively used to determine the carbaryl concentration in samples of fruits, vegetables, and water [54]. Organophosphorus compounds are also one of the class of insecticides used for pest control. Even at small exposure levels, this can still interfere with acetylcholinesterase action and affect human health. A new acetylcholinesterase biosensor based on an Au/3D graphene layer was developed to identify organophosphorus pesticides in water and vegetable samples. This film increased the electro-transfer efficiency and signal of electro-oxidation of thiocholine and offered a particular surface area for acetylcholinesterase adsorption [55]. In another study, a 3D graphene nanocomposite coated with copper oxide nanoparticles served as the foundation for the establishment of a non-enzymatic electrochemical sensing platform to detect malathion. Copper oxide nanoparticles were shown to have a strong affinity for malathion, which could hinder their capacity to undergo a redox reaction and reduce the electrochemical signal. The 3D graphene's huge surface area for loading large numbers of copper oxide nanoparticles and its good conductivity for facilitating electron transmission may be responsible for the great sensitivity [56].

8.13 Food additives

To enhance the flavour, structure, and colour of food, some synthetic or natural dyes, such as Ponceau 4R, Sudan red, and Sunset yellow, are added throughout the manufacturing process. However, consuming too much of most food additives having aromatic ring structures is harmful for human health. Therefore, developing quick, accurate, and simple detection techniques is crucial [57].

Direct electrochemical methods or indirect electrochemical approaches can be used to find those additions. Using Au/SH-cyclodextrin and 3D reduced GO aerogels functionalized with PEI, for instance, sunset yellow is detected electrochemically using a dual-signal approach. As a colourant, Ponceau 4R has been used extensively in both food and soft beverages. To quantify Ponceau 4R,

3D-cyclodextrin functionalized graphene aerogels were developed and used as an improved electrochemical sensing platform. Through host–guest recognition, -cyclodextrin with a unique cavity structure may effectively capture Ponceau 4R, and using graphene aerogels as the substrate material would increase the conductivity and loading of -cyclodextrin [58].

8.14 Detection of foodborne microorganisms

A possible alternative to traditional analysis methods for the detection of harmful bacteria in food samples is electrochemical biosensors. An electrochemical DNA biosensor with high sensitivity and appropriate selectivity for sulfate-reducing bacteria was constructed using a 3D graphene hierarchical framework functionalized with Au nanoparticles [59]. Cyclic voltammetry was used to measure the synthesized material's electrochemical response. The increased performance of this biosensor was due to the 3D graphene–Au hierarchical structure's large surface/volume ratio and strong conductivity. Using 3D graphene-based sensors, harmful microorganisms have only been detected in a few cases so far. So, for the electrochemical detection of pathogenic bacteria in food samples, this would be the research area for 3D graphene composites [60].

8.15 Other compounds

In the food sector, hydrogen peroxide (H_2O_2), an effective oxidising agent, is essential. It is typically used as a sterilising agent on aseptic milk and fruit juice products as well as an antibacterial agent in food. H_2O_2 in excess, however, can cause severe respiratory irritation and neurological reactions that could be fatal [61]. This method of using an electrochemical sensor to calculate H_2O_2 is generally acknowledged. A 3D-structured H_2O_2 sensing material made of a combination of graphene, carbon nanotubes, and MnO_2 was developed [62]. The final electrode had excellent anti-interference performance, quick reaction, high repeatability, long-term storage stability, and high electrocatalytic activity towards H_2O_2 in milk samples. Chloramphenicol, an antibiotic with a broad spectrum has been prohibited in animals raised for food because their excessive use and build-up pose serious risks to the public's health. Zhang *et al* [63], created a 3D reduced GO architecture using a template-free, environmentally friendly method, which has been used as the active electrode material in highly selective electrochemical sensors for chloramphenicol detection. With a linear range of 1–113 M and a detection limit of 0.15 M at physiological pH 7.4, the reported 3D reduced GO sensor showed an impressive analytical performance. Milk was used to test the effectiveness of a 3D-reduced GO-based electrochemical sensor for chloramphenicol detection. The milk did not contain any analytes, and the achieved good relative recoveries ranged from 97.3% to 98.4% [63].

8.16 Conclusion

In conclusion about graphene, we would like to say that graphene is a flawless 2D lattice of carbon atoms. The outlook is changing, though, as new technological paths are being opened by pure science. For example, certain applications can make use of graphene layers that aren't perfectly flat. In that it combines numerous exceptional properties, from mechanical to electrical, graphene is a special crystal. This means that beyond when it is used to replace other materials in existing applications, its full capability will only be realised in unique applications that are explicitly built with this material in mind. Experimental evidence has shown that graphene-made nanomaterials have a significant potential for use in biomedical applications such as photothermal therapy, drug delivery, biosensing, and stem cell technologies. The noble material graphene recently brought about a revolution in the field of nanoelectronics. Its extraordinary contributions extend beyond nanoelectronics to include areas like medical research, nanorobotics, commercial production of products made from synthetic graphene, and more.

References

[1] Yáñez L *et al* 2002 Overview of human health and chemical mixtures: problems facing developing countries *Environ. Health Perspect.* **110** 901–9

[2] Wu G *et al* 2018 Epidemiology of foodborne disease outbreaks from 2011 to 2016 in Shandong Province, China *Medicine (United States)* **97** 1–7

[3] Wu S *et al* 2012 Multiplexed fluorescence resonance energy transfer aptasensor between upconversion nanoparticles and graphene oxide for the simultaneous determination of mycotoxins *Anal. Chem.* **84** 6263–70

[4] Geim A K 2009 Graphene: status and prospects *Science (80-.)* **324** 1530–4

[5] Novoselov K S *et al* 2007 Electronic properties of graphene *Phys. Status Solidi Basic Res.* **244** 4106–11

[6] Meyer J C, Geim A K, Katsnelson M I, Novoselov K S, Booth T J and Roth S 2007 The structure of suspended graphene sheets *Nature* **446** 60–3

[7] Sofo J O, Chaudhari A S and Barber G D 2007 Graphane: a two-dimensional hydrocarbon *Phys. Rev.* B **75** 153401

[8] Hall E O *et al* 2009 Control of graphene's properties by reversible hydrogenation: evidence for graphane *Science (80-.)* **323** 610–4

[9] Schedin F *et al* 2007 Detection of individual gas molecules adsorbed on graphene *Nat. Mater* **6** 652–5

[10] Hernandez Y *et al* 2008 High-yield production of graphene by liquid-phase exfoliation of graphite *Nat. Nanotechnol.* **3** 563–8

[11] Dikin D A *et al* 2007 Preparation and characterization of graphene oxide paper *Nature* **448** 457–60

[12] Neto A C, Guinea F and Peres N M 2006 Drawing conclusions from graphene *Phys. World* **19** 33–7

[13] Berger C *et al* 2004 Ultrathin epitaxial graphite: 2D electron gas properties and a route toward graphene-based nanoelectronics *J. Phys. Chem.* B **108** 19912–6

[14] Novoselov K S *et al* 2006 Unconventional quantum Hall effect and Berry's phase of 2π in bilayer graphene *Nat. Phys.* **2** 177–80

[15] Novoselov K S *et al* 2005 Two-dimensional gas of massless Dirac fermions in graphene *Nature* **438** 197–200

[16] Peres N M R, Guinea F and Castro Neto A H 2006 Electronic properties of disordered two-dimensional carbon *Phys. Rev.* B **73** 125411
[17] Zhang Y, Tan Y-W, Stormer H L and Kim P 2005 Experimental observation of the quantum Hall effect and Berry's phase in graphene *Nature* **438** 201–4
[18] Fröhner M and Wirth M P 2012 Alter *u*nd Komorbidität beim frühen Prostatakarzinom *Urologe* **51** 1356–61
[19] Eizenberg M and Blakely J M 1979 Carbon interaction with nickel surfaces: monolayer formation and structural stability *J. Chem. Phys.* **71** 3467–77
[20] Lang B 1975 A LEED study of the deposition of carbon on platinum crystal surfaces *Surf. Sci.* **53** 317–29
[21] Bhuyan M S A, Uddin M N, Islam M M, Bipasha F A and Hossain S S 2016 Synthesis of graphene *Int. Nano Lett.* **6** 65–83
[22] Novoselov K S, Fal'Ko V I, Colombo L, Gellert P R, Schwab M G and Kim K 2012 A roadmap for graphene *Nature* **490** 192–200
[23] Zhang Y, Small J P, Pontius W V and Kim P 2005 Fabrication and electric-field-dependent transport measurements of mesoscopic graphite devices *Appl. Phys. Lett.* **86** 1–3
[24] Behabtu N *et al* 2010 Spontaneous high-concentration dispersions and liquid crystals of graphene *Nat. Nanotechnol.* **5** 406–11
[25] Park S, Lee K S, Bozoklu G, Cai W, Nguyen S B T and Ruoff R S 2008 Graphene oxide papers modified by divalent ions – enhancing mechanical properties via chemical cross-linking *ACS Nano* **2** 572–8
[26] Xu Y, Bai H, Lu G, Li C and Shi G 2008 Flexible graphene films via the filtration of water-soluble noncovalent functionalized graphene sheets *JACS* **130** 5856–7
[27] Wu Y H, Yu T and Shen Z X 2010 Two-dimensional carbon nanostructures: fundamental properties, synthesis, characterization, and potential applications *J. Appl. Phys.* **108** 071301
[28] Brodie B C 1983 On the atomic weight of graphite *Philos. Trans. R. Soc.* B **303** 1–62
[29] Marcano D C *et al* 2010 Improved synthesis of graphene oxide *ACS Nano* **4** 4806–14
[30] Qian M *et al* 2011 Formation of graphene sheets through laser exfoliation of highly ordered pyrolytic graphite *Appl. Phys. Lett.* **98** 12–5
[31] Li X *et al* 2009 Large-area synthesis of high-quality and uniform graphene films on copper foils *Science (80-.)* **324** 1312–4
[32] Van Bommel A J, Crombeen J E and Van Tooren A 1975 [001] l Si-face [00i] I C-face *Surf. Sci.* **48** 463–72
[33] Kholin N A, Rut'kov E V and Tontegode A Y 1984 The nature of the adsorption bond between graphite islands and iridium surface *Surf. Sci.* **139** 155–72
[34] Yu Q, Lian J, Siriponglert S, Li H, Chen Y P and Pei S S 2008 Graphene segregated on Ni surfaces and transferred to insulators *Appl. Phys. Lett.* **93** 1–4
[35] Li X *et al* 2011 Large-area graphene single crystals grown by low-pressure chemical vapor deposition of methane on copper *J. Am. Chem. Soc.* **133** 2816–9
[36] Reina A *et al* 2009 Growth of large-area single- and Bi-layer graphene by controlled carbon precipitation on polycrystalline Ni surfaces *Nano Res.* **2** 509–16
[37] Li X, Cai W, Colombo L and Ruoff R S 2009 Evolution of graphene growth on Ni and Cu by carbon isotope labeling *Nano Lett.* **9** 4268–72
[38] Hou X, Xu H, Zhen T and Wu W 2020 Recent developments in three-dimensional graphene-based electrochemical sensors for food analysis *Trends Food Sci. Technol.* **105** 76–92

[39] Adusei P K *et al* 2019 A scalable nano-engineering method to synthesize 3D-graphene-carbon nanotube hybrid fibers for supercapacitor applications *Electrochim. Acta* **312** 411–23
[40] Li C, Yang Z, Tang Z, Guo B, Tian M and Zhang L 2019 A scalable strategy for constructing three-dimensional segregated graphene network in polymer via hydrothermal self-assembly *Chem. Eng. J.* **363** 300–8
[41] Goldoni A, Alijani V, Sangaletti L and D'Arsiè L 2018 Advanced promising routes of carbon/metal oxides hybrids in sensors: a review *Electrochim. Acta* **266** 139–50
[42] Chen M *et al* 2018 Three-dimensional electrochemical DNA biosensor based on 3D graphene-Ag nanoparticles for sensitive detection of CYFRA21-1 in non-small cell lung cancer *Sensors Actuators* B **255** 2910–8
[43] Gulzar A *et al* 2017 Bioapplications of graphene constructed functional nanomaterials *Chem. Biol. Interact.* **262** 69–89
[44] Chen J *et al* 2021 Self-assembling porphyrins as a single therapeutic agent for synergistic cancer therapy: a one stone three birds strategy *ACS Appl. Mater. Interfaces* **13** 27856–67
[45] Khatun Z *et al* 2015 A hyaluronic acid nanogel for photo-chemo theranostics of lung cancer with simultaneous light-responsive controlled release of doxorubicin *Nanoscale* **7** 10680–9
[46] Gao H *et al* 2019 Three-dimensional porous cobalt phosphide nanocubes encapsulated in a graphene aerogel as an advanced anode with high coulombic efficiency for high-energy lithium-ion batteries *ACS Appl. Mater. Interfaces* **11** 5373–9
[47] Basavarajappa P S, Seethya B N H, Ganganagappa N, Eshwaraswamy K B and Kakarla R R 2018 Enhanced photocatalytic activity and biosensing of gadolinium substituted $BiFeO_3$ nanoparticles *ChemistrySelect* **3** 9025–33
[48] Paquin F, Rivnay J, Salleo A, Stingelin N and Silva C 2015 Multi-phase semicrystalline microstructures drive exciton dissociation in neat plastic semiconductors *J. Mater. Chem.* C **3** 10715–22
[49] Zuo Y, Xu J, Zhu X, Duan X, Lu L and Yu Y 2019 Graphene-derived nanomaterials as recognition elements for electrochemical determination of heavy metal ions: a review *Microchim. Acta* **186** 171
[50] Shi L *et al* 2017 A facile and green strategy for preparing newly-designed 3D graphene/gold film and its application in highly efficient electrochemical mercury assay *Biosens. Bioelectron.* **89** 871–9
[51] Qiu N, Liu Y and Guo R 2016 *A Novel Sensitive Electrochemical Sensor for Lead Ion Based on Three-Dimensional Graphene/Sodium Dodecyl Benzene Sulfonate Hemimicelle Nanocomposites* (Amsterdam: Elsevier) p 212
[52] Lu M *et al* 2019 Graphene aerogel-metal-organic framework-based electrochemical method for simultaneous detection of multiple heavy-metal ions *Anal. Chem.* **91** 888–95
[53] Patel H, Rawtani D and Agrawal Y K 2019 A newly emerging trend of chitosan-based sensing platform for the organophosphate pesticide detection using acetylcholinesterase – a review *Trends Food Sci. Technol.* **85** 78–91
[54] Pundir C S, Malik A and Preety 2019 Bio-sensing of organophosphorus pesticides: a review *Biosens. Bioelectron.* **140** 111348
[55] Dong P, Jiang B and Zheng J 2019 A novel acetylcholinesterase biosensor based on gold nanoparticles obtained by electroless plating on three-dimensional graphene for detecting organophosphorus pesticides in water and vegetable samples *Anal. Methods* **11** 2428–34

[56] Xie Y *et al* 2018 CuO nanoparticles decorated 3D graphene nanocomposite as non-enzymatic electrochemical sensing platform for malathion detection *J. Electroanal. Chem.* **812** 82–9

[57] Ma J Q, Liu L, Wang X, Chen L Z, Lin J M and Zhao R S 2019 Development of dispersive solid-phase extraction with polyphenylene conjugated microporous polymers for sensitive determination of phenoxycarboxylic acids in environmental water samples *J. Hazard. Mater.* **371** 433–9

[58] Ma X *et al* 2018 Three-dimensional β-cyclodextrin functionalized graphene aerogels: an enhanced electrochemical sensing platform for quantification of Ponceau 4R *J. Electroanal. Chem.* **823** 437–44

[59] Chen S, Frank Cheng Y and Voordouw G 2018 Three-dimensional graphene nanosheet doped with gold nanoparticles as electrochemical DNA biosensor for bacterial detection *Sensors Actuators* B **262** 860–8

[60] Li F, Yu Z, Han X and Lai R Y 2019 Electrochemical aptamer-based sensors for food and water analysis: a review *Anal. Chim. Acta* **1051** 1–23

[61] Kumar J S, Murmu N C and Kuila T 2018 Recent trends in the graphene-based sensors for the detection of hydrogen peroxide *AIMS Mater. Sci.* **5** 422–66

[62] Ye D *et al* 2013 A three-dimensional hybrid of MnO_2/graphene/carbon nanotubes based sensor for determination of hydrogen-peroxide in milk *Electrochim. Acta* **109** 195–200

[63] Zhang X, Zhang Y C and Zhang J W 2016 A highly selective electrochemical sensor for chloramphenicol based on three-dimensional reduced graphene oxide architectures *Talanta* **161** 567–73

IOP Publishing

Graphene-Based Sensors

Jamballi G Manjunatha

Chapter 9

Graphene-based electrodes for determination of neurotransmitters

Sanu K Anand, Manna Rachel Mathew and K Girish Kumar

Neurotransmitters (NTs) are biologically significant endogenous compounds that play a pivotal role in the optimal functioning of central nervous, cardiovascular, renal and hormonal systems. The precise and facile detection and determination of these chemical messengers is of great significance in disease diagnosis and monitoring of various physiological parameters. In this context, the effective determination of these NTs via electrochemical sensing attracts considerable scientific attention as it is one of the simple, rapid and cost-effective screening methods. This chapter explains the electrochemical detection of NTs using the emerging graphene-based-materials. Among the various available materials, graphene-based ones are the recent and promising environmentally benign, biocompatible and user-friendly electrocatalyst. Graphene, a two-dimensional (2D), single-layered carbonaceous nanomaterial having honeycomb lattice, has guaranteed the sensitive determination of target analytes owing to its remarkable and unique optical, electronic, mechanical, thermal and electrochemical properties. As such, its derivatives including graphene oxide and reduced graphene oxide, and its zero-dimensional material graphene quantum dots, etc, individually, as well as in combination with other materials such as amino acids, conducting polymers, metal oxides and nanomaterials, are promising candidates for the designing of electrochemical sensors with improved analytical performances. Fabrication of an electrode using graphene-based-materials as modifier, its electrochemical studies and its reliability for the determination of NTs to point-of-care analysis are highlighted in this chapter.

9.1 Introduction

NTs are chemical messengers in synaptic transmission that decide the behavioral and emotional state of multicellular organisms [1–3]. They play a key role in the

doi:10.1088/978-0-7503-5578-0ch9

proper functioning of central nervous, cardiovascular, renal and hormonal systems that include regulation of sleep, mood, digestion, heart rate, blood pressure, etc [1, 4, 5]. According to their chemical structure, NTs are mainly classified as amines (e.g. dopamine, adrenaline, noradrenaline, serotonin, etc), amino acids (tyrosine, glycine, glutamate, etc) and gaseous molecules (e.g. nitric oxide and H_2S) [1]. Any imbalance in their optimum level is indicative of various psychiatric, neurodegenerative and renal diseases [6, 7]. To name a few examples, NTs are clinically significant biomarkers of Alzheimer's, Parkinson's, dementia, epilepsy, bipolar disorder, depression, anxiety, celiac, irritable bowel syndrome, etc [1, 8]. Their presence can be easily confirmed by analyzing biological matrices such as plasma, urine, serum, cerebral spinal fluids and platelets, thereby acting as a diagnostic tool for the discrimination of healthy and disease conditions [9]. Due to their importance in pathological research, their selective and sensitive determination in real samples is swiftly increasing [4]. Hence, there is a pressing need for non-invasive and economically feasible screening methods for the effective diagnosis and successful treatment of these diseases. Currently available detection strategies of NTs demand experienced personnel, tedious sample pretreatment and lengthy analysis procedure [10–12]. The above-mentioned shortcomings of conventional method have been easily overcome by electrochemical methods that offer ease of analysis, high selectivity and sensitivity, possibility of multicomponent analysis, excellent detection limits, reproducibility in measurements, requirement of small analyte volumes, ability to be used in turbid biofluids, etc [13, 14]. Also, this method can be easily integrated to portable and miniaturized devices for point-of-care applications when compared to other analytical techniques [9]. Since most of the NTs are electrochemically active compounds, development of electrochemical sensors follow the merits of convenient analysis, rapidity in measurements and sensitivity in determination.

An electrochemical sensor comprises a receptor layer and an electrochemical transducer. The receptor unit interacts with the analyte of interest and this binding event is transduced into an electrochemical signal [4, 15]. An electrochemical setup, mainly voltammetry, comprises of a three-electrode cell system involving the working electrode, reference electrode and auxiliary electrode immersed in an electrically conducting solution. High overpotential, surface fouling, sluggish kinetics and poor sensitivity of the bare working electrode can be easily rectified using chemically modified electrodes [16, 17]. A proper modifier often increases the surface area of the electrode so as to provide more reaction sites to give a favorable reaction kinetics [17]. The unique and well-defined electrocatalytic properties of nanomaterials have culminated it as an effective electrode modifier to tackle the challenges in electroanalytical applications [18, 19]. Noble metal nanoparticles (MNPs) have been already established as one of the potent electrocatalysts and contributed remarkably for the scientific progress in this field [19, 20]. Today, carbon-based nanomaterials, especially graphene-based ones, have been considered as one of the most explored materials as electrode modifiers due to their impressive success in the enhancement of electron transfer rate in electrochemical reactions along with easily available and cost effective graphitic raw materials [19].

This chapter covers the use of environmentally benign, biocompatible and cost effective graphene-based-materials as the electrode modifier for the sensitive determination of NTs.

The International Union for Pure and Applied Chemistry (IUPAC) defines graphene as a 'single carbon layer of graphite structure describing its nature by analogy to a polycyclic aromatic hydrocarbon of quasi-infinite size' [21, 22]. Graphene, a 2D, single-layered nanosheet of sp^2 hybridized carbon atoms having a honeycomb lattice, is a well-accepted nanomaterial as an electrode modifier owing to its remarkable chemical/thermal stability and outstanding electronic, thermal and mechanical properties [6, 22]. This sp^2 hybridized network is responsible for the characteristic properties of this material [22]. It can easily bind with the oligonucleotides via π–π stacking interaction, and its outstanding electronic properties guarantee the development of highly sensitive biosensors [6]. Graphene on oxidation results in graphene oxide (GO), which is the oxygenated derivative having both sp^2 and sp^3 hybridized carbon atoms [23–25]. These oxygen moieties such as epoxy and hydroxyl groups in the basal plane as well as carboxyl and carbonyl groups in the edge plane of the graphitic layer, result in better biocompatibility and hydrophilicity. Additionally, it enhances the immobilization of bioreceptors and also surface functionalization compared to pristine graphene [23, 24]. On the other hand the excess oxygenated groups of GO reduce the conductivity, which depends on the degree of oxidation [23, 26]. The limitations of pristine graphene and GO are mitigated by the introduction of reduced GO (RGO), where a large number of oxygen functional groups have been removed by employing thermal, hydrothermal, electrochemical methods, etc [23, 27]. By this reduction process, the electrical conductivity of RGO will be restored and at the same time the presence of remaining oxygen functionalities makes the electrode surface modification much easier and thereby assures the sensitive determination of targets [23, 28].

Apart from graphene and its derivatives, graphene quantum dots (GQDs), a zero-dimensional nanostructured fragment of graphene, due to its peculiar optical and electronic properties, has made great impact on researchers as a potential catalyst support for electroanalytical applications [14, 29]. It can be prepared either from graphene-based-materials or graphene-like organic molecules [30]. Their surface is rich in oxygen functionalities and exhibits high surface-to-volume ratio due to their quantum confinement and edge effects, that endow the surface functionalization much more easily [29, 30]. Moreover, recent findings have shown that the heteroatom doping of GQDs tunes their intrinsic characteristics including electronic and optical properties, providing more active sites and thereby widening their window of applications [30, 31]. The excellent biocompatibility, low cytotoxicity, and water solubility, along with outstanding physicochemical properties, make this emerging material excellent for biomedical applications compared to conventional semiconductor quantum dots [30, 32].

Apart from these single materials, their synergistic effects when combined with other electroactive materials accelerate the analytical performances of the sensor in terms of wide dynamic range, high selectivity and sensitivity with low detection limits [1, 33]. The exponential increase in the number of research publications of graphene-based-materials, especially for biomedical applications shows their relevance in current sensor research. This chapter discusses the fabrication of an electrode using different graphene-based-materials as modifiers, especially graphene, GO, RGO and GQDs, their electrochemical studies and reliability for the selective, sensitive and effective determination of NTs for biomedical analysis, and it is structured accordingly.

9.2 Graphene-based electrochemical sensors for neurotransmitters

9.2.1 Graphene, graphene oxide and reduced graphene oxide as electrocatalyst

Graphene is considered as a basic building block for graphitic materials [34]. It can be synthesized via chemical reduction of GO, which is produced by any of the oxidative treatments of graphite reported by Brodie [35], Hummers [36], Hofmann [37] and Staudenmaier [38]. The electrochemical activity and sensitivity of graphene-based sensors towards the detection of many NTs are greatly dependent on the π–π interaction between the graphene and the analyte molecule.

A major concern regarding the fabrication of graphene nanomaterial-based sensors is the selection of appropriate non-toxic and biocompatible dispersing base. An interesting choice was reported, where gelatin, a soluble polypeptide, was used as both the dispersing agent and the stabilizing agent for graphite dispersion [39]. Gelatin is a commonly used biopolymer in biomedical applications due to its excellent biocompatibility and biodegradability. In another approach, cyclodextrin is used as dispersing agent, where cyclodextrin entrapped graphite composite probe is fabricated for the detection of dopamine (DA) [40].

Additionally, the existence of large number of defects at the edge plane of graphene assists the rapid and sensitive detection of NTs [41, 42]. For instance, functionalized graphene nanoribbon, having defects at the edge plane, have been modified on a screen-printed electrode and employed for the determination of DA and serotonin hydrochloride (ST) with good sensitivity and reproducibility [43]. In another approach, the synthesis of multilayer graphene nanoflake films onto Si substrates has been achieved via microwave plasma enhanced chemical vapor deposition method [44]. These graphene nanoflakes comprise high graphitized knife-edge structure and exhibit faster electron transfer kinetics. The sensor has demonstrated good sensitivity and excellent electrocatalytic activity towards the simultaneous detection of ascorbic acid (AA), DA, and uric acid (UA) with a very low detection limit. A pristine three-dimensional (3D) graphene-based sensor has been developed via chemical vapor deposition method [45]. Here, the 3D graphene was made on macroporous Ni foam scaffolds, which were finally removed by thermal annealing and etching process. By this approach, the researchers achieved macromolar detection limit for DA and were able to separate the signals of AA, DA and UA.

The possibility of functionalization of graphene further extends its properties, lending promising applications for the determination of NTs [46]. Although GO is an electrically insulating material, the presence of carboxyl and hydroxyl functionalities can form hydrogen bond interactions with the hydroxyl and amine groups of

NTs, which makes the sensing facile. Moreover, the aromatic structure of graphene offers π–π stacking between the substrate and the NTs. It is reported that GO-modified sensors can suppress the electro-oxidation peak of AA due to the electrostatic repulsion between the negatively charged carboxylate functional groups of GO with the negatively charged AA. For instance, Gao *et al* developed a GO-modified glassy carbon electrode (GCE), which demonstrated enhanced sensitivity for DA detection, in the presence of AA [47]. A functionalized graphene sheets-modified graphite electrode is also reported for the selective determination of DA in the presence of AA and UA [48]. Here, a graphite electrode is modified with functionalized graphene, which is prepared by solvothermal reduction of colloidal suspension of graphite oxide. A sensing approach with a GO-modified gold electrode as the working electrode is proposed for the sensitive enzymeless determination of γ-aminobutyric acid. Since it is a nonelectrooxidizable NT, a mixture of orthophthalaldehyde and an alkylthiol is used as a ligand to interact with γ-aminobutyric acid, which becomes an electroactive compound. Nanomolar determination was attained and the study was extended to analysis in human urine and serum samples [49].

Considering the potential benefits of RGO such as high electrical conductivity, high surface area and fascinating physicochemical properties compared to pristine graphene and its oxygenated derivative GO, it has been widely exploited for multifarious applications [50]. It is prepared by the reduction of GO, where oxygen functional groups decorated at the edge (carboxylic acid and hydroxyl) and basal plane (hydroxyl, epoxy and carbonyl) of GO have been reduced and thereby its conductivity has been amplified by four to six times over that of GO due to the restoration of the π conjugated system [51–53]. The residual functional groups after the reduction process impart hydrophilicity in RGO and also the chemical reaction with the organic molecules make it organophilic [54, 55]. Furthermore, these oxygen moieties make the electrode surface negatively charged which enriches the positively charged target molecules via electrostatic interaction, and the honeycomb lattice of RGO provides π–π stacking interaction with the phenyl moieties of NTs, thereby guaranteeing the sensitive determination of targets [56–58].

Chemical oxidation of graphite to GO via the well-accepted Hummers' method, followed by its reduction to RGO is the common procedure for the preparation of RGO [59, 60]. The reduction can be accomplished by chemical, thermal, photochemical or electrochemical treatments. Even though chemical reduction demands the use of toxic reducing agents such as hydrazine hydrate, dimethyl hydrazine, hydroquinone and $NaBH_4$, it is widely employed due to its relatively high reduction efficiency [61, 62]. The electrode is modified by casting the resultant RGO suspension onto its surface and has been successfully used in electrochemical and biosensing platforms [63, 64]. Thermal reduction of GO is performed at very high temperature that mainly removes the oxygen functionalities of basal plane and partially from the edges, where the reduction temperature, heating rate, etc, determine the quality of the produced RGO [65, 66]. It is reported that a thermally RGO-modified GCE in physiological pH allowed the simultaneous determination of DA and UA in micromolar ranges in the presence of highly electroactive AA [66].

RGO can also synthesized hydrothermally via the one-pot synthetic method that provides pure and homogeneous products, and hence it is considered as an efficient method to prepare RGO and its composites [67]. Even though it is a tedious and lengthy synthetic procedure, numerous reports are available for the determination of various NTs in good analytical performances [59, 68–71]. Photochemical reduction requires the dispersion of GO in solvents such as polyvinyl pyrrolidone, *N,N*-dimethylformamide, *N*-methyl-2-pyrrolidone, etc, followed by the continuous irradiation of UV for several hours where hydrated electrons act as reducing agents [60, 72]. RGO, thus produced by the aforementioned methods has enough catalytically active defect sites, which can be effectively used for the loading of target species [63, 73]. Even though it exhibits excellent analytical applications, the unpleasant reaction conditions of these methods hinder its point-of-care applications.

Hence, in search of a green method, electrochemical reduction methods, mainly potentiostatic methods find tremendous interest since they are an efficient and cost effective method for the preparation of RGO at room temperature in a very short time without the use of any toxic chemicals [74–77]. Moreover, they can exhibit a definite control over the reduction process by regulating the experimental parameters and display higher electrical conductivity than their precursor GO [56, 74, 78]. Thus formed graphene is called electrochemically reduced graphene oxide (ERGO) and its formation is confirmed by the cathodic peak nearly at -1.3 V, which is attributed to the reduction of oxygen functional groups [57]. The thickness of this ERGO layer can be maintained by varying the experimental parameters such as scan rate, number of sweep segments, potential, etc [79]. Thus the formed RGO layer endows stable and reproducible film with a definite control over its morphology and thickness compared to other reduction techniques in which synthesized RGO is coated on the electrode surface by dip-casting or drop-casting method [80].

Geometrical changes after the reduction process alter their properties such as reducing their surface energy and enhancing the surface roughness, which simultaneously enhances the loading of the target species and amplifes the electrochemical responses [74, 81]. Determination of targets with intensified peak current intensity and decreased overpotential is the prerequisite for the development of electrochemical sensors. Catecholamines, one of the important class of NTs, whose determination is achieved via the two-electron oxidation to corresponding *ortho* quinone structure in physiological pH [9]. These oxidized species may passivate the electrode surface and minimize the electrochemical response, which can be overcome by this modified electrode by altering the electrode kinetics [9]. Reports show that, compared to single-walled carbon nanotubes (SWCNTs)- and multi-walled carbon nanotubes (MWCNTs)-modified electrodes, predominant voltammetric responses are shown by DA in the presence of excess AA on RGO-modified electrodes in cyclic voltammetric mode, which is attributed to the greater π–π interaction between graphene and DA [82, 83]. Similarly, Goyal *et al* successfully demonstrated the usefulness of an RGO-modified Pd electrode for the voltammetric sensing of norepinephrine (NE) in human urine samples where the detection limit is found at the nanomolar level [84].

Furthermore, chemical doping of RGO with heteroatom such as N, P, S, B, etc attracts the utmost attention as it tailors the physicochemical properties, thereby extending its applications in various fields [53, 85, 86]. These heteroatoms create unbalanced charge centers called active sites in graphene due to the difference in electronegativity between a carbon atom and the dopant [87, 88]. In addition, the difference in the atomic size between them creates a strain on the lattice and forms structural defects which also exist as active sites [87, 89]. These catalytic sites multiply the properties of bare RGO which reflects in the improved sensing performances in terms of selectivity, wide linear ranges and lower detection limits [90–92]. Nitrogen doping is reported as the best choice because it can make a strong bond with carbon atoms as its atomic size and valance electrons are comparable [86]. Moreover, N doping results in n-type conductivity which makes catalytic sites beneficial for sensing applications. Here, the electrode is modified by drop-casting the solution or electrodeposition on the electrode surface [86, 93]. The π–π stacking interaction between the RGO and the phenyl group of DA/NE could be the driving force to load these molecules on the electrode surface so as to produce a noticeable Faradaic current and at the same time doping enhances the conductivity, where a nanomolar detection limit has been achieved for their determination [86, 94–96]. Reports suggest that N-doped RGO-modified electrodes have the efficiency to separate the overlapped electrochemical responses of AA and UA from DA, thereby making the analysis free from the interference of these major co-existing species [91, 97].

As a well-known electrocatalyst, RGO and its composites are widely used as a supporting material in specific sensing platforms such as biosensors and molecularly imprinted polymer (MIP)-based sensors for NTs, thereby improving the performance of the sensor [98]. As the name indicates, specific sensors specifically capture their targets onto the electrode surface and facilitate the analysis free from the interference of co-existing electroactive species, thus broadening its applications in every sector [98, 99]. The conjugated backbone as well as the residual oxygen functionalities of RGO act as the anchoring sites for the attachment of these biocomponents via π–π stacking, hydrogen bonding and covalent interactions, thus ensuring highly selective and sensitive determinations of NTs [100, 101].

9.2.2 Graphene nanocomposites as electrocatalyst

Recent studies have demonstrated that graphene nanohybrid materials show excellent dispersion compared to graphene alone due to uniform distribution and intercalation of materials in between the graphene layers [102]. Furthermore, these graphene nanohybrids enhance the electron transfer kinetics and the electrocatalytic property towards the oxidation of NTs. Hence, the sensing ability has greatly improved compared to graphene only modifications. Moreover, the low toxicity and low cost of these materials has expedited research to fabricate ecofriendly and less expensive electrochemical sensors with good analytical performance [102–105]. Metallic and polymeric nanocomposites have been gaining much interest for the electro-oxidation of NTs. Not only do such composites prevent their self-agglomeration, but also their synergistic catalytic contributions minimize their individual limitations, which ensures enhanced sensing behaviors [56, 106]. Furthermore, composite materials can efficiently distinguish the overlapped oxidation peaks of NTs making simultaneous determination feasible. Superior electrocatalytic properties of their individual counterparts have led to well-defined and well-separated analytical signals and hence an innumerable number of sensing platforms with excellent sensing abilities have been evolved for various combinations of NTs [107, 108].

9.2.2.1 Graphene nanocomposites with metal oxides

Different nanocomposites of RGO with nanoparticles of ZnO, $FeWO_4$ and AuNPs with enlarged surface area, are effective for enhancing the electro-oxidation of DA, epinephrine (EP) and NE, respectively, to their corresponding quinone form. These modifications have resulted in enhanced selectivity, wide linear ranges and low detection limits [50, 56]. Also, their simple and cost effective method of preparation, biocompatibility and catalytic activity along with reproducibility in measurements offer great opportunities in real-time analysis [46]. Likewise, RGO/Co_3O_4 composite has provided the amperometric sensing of ST in which the sensor has effectively distinguished the peak of ST from DA, AA and UA [109]. The RGO/$CuAlO_2$ nanocomposite was found to outperform the detection of DA with enhanced electrocatalytic performances. The effective interaction between graphene sheets and $CuAlO_2$ nanostructures has contributed to the increase in conductivity and enhanced electron transfer ability. The sensor demonstrated excellent analytical performances with limit of detection in the nanomolar range with less interference from UA and AA [110]. In another approach, Wang and coworkers have proposed a magnetic graphene nanocomposite platform (Fe_2O_3–NiO@GO), which offered broader linear range and nanomolar detection limit [111]. The synergistic effect between GO and Fe–Ni bimetallic oxides has contributed to the enhanced electrocatalytic ability of GO and magnetic nanoparticles. Another example is the simultaneous detection of DA and UA using the nanocomposite of RGO and Mn_3O_4 (RGO/Mn_3O_4) [112]. Well-separated oxidation peaks with enhanced peak current were observed for both DA and UA, which was due to the synergistic effect arising from the controlled structure of Mn_3O_4 and larger surface area of graphene. This contributes to the enhanced electrocatalytic activity towards the electro-oxidation of DA and UA. Like graphene, MoS_2 is one of the most exploited 2D materials belonging to the transition metal dichalcogenides family, owing to its unique optoelectronic properties [113]. The combined effect of MoS_2 and RGO is widely reported for its excellent sensing applications [114]. A ternary nanocomposite of MoS_2, RGO and Fe_3O_4 has shown about 61% enhancement in peak current with reduction in peak potential from 0.3 to 0.2 V over the bare electrode for EP in physiological conditions [115]. Also, the synergy of MoS_2 and N-doped RGO ensures the EP determination in good analytical figures of merit [116].

9.2.2.2 Graphene nanocomposites with electropolymers

Nanocomposites of RGO with various electropolymers also show extensive sensing opportunities for the determination of NE, DA, EP, etc [10, 117–122]. On association with the polymers, the nanocomposite acquires new properties, which improves the electrocatalytic activity of the electrode towards NTs. For the detection of DA, Teng *et al* have employed nitrogen-doped graphene and poly (3,4-ethylenedioxythiophene) (PEDOT) hybrid modified GCE [123]. The PEDOT component in the nanohybrid increases its conductivity and inhibits the stacking of nitrogen-doped graphene, resulting in the increase in surface area of the nanohybrids. This in turn increases the electron transfer rate and the electrocatalytic activity towards the oxidation of DA, with a limit of detection of 54 nM. Polyaniline (PANI) has also been utilized along with thermally reduced graphene (TRGO) for DA detection [124]. In the PANI/TRGO composite, the appropriate amount of oxygen groups and homogenous distribution of PANI facilitates the attraction of cationic DA towards the electrode–electrolyte interface, which was responsible for the remarkable sensor performance, having a detection limit of 430 nM. A nanohybrid-based on poly(*para*-phenylene) and chemically reduced GO has demonstrated to be an excellent platform for the simultaneous determination of DA, AA and UA [125]. The redox sites in the polymer provide electrocatalytic activity and its anionic sites prevent non-specific interactions. These properties of the polymer and the larger surface area of RGO synergistically contribute to the improved electron transfer and enhanced electrocatalytic properties. In an alternative approach, coupling of positively charged polymer film, poly(L-lysine) and negatively charged ERGO has been utilized for the fabrication of the electrode [126]. This stable film formed by the electrostatic interaction has proved to be a superior hybrid material for the determination of DA, AA and UA with good analytical features [126]. Gold nanoparticle decorated polypyrrole/GO nanosheets are demonstrated as an effective strategy for the simultaneous determination of AA, DA and UA as well [127]. Moreover, Ghanbari *et al* studied the combined effect of RGO, PtNPs and polypyrrole for the simultaneous electro-oxidation of DA and 5-hydroxytrytamine (ST) at higher concentration of AA [33].

9.2.2.3 Graphene nanocomposites with metal nanoparticles

It was found that, in association with metal nanoparticles, the catalytic properties of graphene have been greatly improved and have allowed the selective determination of NTs. For instance, Inguanta *et al* effectively discriminated the overlapped oxidation peaks of DA, AA and UA using the composite of RGO with Au and Pt nanoparticles [128]. In another approach, Au nanoparticles were electrodeposited onto the surface of GCE/N-doped RGO and employed as the sensing platform for the simultaneous detection of DA, AA, and UA [129].

9.2.2.4 Graphene nanocomposites with chitosan

Chitosan is a natural polycationic linear polysaccharide which is derived from chitin. Owing to its non-toxicity, biocompatibility, biodegradability and its film-forming ability, Chitosan is believed to be an excellent natural biopolymer for the development of electrochemical sensors [130]. However, its low solubility in neutral and alkaline media and relatively poor conductivity limit its application. Nevertheless, chemical modification of chitosan into composites shows new functional properties and it can be used for different applications [46]. In addition, the vast number of hydroxyl and amino groups in the chitosan facilitates the modification of its surface with graphene, conductive polymers or metal nanoparticles, which thereby increases the effective surface area of the nanohybrid and the electron transfer rate at the interface [131]. Chitosan is most likely to be associated with highly conductive graphene sheets for developing highly sensitive electrochemical sensors. This association improves the dispersion and other chemical properties of individual compounds. The interaction may be either due to the covalent bond formation or the electrostatic interaction between the amino or hydroxyl group of the chitosan and the carboxyl group of the graphene [102]. Moreover, the presence of amino groups in the chitosan can be easily protonated and cationic centers can be created, which helps to easily interact with negatively charged groups of the desired biomolecule [132].

Kafi *et al* developed a bendable biostrip with mesoporous-chitosan–GO (m-Chit–GO) composite-based sensing electrode for label-free electrochemical detection of DA [133]. The use of a chitosan–GO composite working electrode lowered the limit of detection to 10 pM, which permitted a biostrip appropriate for *in vivo* DA detection. It also offered high selectivity towards DA detection in the presence of AA and UA. Since all the materials used are biocompatible and the fabrication methods are ecofriendly, the authors have mentioned that their work aligns well with green technology. An RGO/chitosan/chromium oxide (GCE/RGO/CS/Cr_2O_3) nanocomposite-modified GCE is proposed for the synchronous determination of DA, UA, xanthine and hypoxanthine [134]. Here, it was proven that the large surface area and good electrical conductivity of graphene, good electrostatic interaction of chitosan and graphene and higher catalytic activity of Cr_2O_3 nanoparticles, synergistically contributed to the enhanced electrocatalytic property of the sensor. This nanohybrid-based sensor has been successfully applied for the analysis of DA, UA, xanthine and hypoxanthine in fish meat samples. A new approach for porous chitosan–graphene matrix preparation is proposed for synergic detection of DA and UA [135]. The authors demonstrated that a higher degree of amide functionalization of chitosan–graphene matrix enhances the electrocatalytic performance. Moreover, the matrix demonstrated good selectivity towards the detection of DA and UA in excess concentration of AA and glucose. Another strategy has been established, where tyrosinase, chitosan and RGO are sequentially used to modify a screen-printed carbon electrode (SPCE) for the detection of DA [136]. The approach was enzyme-based, which enables specific detection of DA without interference from UA or AA with detection limit in the nanomolar range. The presence of chitosan helps to enhance the mechanical as well as the electrical stability of graphene and the effective immobilization of tyrosinase enzyme. Moreover, application of the strategy in human urine samples gave satisfactory results, thus proving their feasibility for analysis in physiological samples.

9.2.3 Graphene quantum dots as electrocatalyst

GQDs, irrespective of their synthetic route, are zero-dimensional nanomaterials having sizes lower than 100 nm [137–139]. Compared with other electrode modifying materials, GQDs and graphene oxide quantum dots (GOQDs) have received tremendous attention as they exhibit unique features including remarkable conductivity, biocompatibility, high stability, chemical inertness, low cytotoxicity, readily available precursors, simple and economical synthesis. They are often synthesized from graphene and rarely from carbon black [138] and some natural biological sources. The conductivity of GQDs is higher than that of GO and, therefore, GQDs are widely used in designing electrochemical sensors.

A GCE modified with GQDs prepared from GO was developed by Pang and coworkers, by simply drop-casting GQDs suspension on the cleaned GCE [140]. The selective DA detection in the presence of interferents could be due to the possible electrostatic and π–π interactions between DA and GQDs. In another approach, GQDs prepared via carbonization of a citric acid precursor are dropped on the surface of a GCE functionalized with ethylenediamine [141]. This sensor has exhibited good electrical conductivity and favorable electrocatalytic activity towards the determination of DA. Apart from drop-casting GQDs, the electrode with electrodeposited GQDs have also been applied towards the selective determination of DA. Electrodeposition can be done by applying a potential sweep from −1.5 to 2.0 V at a scan rate of 100 $mV \cdot s^{-1}$ for 30 cycles after dipping GCE into the GQDs solution [142]. The literature suggests that the remarkable electrocatalytic effect is mainly due to the π–π interaction between GQDs and DA. Moreover, the electrostatic interaction between the cationic DA and anionic groups on GQDs also facilitates its electrocatalytic capacity.

It can be seen that the combined effect of GQDs and acid-functionalized MWCNTs favors the sensitive detection of DA with detection limit in the nanomolar range [137, 143]. GQDs increase the active surface area of the electrode and MWCNTs offer good conductivity. Here, MWCNTs in HCl were added to GQDs prepared from glucose to form a homogeneous suspension of GQDs-MWCNTs, which was further drop-casted on GCE. The presence of large conjugated π bonds in the GQDs-MWCNTs composite offers electron transfer of DA through π–π stacking interaction. Furthermore, the anionic groups in GQDs could attract cations, which leads to the enhanced selectivity of the sensor. Another strategy of combining nanoparticles and GQDs is the mixing of AuNPs and GQDs made from the hydrothermal treatment of carbon black and it was later drop-casted on GCE [138]. The electrode after coating with Nafion has been successfully employed for the selective detection of DA in the presence of UA and AA. N-doped GQDs have also been utilized along with SnO_2 nanoparticles and PANI as modifiers on GCE for the determination of DA in the presence of a mixture of AA and UA [144]. The N-GQDs were synthesized by heating a mixture of citric acid and urea and the SnO_2/PANI composite was prepared from the oxidative polymerization of aniline in the presence of SnO_2 nanoparticles. The SnO_2/PANI composite with positive charge interacts with negatively charged N-GQDs through electrostatic attraction.

The molecularly imprinted approach based on GQDs incorporated 2D hexagonal boron nitride (2D-hBN) nanosheets was demonstrated by Yola and coworkers [145]. This nanocomposite-based electrode has been synthesized by ultrasonicating GQDs and 2D-hBN suspension for 10 h and then drop-casted on GCE. ST imprinted polymer was developed on GQDs/2D-hBN nanocomposite and it was applied for the selective determination of ST.

The electrode modification with GQDs preserves the biological activity of the enzyme immobilized, and it also facilitates enhancing the electrocatalytic efficiency of the enzyme-based electrochemical sensor. GQDs are strong and stable materials, to which the enzyme binds effectively, and this considerably improves the electrical contact between the enzyme and the electrode. Baluta *et al* presented a GQDs-based electrochemical biosensor for the determination of epinephrine, based on the oxidation of catecholamine in the presence of the enzyme—laccase [146]. The fabrication of the sensor and its sensing strategy is simple and the results are selective and sensitive making the biosensor a good tool for EP detection.

A triple nanocomposite-based on chitosan–GQDs–graphite mixed with paraffin has been utilized as the modifier to construct bulk-modified carbon paste electrode via simple ultrasonication [132]. This nanocomposite holds the excellent biocompatibility of chitosan and conductivity of GQDs and graphite. The resulting electrode has been applied for the detection of EP in real samples. GQDs and ionic liquid composite for carbon paste electrode modification have also been demonstrated by simply mixing GQDs, graphite powder and the ionic liquid in a mortar [147]. The modified electrode has been used for the simultaneous analysis of NE and acetylcholine. A similar approach with a room temperature ionic liquid/GQDs-modified carbon paste electrode has also been proposed for nanomolar determination of levodopa in the presence of ST [148]. The high electrical conductivity of GQDs and ionic liquid favors the enhanced electrocatalytic activity of this composite electrode.

GQDs have also been used in the fabrication of sensors using screen-printed electrodes. They are economical signal transducers and an important component for developing miniaturized electrochemical sensors. GQDs-modified screen-printed electrodes have been proposed for the determination of DA in the presence of tyrosine [149]. GQDs were prepared via direct pyrolysis of citric acid and the suspension was drop-casted on a screen-printed electrode. The simultaneous determination of DA and tyrosine was successfully performed with good sensitivity and selectivity. A nanocomposite comprising chitosan and microwave-assisted N-doped GQDs, on an SPCE is presented for the sensitive and selective determination of DA [150]. The N-GQDs–chitosan nanocomposite has been prepared by mixing and stirring the microwave-assisted N-doped GQDs and chitosan for one hour. This modified electrode has proved to be selective towards the determination of DA. The authors proposed that the chitosan present in the nanocomposite plays a significant role towards the selectivity of the sensor, as it hinders the interference caused by AA and increases the peak separation between DA and UA.

9.3 Conclusion

In conclusion, both RGO and GQDs are understood to be worthwhile in terms of selectivity and sensitivity, based on the fact that selective and sensitive electrodes can be fabricated from both RGO and GQDs with low detection limits. In both cases, the functionalization with certain groups is very interesting and the corresponding electrode exhibits lower detection limits. The application of graphene in the sensing approach with molecular recognition has only a few reports. It could be observed that the graphene modified molecular imprinted electrodes yield determination of NTs in picomolar level due to the specificity and the selectivity of molecular imprinted polymers. Finally, the miniaturization of the sensors is commanding for point-of-care diagnosis of NTs, since *in vivo* studies generally demand convenient, portable and integrative equipment.

References

[1] Madhurantakam S, Karnam J B, Brabazon D, Takai M, Ahad I U, Rayappan J J B and Krishnan U M 2020 *ACS Chem. Neurosci.* **11** 4024–47
[2] Durairaj S, Sidhureddy B, Cirone J and Chen A 2018 *Appl. Sci.* **8** 1504
[3] Ou Y, Buchanan A M, Witt C E and Hashemi P 2019 *Anal. Methods* **11** 2738–55
[4] Kaur H, Siwal S S, Saini R V, Singh N and Thakur V K 2022 *ACS Nanosci. Au* **3** 1–27
[5] Dăscălescu D and Apetrei C 2021 *Chemosensors* **9** 1–14
[6] Jiao L, Lu G, Zhang C, Yuan Y, Wang J, Zhang Q and Li P 2013 *Micro Nano Lett.* **8** 903–5
[7] Khoshnevisan K, Maleki H, Honarvarfard E, Baharifar H, Gholami M, Faridbod F, Larijani B, Majidi R F and Khorramizadeh M R 2019 *Microchim. Acta* **186** 49
[8] Abbaspour A and Noori A 2011 *Biosens. Bioelectron.* **26** 4674–80
[9] Ribeiro J A, Fernandes P M V, Pereira C M and Silva F 2016 *Talanta* **160** 653–79
[10] Anand S K, Mathew M R, Redecki J, Radecka H and Girish Kumar K 2020 *J. Electroanal. Chem.* **878** 114531
[11] Jianxiu W, Meixian L, Zujin S, Nanqiang L and Zhennan G 2002 *Electroanalysis* **14** 225–30
[12] Sam S, Mathew M R and Girish Kumar K 2022 *J. Electrochem. Soc.* **169** 027511
[13] Sanghavi B J, Wolfbeis O S, Hirsch T and Swami N S 2015 *Microchim. Acta* **182** 1–41
[14] Xi J, Xie C, Zhang Y, Wang L, Xiao J, Duan X, Ren J, Xiao F and Wang S 2016 *ACS Appl. Mater. Interfaces* **8** 22563–73
[15] Mohammadpour-Haratbar A, Zare Y and Rhee K Y 2022 *Adv. Colloid Interface Sci.* **309** 102795
[16] Sajid M, Nazal M K, Mansha M, Alsharaa A, Jillani S M S and Basheer C 2016 *Trends Anal. Chem.* **76** 15–29
[17] Sajid M, Baig N and Alhooshani K 2019 *Trends Anal. Chem.* **118** 368–85
[18] Rezaei B, Havakeshian E and Ensafi A A 2013 *Biosens. Bioelectron.* **48** 61–6
[19] Chen A and Chatterjee S 2013 *Chem. Soc. Rev.* **42** 5425–38
[20] Kwon S J and Bard A J 2012 *J. Am. Chem. Soc.* **134** 10777–9
[21] Fitzer E, Kochling K H, Boehm H P and Marsh H 1995 *Pure Appl. Chem.* **67** 473–506
[22] Reddy Y V M, Shin J H, Palakollu V N, Sravani B, Choi B H, Park K, Kim S K, Madhavi G, Park J P and Shetti N P 2022 *Adv. Colloid and Interface Sci.* **304** 102664

[23] Taniselass S, Md Arshad M K and Gopinath S C B 2019 *Biosens. Bioelectron.* **130** 276–92

[24] Stankovich S, Dikin D A, Dommett G H B, Kohlhaas K M, Zimney E J, Stach E A, Piner R D, Nguyen S T and Ruoff R S 2006 *Nature* **442** 282–6

[25] Stankovich S, Dikin D A, Piner R D, Kohlhaas K A, Kleinhammes A, Jia Y, Wu Y, Nguyen S T and Ruoff R S 2007 *Carbon* **45** 1558–65

[26] Wilson N R *et al* 2009 *ACS Nano* **3** 2547–56

[27] Lightcap I V, Kosel T H and Kamat P V 2010 *Nano Lett.* **10** 577–83

[28] Gilje S, Han S, Wang M, Wang K L and Kaner R B 2007 *Nano Lett.* **7** 3394–8

[29] Xu Q, Yuan H, Dong X, Zhang Y, Asif M, Dong Z, He W, Ren J, Sun Y and Xiao F 2018 *Biosens. Bioelectron.* **107** 153–62

[30] Li M, Chen T, Gooding J J and Liu J 2019 *ACS Sens.* **4** 1732–48

[31] Ju J and Chen W 2015 *Anal. Chem.* **87** 1903–10

[32] Ruan J, Wang Y, Li F, Jia R, Zhou G, Shao C, Zhu L, Cui M, Yang D P and Ge S 2018 *ACS Appl. Mater. Interfaces* **10** 14342–55

[33] Ghanbari K and Bonyadi S 2020 *Int. J. Polym. Anal.* **25** 105–25

[34] Yan F, Zhang M and Li J 2014 *Adv. Healthc. Mater.* **3** 313–31

[35] Brodie B C 1859 *Philos. Trans. R. Soc. London* **149** 249–59

[36] Hummers W S and Offeman R E 1958 *J. Am. Chem. Soc.* **80** 1339

[37] Hofmann U and Kçnig E 1937 *Z. Anorg. Allg. Chem.* **234** 311–36

[38] Staudenmaier L 1898 *Ber. Dtsch. Chem. Ges.* **31** 1481–7

[39] Rajkumar C, Thirumalraj B, Chen S-M and Chen H-A 2017 *J. Colloid Interface Sci.* **487** 149–55

[40] Palanisamy S, Sakthinathan S, Chen S M, Thirumalraj B, Wu T H, Lou B S and Liu X 2016 *Carbohydr. Polym.* **135** 267–73

[41] Guell A G, Cuharuc A S, Kim Y-R, Zhang G, Tan S-Y, Ebejer N and Unwin P R 2015 *ACS Nano* **9** 3558–71

[42] Rahman M M and Lee J-J 2019 *J. Electrochem. Sci. Technol.* **10** 185–95

[43] Valentini F, Romanazzo D, Carbone M and Palleschi G 2012 *Electroanalysis* **24** 872–81

[44] Shang N G, Papakonstantinou P, McMullan M, Chu M, Stamboulis A, Potenza A, Dhesi S S and Marchetto H 2008 *Adv. Funct. Mater.* **18** 3506–14

[45] Xiao X, Miller P R, Narayan R J, Brozik S M, Wheeler D R, Brener I, Wang J, Burckel D B and Polsky R 2014 *Electroanalysis* **26** 52–6

[46] Cernat A, Ştefan G, Tertis M, Cristea C and Simon I 2020 *Bioelectrochemistry* **136** 107620

[47] Gao F, Cai X, Wang X, Gao C, Liu S, Gao F and Wang Q 2013 *Sensors Actuators* B **186** 380–7

[48] Mallesha M, Manjunatha R, Nethravathi C, Suresh G S, Rajamathi M, Melo J S and Venkatesha T V 2011 *Bioelectrochemistry* **81** 104–8

[49] Alamry K A, Hussein M A, Choi J and El-Said W A 2020 *J. Electroanal. Chem.* **879** 114789

[50] Babulal S M, Karuppiah C, Chen S M, Babulal L M, Palanisamy S, Yang C C and Chiesa M 2022 *FlatChem.* **36** 100445

[51] Iman Sengupta I, Sharat Kumar S S S, Pal S K and Chakraborty S 2020 *J. Mater. Res.* **35** 1197–204

[52] Zhan D, Ni Z, Chen W, Sun L, Luo Z, Lai L, Yu T, Wee A T S and Shen Z 2011 *Carbon* **49** 1362–6

[53] Guo H L, Su P, Kang X and Ning S K 2013 *J. Mater. Chem.* A **1** 2248–55

[54] Becerril H A, Mao J, Liu Z, Stoltenberg R M, Bao Z and Chen Y 2008 *ACS Nano* **2** 463–70

[55] Sharma D, Kanchi S, Sabela M I and Bisetty K 2016 *Arab. J. Chem.* **9** 238–61
[56] Li F, Ni B, Zheng Y, Huang Y and Li G 2021 *Surf. Interfaces* **26** 101375
[57] Yang L, Liu D, Huang J and You T 2014 *Sensors Actuators* B **193** 166–72
[58] Lee E J, Choi J H, Um S H and Oh B K 2017 *Korean J. Chem. Eng.* **34** 1129–32
[59] Shahid M M, Rameshkumar P, Pandikumar A, Lim H N, Ng Y H and Huang N M 2015 *J. Mater. Chem.* A **3** 14458–68
[60] Soni M, Kumar P, Kumar R, Sharma S K and Soni A 2017 *J. Phys. D: Appl. Phys.* **50** 124003
[61] Borowiec J, Wang R, Zhu L and Zhang J 2013 *Electrochim. Acta* **99** 138–44
[62] Gao F, Wang Q, Gao N, Yang Y, Cai F, Yamane M, Gao F and Tanaka H 2017 *Biosens. Bioelectron.* **97** 238–45
[63] Zhou M, Zhai Y and Dong S 2009 *Anal. Chem.* **81** 5603–13
[64] Li W, Ma L, Wu B, Zhang Y and Li Z 2019 *Anal. Methods* **9** 3819–24
[65] Sengupta I, Chakraborty S, Talukdar M, Pal S K and Chakraborty S 2018 *J. Mater. Res.* **33** 4113–22
[66] Sengupta I, Sharat Kumar S, Pal S and Chakraborty S 2020 *J. Mater. Res.* **35** 1197–204
[67] Lencka M M, Oledzka M and Riman R E 2000 *Chem. Mater.* **12** 1323–30
[68] Cheng C, Zhang C, Gao X, Zhuang Z, Du C and Chen W 2018 *Anal. Chem.* **90** 1983–91
[69] Ramu A G, Umar A, Ibrahim A A, Algadi H, Ibrahim Y S A, Wang Y, Hanafiah M M, Shanmugam P and Choi D 2021 *Environ. Res.* **200** 111366
[70] Fan Y, Lu H T, Liu J H, Yang C P, Jing Q S, Zhang Y X, Yang X K and Huang K J 2011 *Colloids Surf.* B **83** 78–82
[71] Kokulnathan T, Chen T W, Chen S M, Kumar J V, Sakthinathan S and Nagarajan E R 2019 *Compos. B. Eng.* **169** 249–57
[72] Ding Y H, Zhang P, Zhuo Q, Ren H M, Yang Z M and Jiang Y 2011 *Nanotechnology* **22** 215601
[73] Wang H, Zhang S, Li S and Qu J 2018 *Talanta* **178** 188–94
[74] Liu L, Gou Y, Gao X, Zhang P, Chen W, Feng S, Hu F and Li Y 2014 *Mater. Sci. Eng.* C **42** 227–33
[75] You J M, Kim D, Kim S K, Kim M S, Han H S and Jeon S 2013 *Sensors Actuators* B **178** 450–7
[76] Hu X, Dou W C, Fu L L and Zhao G Y 2013 *Anal. Biochem.* **434** 218–20
[77] Wang J, Yang B, Zhong J, Yan B, Zhang K, Zhai C, Shiraishi Y, Du Y and Yang P 2017 *J. Colloid Interface Sci.* **497** 172–80
[78] Sun W, Wang Y H, Lu Y X, Hu A H, Shi F and Sun Z F 2013 *Sensors Actuators* B **188** 564–70
[79] Toh S Y, Loh K S, Kamarudin S K and Daud W R W 2014 *Chem. Eng. J.* **251** 422–34
[80] Alanyalıoğlu M, Segura J J, Oró-Solè J and Casañ-Pastor N 2012 *Carbon* **50** 142–52
[81] Amal Raj M and Abraham John S 2013 *Anal. Chim. Acta* **771** 14–20
[82] Alwarappan S, Erdem A, Liu C and Li C Z 2009 *J. Phys. Chem.* C **113** 8853–7
[83] Wang Y, Li Y, Tang L, Lu J and Li J 2009 *Electrochem. Commun.* **11** 889–92
[84] Rosy , Yadav S K, Agrawal B, Oyama M and Goyal R N 2014 *Electrochim. Acta* **125** 622–9
[85] Lee S U, Belosludov R V, Mizuseki H and Kawazoe Y 2009 *Small* **5** 1769
[86] Soni R, Palit K, Soni M, Kumar R and Sharma S K 2020 *Bull. Mater. Sci.* **43** 175
[87] Putri L K, Ong W J, Chang W S and Chai S P 2015 *Appl. Surf. Sci.* **358** 2–14

[88] Yang L, Jiang S, Zhao Y, Zhu L, Chen S, Wang X, Wu Q, Ma J, Ma Y and Hu Z 2011 *Angew. Chem.* **50** 7132–5
[89] Jin Z, Nie H, Yang Z, Zhang J, Liu Z, Xu X and Huang S 2012 *Nanoscale* **4** 6455–60
[90] Gaidukevic J, Aukstakojyte R, Barkauskas J, Niaura G, Murauskas T and Pauliukaite R 2022 *Appl. Surf. Sci.* **592** 153257
[91] Li S M *et al* 2013 *Carbon* **59** 418–29
[92] Manna B and Raj C R 2018 *ACS Sustain. Chem. Eng.* **6** 6175–82
[93] Dinesh B, Veeramani V, Chen S M and Saraswathi R 2017 *J. Electroanal. Chem.* **786** 169–76
[94] Ghanbari M H, Norouzi Z and Ghanbari M M 2020 *Microchem. J.* **156** 104994
[95] Huang Y, Zhu C, Pan H, Xu D, Lu T, Mao L, Meng X, Chen Z, Zhang D and Zhu S 2017 *RSC Adv.* **7** 36000–6
[96] Chen S, Shi M, Xu Q, Xu J, Duan X, Gao Y, Lu L, Gao F, Wang X and Yu Y 2021 *Nanotechnology* **32** 265501
[97] Sheng Z H, Zheng X Q, Xu J Y, Bao W J, Wang F B and Xia X H 2012 *Biosens. Bioelectron.* **34** 125–31
[98] Xue C, Wang X, Zhu W, Han Q, Zhu C, Hong J, Zhou X and Jiang H 2014 *Sensors Actuators* B **196** 57–63
[99] Chen T, Tang L, Yang F, Zhao Q, Jin X, Ning Y and Zhang G J 2016 *Anal. Lett.* **49** 2223–33
[100] Wei B, Zhong H, Wang L, Liu Y, Xu Y, Zhang J, Xu C, He L and Wang H 2019 *Int. J. Biol. Macromol.* **135** 400–6
[101] Adumitrăchioaie A, Tertiş M, Suciu M, Graur F and Cristea C 2019 *Electrochim. Acta* **311** 50–61
[102] Hsine Z, Mlika R, Jaffrezic-Renault N and Korri-Youssoufi H 2022 *Chemosensors* **10** 249
[103] Balandin A A, Ghosh S, Bao W, Calizo I, Teweldebrhan D, Miao F and Lau C N 2008 *Nano Lett.* **8** 902–7
[104] Xu Z and Gao C 2015 *Mater. Today* **18** 480–92
[105] Yang Y, Asiri A M, Tang Z, Du D and Lin Y 2013 *Mater. Today* **16** 365–73
[106] Liu B, Ouyang X, Ding Y, Luo L, Xu D and Ning Y 2016 *Talanta* **146** 114–21
[107] Khan M Z H, Liu X, Tang Y, Zhu J, Hu W and Liu X 2018 *Microchim. Acta* **185** 439
[108] Liang W *et al* 2019 *Microchim. Acta* **186** 751
[109] Shahid M M, Rameshkumar P, Numan A, Shahabuddin S, Alizadeh M, Khiew P S and Chiu W S 2019 *Mater. Sci. Eng.* C **100** 388–95
[110] Subramaniam T, Kesavan G and Venkatachalam G 2020 *ACS Appl. Bio. Mater.* **3** 7769–78
[111] Wang W, Wei F and Han B 2022 *Int. J. Electrochem. Sci.* **17** 220232
[112] Vinodhkumar G, Jose S P, Lokeswarareddy S, Sekar C, Potheher I V and Peter A C 2021 *Synth. Met.* **280** 116859
[113] Ganatra R and Zhang Q 2014 *ACS Nano* **8** 4074–99
[114] Xing L and Ma Z 2016 *Microchim. Acta* **183** 257–63
[115] Kalia S, Rana D S, Thakur N, Singh D, Kumar R and Singh R K 2022 *Mater. Chem. Phys.* **287** 126274
[116] Kumar S, Singh D, Pathania D, Awasthi A and Singh K 2023 *Mater. Chem. Phys.* **297** 127446
[117] Mathew M R, Anand S K and Girish Kumar K 2022 *Ionics* **28** 3023–36

[118] Huang K J, Jing Q S, Wu Z W, Wang L and Wei C Y 2011 *Colloids Surf. B Biointerfaces* **88** 310–4
[119] Qin Q, Bai X and Hua Z 2016 *J. Electroanal. Chem.* **782** 50–8
[120] Hu Y, Yang H, Li W, Du K, Song S, Zhang W, Wei Z and Li S 2021 *Int. J. Electrochem. Sci.* **16** 211224
[121] Swathy S, Mathew M R and Girish Kumar K 2022 *J. Electrochem. Soc.* **169** 087519
[122] Chandran A, Sam S and Girish Kumar K 2022 *J. Electrochem. Soc.* **169** 127507
[123] Teng H, Song J, Xu G, Gao F and Luo X 2020 *Electrochim. Acta* **355** 136772
[124] Minta D, González Z, Melendi-Espina S and Gryglewicz G 2022 *Surf. Interfaces* **28** 101606
[125] Hsine Z, Bizid S, Mlika R, Sauriat-Dorizon H, Haj Said A and Korri-Youssoufi H 2020 *Sensors* **20** 1256
[126] Zhang D, Li L, Ma W, Chen X and Zhang Y 2017 *Mater. Sci. Eng.* C **70** 241–9
[127] Tan C, Zhao J, Sun P, Zheng W and Cui G 2020 *New J. Chem.* **44** 4916–26
[128] Patella B, Sortino A, Mazzara F, Aiello G, Drago G, Torino C, Vilasi A, O'Riordan A and Inguanta R 2021 *Anal. Chim. Acta* **1187** 339124
[129] Minta D, González Z, Wiench P, Gryglewicz S and Gryglewicz G 2020 *Sensors* **20** 4427
[130] El Khamlichi R *et al* 2018 *J. Solid State Electrochem.* **22** 1983–94
[131] Annu R A N 2020 *Int. J. Biol. Macromol.* **164** 4231–44
[132] Tashkhourian J, Nami-Ana S F and Shamsipur M 2018 *J. Mol. Liq.* **266** 548–56
[133] Kafi M A, Paul A, Vilouras A and Dahiya R 2020 *Biosens. Bioelectron.* **147** 111781
[134] Ghanbari K and Nejabati F 2020 *Anal. Methods* **12** 1650–61
[135] Begum H, Ahmed M S and Jeon S 2017 *ACS Omega* **2** 3043–54
[136] Liu C-Y, Chou Y-C, Tsai J-H, Huang T-M, Chen J-Z and Yeh Y-C 2019 *Appl. Sci.* **9** 622
[137] Huang Q, Lin X, Tong L and Tong Q-X 2020 *ACS Sustain. Chem. Eng.* **8** 1644–50
[138] Jang H-S, Kim D, Lee C, Yan B, Qin X and Piao Y 2019 *Inorg. Chem. Commun.* **105** 174–81
[139] Liu H, Li N, Zhang H, Zhang F and Su X 2018 *Talanta* **189** 190–5
[140] Pang P, Yan F, Li H, Li H, Zhang Y, Wang H, Wu Z and Yang W 2016 *Anal. Methods* **8** 4912–8
[141] Li Y, Jiang Y, Mo T, Zhou H, Li Y and Li S 2016 *J. Electroanal. Chem.* **767** 84–90
[142] Zheng S, Huang R, Ma X, Tang J, Li Z, Wang X, Wei J and Wang J 2018 *Int. J. Electrochem. Sci.* **13** 5723–35
[143] Arumugasamy S K, Govindaraju S and Yun K 2020 *Appl. Surf. Sci.* **508** 145294
[144] Hsu W-F and Wu T-M 2019 *J. Mater. Sci., Mater. Electron.* **30** 8449–56
[145] Yola M L and Atar N 2018 *Appl. Surf. Sci.* **458** 648–55
[146] Baluta S, Lesiak A and Cabaj J 2018 *Electroanalysis* **30** 1781–90
[147] Jahani P M 2020 *Int. J. Electrochem. Sci.* **15** 947–58
[148] Sanati A L, Faridbod F and Ganjali M R 2017 *J. Mol. Liq.* **241** 316–20
[149] Beitollahi H, Dourandish Z, Ganjali M R and Shakeri S 2018 *Ionics* **24** 4023–31
[150] Ben A S 2017 *R. Soc. Open Sci.* **4** 171199

IOP Publishing

Graphene-Based Sensors

Jamballi G Manjunatha

Chapter 10

Graphene-modified electrodes for detection of vitamins

Nida Aydogdu Ozdogan, Ersin Demir and Murat Misir

Vitamins are organic compounds found in various food products and required in specific amounts for the proper functioning of cells and organs in living things. Vitamins have a unique molecular structure and are classified into 'fat-soluble' and 'water-soluble' vitamins according to their dissolution properties. There are 13 vitamins, mainly ascorbic acid, retinol, α-tocopherol, and folic acid. These biologically necessary substances cannot be synthesized by human and animal cells and must be taken into the body by foodstuffs. The deficiency of these vitamins, which should be consumed frequently, causes many diseases such as deterioration in DNA synthesis, visual impairment, characteristic changes in developing blood cells, anemia, but also fatigue, and irritability. It is important to determine vitamins in biological samples with accurate, sensitive, reliable, and fast analysis methods. Graphene and its derivatives have an important place in the development of electrochemical sensors, as they have many advantages such as conductivity, catalytic activity, fast electron transfer kinetics, large surface area, high sensitivity, and good mechanical and electrical properties. According to the literature review, traditional analytical methods such as high-performance liquid chromatography (HPLC), micellar electrokinetic chromatography method (MEKC), spectrophotometry and capillary electrophoresis, colorimetric, spectrofluorometric have been widely used in the determination of target analytes. Although these methods have advantages such as high selectivity and sensitivity, they have shortcomings such as long-term analysis, complex sample preparation, large sample amount and expensive equipment to be used. For this chapter, electroanalytical methods, which are easy to apply, are preferred as a good alternative to other methods because they are simple, sensitive, and selective. Consequently, the physical and chemical properties of graphene-modified sensors used in the electrochemical analysis of vitamins in the last 10 years are examined in detail.

10.1 Introduction

Vitamins are a group of organic chemical compounds found in very low amounts in natural foods. Casimir Funk first coined the term 'vitamin' in 1912. The principal period of discovery began in the early nineteenth century and ended in the middle of the twentieth century [1]. Vitamins are essential for functioning in vital biological processes such as digestion and the immune system. For example, vitamins can be used as cofactors to help them complete various biochemical reactions and activate enzymes [2]. According to their solubility, vitamins are divided into water-soluble vitamins and fat-soluble vitamins (figure 10.1). The body cannot synthesize most vitamins, so they must be taken in through the diet. While fat-soluble vitamins can be stored in adipose tissue, water-soluble vitamins cannot be stored in the body. Insufficient intake of vitamins that cannot be stored in the body or excessive intake of stored vitamins leads to the formation of various diseases [2]. Therefore, the determination of these substances is important in every respect.

With electroanalytical methods, analysis is performed based on the electrochemical (reduction or oxidation) properties of the analyte in the electrochemical cell. Electroanalytical methods are named voltammetry, amperometry, potentiometry, and polarography based on the relationship between the concentration of the analyte and electrical parameters (such as potential, current, or conductivity) (figure 10.2). Cyclic voltammetry (CV) and linear sweep voltammetry (LSV), square-wave voltammetry (SWV) and differential pulse voltammetry (DPV), and absorptive stripping voltammetry techniques are types of various voltammetry methods. Electrochemical sensors are transducers that selectively generate electrical signals based on the concentration of chemical compounds or ions. A triple electrode system is generally used: reference electrode, working electrode, and counter electrode in electrochemical techniques like voltammetry. Various indicator

Figure 10.1. Classification of vitamins.

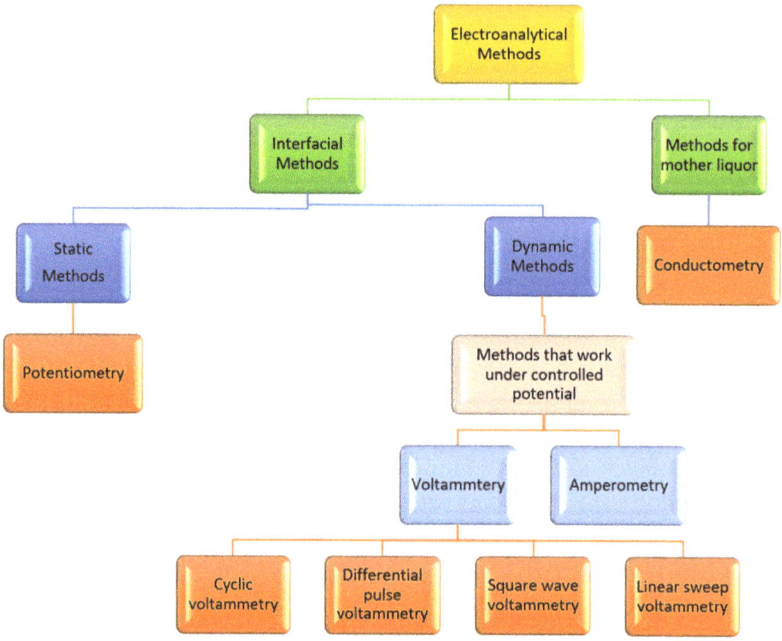

Figure 10.2. Classification of electroanalytical methods.

electrodes such as glassy carbon electrode (GCE), hollow cathode electrode, gold electrode, and boron-doped diamond electrode have been used for analysis of substances. With the modification of different nanomaterials, a more sensitive, selective, and stable electrochemical sensor can be developed by the interaction between the analyte and the electrode surface [3].

While graphene generally refers to a single layer of sp^2-bonded carbon atoms, there are also studies on graphene with two or more layers (figure 10.3). In 2004, graphene was produced from graphite sheets by micromechanical fission by work teams of Andre Geim and Konstantin Novoselov. Since then, much progress has been made in graphene synthesis, and several methods have been developed to prepare monolayer graphene and several-layer graphene. Since 2004, graphene has attracted significant attention in the scientific and technological fields since its many superior electrical, mechanical, thermal, and optical properties, thus attracting the scientific world. Graphene oxide (GO) is highly electroactive like graphene due to its low atomic number and two-dimensional structure. While GO which is containing oxygenated groups (epoxy, hydroxyl) has the electrochemical advantage of increasing the surface area, its disadvantage is reduced electrical conductivity. The decrease in electrical conductivity is due to oxygenated groups forming an insulation-like barrier [4–6]. Due to the reduced electrical conductivity, reduced GO (rGO) has replaced GO in many electrochemical sensors. It has made the electrochemical properties better by reducing the oxygen-containing groups using many approaches, such as electrochemical and hydrothermal [7–10]. Graphene quantum dots (GQDs)

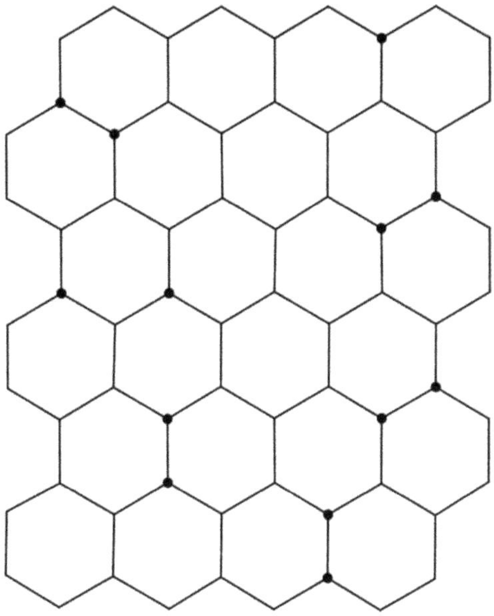

Figure 10.3. The graphene structure.

consist of one or a few graphene layers and are smaller than 100 nm [11, 12]. They are chemically and physically stable and have a sizeable surface-to-mass ratio. Studies in the literature show that since the discovery of graphene, various graphene derivatives (Gr, GQDs, GO, rGO, etc) have been widely used in electrochemical analyses, given their advantages [13–17]. Graphene-based electrodes are frequently used for the determination of not only vitamins but also numerous substances such as antioxidants, plant extracts, proteins, pesticides, and pesticides [18–25].

In the literature, traditional analytical methods such as high-performance liquid chromatography (HPLC) [26–29], micellar electrokinetic chromatography method (MEKC) [30, 31], spectrophotometry [32, 33], capillary electrophoresis [34–36], colorimetry [37], and spectrofluorimetry [38] have been widely used for the determination of analytes of vitamins. While these methods have advantages such as high selectivity and sensitivity, they have shortcomings such as long analysis time, complex sample preparation, large sample amount and expensive equipment, and trained personnel requirement. Therefore, electrochemical sensors are widely applied in various fields, including pharmaceuticals, food, and clinical analysis. Since most vitamins are electrochemically active, they can be detected by electrochemical sensors. Recently, many electrochemical sensors have been proposed for the detection of vitamins. This review has investigated the analysis of vitamins by various electroanalytical methods using graphene-based electrochemical sensors in the last 10 years (table 10.1).

Zhang and Zhang, in their study published in 2018, developed a new-generation composite electrode for the simultaneous determination of vitamin B_2 (VB_2), also known as riboflavin and dopamine (DA) [17]. The GCE was treated with cystine (Cys),

Table 10.1. Determination with graphene-based electrodes of water-soluble vitamins.

ss	Analyte	Method	Linear range (μM)	LOD (nM)	LOQ (nM)	Samples	Interference	Ref
PIXCoCl/AuNPsnGr	Vitamin B_1	CV	1×10^{-5}–0.1	0.003	0.01	Water, multivitamin tablet, blueberry syrup, urine	Ascorbic acid, CH_3COONa, $FeSO_4$, fructose, glucose, maltodextrin, NH_4Cl, vitamin B_{12}	[39]
PIX/AuNPsnGr	Vitamin B_1	CV	1×10^{-6}–10.0	0.0003	0.001	Water, multivitamin tablet, blueberry syrup, urine	Ascorbic acid, CH_3COONa, $FeSO_4$, fructose, glucose, maltodextrin, NH_4Cl, vitamin B_{12}	[39]
Cystine/MWCNT/GO/GCE	Vitamin B_2	CV	0.002–300	1.2	—	Urine, blood samples	—	[17]
PApA/GNpPE	Vitamin B_2	CV	10–100	0.33	—	—	—	[40]
rGO/A_{32}/AuE	Vitamin B_2	DPV	0.025–2.75	15	—	Human urine	—	[41]
GO/Au/polyE AmVS/GCE	Vitamin B_2	DPV	1.0–100.0	72	—	Non-alcoholic beverage, energy drink, vitamin C drink	Glucose, fructose, galactose, ascorbic acid, maltose, folic acid	[42]
WO_3/TiO_2/Gr/GCE	Vitamin B_2	DPV	0.02–2.5	25.24	—	Capsule, tablet, egg white, egg yolk, whole egg, milk	Chloramphenicol, glucose, uric acid, ascorbic acid, vitamin B complex, vitamin E, vitamin A tablets	[43]
PApA/GNpPE	Vitamin B_2	DPV	2–100	0.02	—	Pharmaceutical tablets, blood serum, human urine	Zn^{2+}, Na^+, Mg^{2+}, K^+, Fe^{2+}, Hg^{2+}, glycine, glutamic acid, galactose, sucrose, tyrosine, fructose	[40]

(*Continued*)

Table 10.1. (*Continued*)

ss	Analyte	Method	Linear range (μM)	LOD (nM)	LOQ (nM)	Samples	Interference	Ref
AuNPs@PDA–rGO	Vitamin B_2	DPV	0.02–60	9.6	—	Pharmaceutical formulation	—	[44]
PGBHA–afGQDs–MnO_2/GCE	Vitamin B_2	DPV	0.1–100	40	—	Injection samples	—	[45]
ERGO/GCE	Vitamin B_2	DPV	0.1–1.0; 1.0–40	16	—	Pharmacological specimens	Ascorbic acid, acetic acid, vitamin B_1, vitamin B_9, D-glucose, dopamine, H_2O_2, urea, K^+, HCO_3^-, Ca^{2+}, Cu^{2+}, Mg^{2+}, Zn^{2+}	[46]
Poly-amino acid/GQD/GCE	Vitamin B_2	DPV	1–1000	220	—	Commercial multivitamin	Niacin, thiamin, Mg^{2+}, Zn^{2+}, Fe^{2+}, Ca^{2+}	[47]
Sm_2O_3/rGO/CPE	Vitamin B_2	FFTSWV	0.01–0.4	0.86	—	Human plasma	—	[48]
S_2-MIPs/GO-SWCNTs/GCE	Vitamin B_2	LSV	0.002–0.9	0.7	—	Pharmaceutical solutions	Vitamin B_1, vitamin B_3, vitamin B_6, vitamin B_{12}, vitamin C, uric acid, glucose, and 4 acetaminophen	[13]
N,*S*-GQD–CS/GCE	Vitamin B_2	SWASV	0.001–8.0	0.30	1.01	Energy drink	Niacin, pantothenic acid, vitamin B_{12}, pyridoxine, caffeine, taurine	[49]
SnO_2/rGO/GCE	Vitamin B_2	SWV	0.1–150	34	—	Pharmaceutical compounds and energy drinks	L-Tryptophan, L-proline, L-cysteine, L-lysine, L-glutamic acid, L-ascorbic acid, glucose, cholesterol, dopamine HCl, uric acid, sodium sulfate, ammonium chloride, thiamine	[50]

HMS@GO/GCE	Vitamin B_2	SWV	0.001–4.0	0.26	—	Egg yolk, shrimp, milk powder, honey	K^+, Mg^{2+}, Ca^{2+}, Fe^{3+}, Cd^{2+}, Cr^{3+}, Pb^{2+}, Cu^{2+}, Co^{2+}, Zn^{2+}, Cl^-, NO_3^-, CO_3^{2-}, SO_4^{2-}, vitamins C, B_1, B_3, B_6, B_9, B_{12}, K_1	[51]
Graphene sheets modified GCE	Vitamin B_6	Ampero metry	0.1–100	56		Blood serum, pharmaceutical formulations	L-3,4-dihydroxy phenyl alanine, theophylline	[52]
ERGO/GCE	Vitamin B_6	DPV	20–1000	14	—	Pharmacological specimens	Ascorbic acid, vitamin B_1, vitamin B_9, vitamin B_{12}, dopamine, H_2O_2, caffeine, L-cystine, citric acid, Fe^{2+}	[46]
Pt/β-CD-GR/PGE	Vitamin B_6	DPV	0.002–0.205	1.2	—	Orange juice and energy drink	Li^+, K^+, Ca^{2+}, F^-, methionine, glycine, valine, glucose	[53]
AGCE	Vitamin B_6	DPV	25–454	10 000	32 000	Pharmaceutical tablets	—	[54]
AuNPs@PDA–rGO	Vitamin B_6	DPV	0.03–600	25	—	Pharmaceutical formulation	—	[44]
GO/CMC–GCE	Vitamin B_6	DPV	5–200	100	—	—	Lactic acid, uric acid, vitamins (A, C, E, B_6, B_1, B_2, B_{12})	[55]
CuO–PLL/Gr electrode	Vitamin B_6	DPV	3.0–2076	2300	—	Orange	Cysteine, ascorbic acid, tyrosine, glucose, tryptophan, serine, abscisic acid, indole-3-acetic acid	[56]

(*Continued*)

Table 10.1. (*Continued*)

ss	Analyte	Method	Linear range (μM)	LOD (nM)	LOQ (nM)	Samples	Interference	Ref
GO/Fe_3O_4@SiO_2/SPE	Vitamin B_6	DPV	1.0–900	520	—	Urine	—	[57]
GQDs/SPCE	Vitamin B_6	DPV	250–1 × 10^4	830	—	—	—	[58]
GO–SiO_2–CMC	Vitamin B_6	DPV	0.5–200	100	—	—	Lactic acid, uric acid	[14]
Gr/MWCNT/GCE	Vitamin B_6	DPV	0.05–200	20	—	vitamin B_6 tablets	—	[59]
N,S-GQD–CS/GCE	Vitamin B_6	SWASV	0.1–18.0	30.1	100.1	Energy drink	Niacin, pantothenic acid, riboflavin, vitamin B_{12}, caffeine and taurine	[49]
Cu-P(1,8DAN)/Gr/GCE	Vitamin B_6	SWV	0.58–24.51	300	—	Pharmaceutical Tablet	Thiamine HCl, riboflavin, ascorbic acid, dopamine	[60]
Gr/ZnO/GCE	Vitamin B_6	SWV	1.0–1600	720	—	—	—	[61]
MoS_2/rGO paper electrode	Vitamin B_9	Amperometry	0.1–1240	37	—	Serum	Glucose, urea, oxalate, H_2O_2, $CuCl_2$, $MgSO_4$, K_2CO_3, NaCl	[62]
GONR/SPCE	Vitamin B_9	Amperometry	0.1–1600	20	—	Human urine	—	[63]
GNSs–MoS_2–AuNPs/SPCE	Vitamin B_9	Amperometry	0.05–1150	38.5	—	Human urine	—	[64]
Au-rGO/PEI/CD	Vitamin B_9	DPV	0.005–10	1	—	Serum	Dopamine, ascorbic acid	[65]
$(PEI/P_2Mo_{16}V_2\text{-rGO})_4$/GCE	Vitamin B_9	DPV	0.01–800	0.284	—	Human serum	NaCl, K_2CO_3, NH_4Cl, $CaCl_2$, ascorbic acid (AA), dopamine, epinephrine, NADH, glucose	[66]
Gr/ZnO NWAs/GFE	Vitamin B_9	DPV	1–60	1000	—	Urine, tablet	Uric acid	[67]
Fe_3O_4@SiO_2/GR/SPE	Vitamin B_9	DPV	0.1–600	50	—	Urine, tablet	—	[68]
MoS_2–rGO/GCE	Vitamin B_9	DPV	0.01–100	10	—	Human serum	Ascorbic acid, dopamine	[69]

Methylene blue/ ERGO/GCE	Vitamin B_9	DPV	4–164	500	—	Tablet	$MgCl_2$, $CaCl_2$, $MgSO_4$, NaCl, KNO_3, Na_2SO_4, $(NH_4)_2C_2O_4$, L-lysine, glucose, glycine, uric acid, ascorbic acid	[70]
PMo–PPy/Gr/GCE	Vitamin B_9	DPV	0.001–0.2	0.033	—	Human blood serum	Uric acid, ascorbic acid, dopamine, glucose, L-cysteine, Ca^{2+}, Mg^{2+}, Al^{3+}, Fe^{2+}, Fe^{3+}, NH_4^+, SO_4^{2-}	[71]
ZnO@GO/GCE	Vitamin B_9	DPV	1–400	40		Apple juice	Folic acid, Ca^{2+}, CO_3^{2-}, K^+, Na^+, PO_4^{3-}, SO_4^{2-}, ascorbic acid, uric acid, citric acid, glucose, glutamic acid, lactose, fructose, gallic acid	[72]
Poly(*o*-aminophenol)/ rGO–AuNPs/MIP	Vitamin B_9	DPV	0.8–10	2.8	—	Infant formula milk, multivitamin tablets, and blood serum	—	[73]
Poly-amino acid/GQD/ GCE	Vitamin B_9	DPV	0.22–100	100	—	Commercial multivitamin	Niacin, thiamin, Mg^{2+}, Zn^{2+}, Fe^{2+}, Ca^{2+}	[47]
AuNCs/AGr/ MWCNT/GCE	Vitamin B_9	SWV	10–170	90	—	Human urine	Uric acid, ascorbic acid, dopamine, folic acid	[74]
MB-rGO/f-MWCNT/ Au-MIP@PGE	Vitamin B_{12}	DPV	1.54×10^{-5}–0.001 06	0.004 13	—	Pharmaceutical, cerebrospinal fluid, urine, blood serum	Cyanocobalamin, methylcobalamine, thiamine, riboflavin, pyridoxine, nicotinamide,	[75]

(*Continued*)

Table 10.1. (*Continued*)

SS	Analyte	Method	Linear range (μM)	LOD (nM)	LOQ (nM)	Samples	Interference	Ref
							nicotinic acid, calcium D-pantothenate, biotin, folic acid, ascorbic acid	
N,S-GQD–CS/GCE	Vitamin B_{12}	SWASV	0.001–8.0	0.32	1.08	Energy drink	Niacin, pantothenic acid, riboflavin, pyridoxine, caffeine, and taurine	[49]
$Ce_2(SO_4)_3$/rGO/SPCE	Vitamin C	—	10–1000	9×10^5	—	—	—	[76]
PMo_{12}–Gr/GCE	Vitamin C	Amperometry	1–8000	500	—	Vitamin C tablets	Uric acid, glucose, citric acid, sodium chloride	[77]
Gr–NiHCF composite	Vitamin C	Amperometry	1–16 280	250	—	Natural fruit juices (orange, lemon, sweet lemon)	Uric acid, glucose, sucrose, urea	[78]
HSE–FLG/GCE	Vitamin C	Amperometry	2–1600	3.2	—	Natural citrus fruit juice	Glucose, fructose, galactose, nitrate, H_2O_2, urea	[79]
TiO_2/rGO/GCE	Vitamin C	Amperometry	25–725	1190		Vitamin C supplement	Dopamine, uric acid	[80]
AuNPs/ERGO/Nafion/GCE	Vitamin C	Amperometry	5–30	2760	—	Pharmaceutical products	Uric acid, dopamine	[81]
GNP/AgNR/poly (AY)/GCE	Vitamin C	Amperometry	1–200	420	—	Human serum, urine	Dopamine	[82]
HNONS@Tr-rGO/GCE	Vitamin C	Amperometry	0.25–13 100	250	—	—	—	[83]
$NiCo_2O_4$/N-rGO-1/GCE	Vitamin C	Amperometry	2.0–6000	390	—	Urine, vitamin C tablet, vitamin B_4 tablet, vitamin C drink	Dopamine, glucose	[84]
pNR/rGO–GCE	Vitamin C	Amperometry	50–750; 900–24 900	1400	—	Human serum, vitamin C beverage	—	[85]

AuNPs–GO/Au-IDA	Vitamin C	Amperometry	4.6–193	1400	4600	Human urine	Glucose, dopamine, epinephrine	[86]
rGO–TiO_2/GCE	Vitamin C	Amperometry	1–1500	500	—	Vitamin C tablets, spiked fruit juice	—	[87]
PFA/GO-3.88 W	Vitamin C	Chronoamperometry	50–5000	1000	—	Tablet	—	[88]
NiO/Gr/graphite electrode	Vitamin C	CV	1000–5000	—	—	—	—	[89]
Pd/rGO/MWCNT/GCE	Vitamin C	CV	12 700–2000	—	—	—	—	[90]
PDDA-RGO/Pt NPs/GCE	Vitamin C	CV	1.0–10 000	0.5	—	Tomato samples	Dopamine, glucose, hydrogen peroxide	[91]
rGO–AuNP/SPE	Vitamin C	DPV	20–380	1100	—	—	Dopamine, uric acid	[92]
GO/MWCNT/AuNR/GCE	Vitamin C	DPV	0.001–0.5; 1–800	0.85	—	Serum	Dopamine, uric acid, glucose, L-glutatione	[93]
GO–Ag–Fe_3O_4/GCE	Vitamin C	DPV	0.2–60	74	—	Human plasma	Glucose, fructose, sucrose, lactose	[94]
AuNR/NG	Vitamin C	DPV	0.005–1.0; 1.0–8000	2.6	—	Human serum	—	[95]
TiO_2NP@CuO–rGO/GCE	Vitamin C	DPV	0.000 01–10	0.003	—	Human synthetic blood serum	Na^+, K^+, Cl^-, Mg^{2+}, Ca^{2+}, SO_4^{2-}, NO_3^-, glucose, fructose, l-cysteine, l-glutamic acid, citric acid, uric acid, glycine, valine	[96]
Pt/IMo_6/GO–GCE	Vitamin C	DPV	50–4000	6400	—	Human blood serum	Glucose, NH_4Cl, KCl, $FeCl_3$, and NaCl	[97]
CuO–CeO_2–rGO–MWCNTs/GCE	Vitamin C	DPV	0.01–28	9.0	—	Blood serum, urine	K^+, Cl^-, Ca^{2+}, Zn^{2+}, Al^{3+}, glucose, lactose, sucrose, tyrosine, uric acid, dopamine	[98]

(*Continued*)

Table 10.1. (*Continued*)

ss	Analyte	Method	Linear range (μM)	LOD (nM)	LOQ (nM)	Samples	Interference	Ref
CTAB/rGO/ZnS/GCE	Vitamin C	DPV	50–1000	30 000	—	Urine	—	[99]
Gr/NiOH/GCE	Vitamin C	DPV	50–450	30 000	—	—	—	[100]
MWCNT/Gr foam/GCE	Vitamin C	DPV	2–64	18 280	—	—	Na^+, K^+, Ca^{2+}, Mg^{2+}, glucose	[101]
Hemin/GO/GCE	Vitamin C	DPV	5–100	300	—	Human serum and urine	Na^+, K^+, Mg^{2+}, Zn^{2+}, Al^{3+}	[102]
Graphene flower/carbon fiber	Vitamin C	DPV	45.4–1489.23	24 700	—	Human serum and urine	—	[103]
MgO/Gr/Ta wire electrode	Vitamin C	DPV	5–50; 50–350	30		Human serum	L-Cysteine, glucose, L-lysine, L-serine, L-tryptophan, L-tyrosine, aniline, catechol, resorcinol, hydroquinone, phenol, $CaCl_2$, CH_3COONa, K_2HPO_4, KCl, $Mg(NO_3)_2$, Na_2S, NaH_2PO_4, $Zn(NO_3)_2$	[104]
ERGO/carbon fiber electrode	Vitamin C	DPV	8–2016.45	4500	—	—	—	[105]
CTAB-GO/MWCNT/GCE	Vitamin C	DPV	5–300	1000	—	Urine	Glucose, cysteine, citric acid	[106]
GO/$TmPO_4$/GCE	Vitamin C	DPV	100–1000	39 000	—	Human urine	—	[107]
Au/rGO/GCE	Vitamin C	DPV	240–1500	51 000	—	Human serum	NaCl, KCl, $NaNO_3$, $CaCl_2$, glucose, L-cysteine, citric acid	[108]
PtNCs–GrNPs–MWCNTs/GCE	Vitamin C	DPV	100–1200	10 000	—	Human serum, vitamin C tablets	Glucose, KCl, NaCl, phenylalanine, histidine	[15]

Pt–Pd NPs/chitosan/N-Gr/GCE	Vitamin C	DPV	2–400	970	—	Vitamin C Injection	—	[109]
ERGO-poly(PR)/AuNPs/GCE	Vitamin C	DPV	0.4–110	0.48	—	Vitamin C tablets, and human urine	—	[110]
Cys/GO/SPE	Vitamin C	DPV	500–5000	0.26×10^6	—	—	Dopamine, uric acid	[111]
rGO–SnO_2/GCE	Vitamin C	DPV	400–1600	38 700	—	Pharmaceutical, food industries, clinical diagnostics	Glucose, urea, uric acid, Na^+	[112]
ERGO-pEBT/AuNPs/GCE	Vitamin C	DPV	10–900	530	—	Vitamin C tablets, urine	Ca^{2+}, Fe^{3+}, L-cystine, Na^+, K^+, glucose, glycine, Cu^{2+}	[113]
Ag NPs–GO/SPCE	Vitamin C	DPV	1–210	25	—	Human urine, vitamin C tablet	Hydroquinone, epinephrine, uric acid, dopamine, catechol	[114]
Poly-amino acid/GQD/GCE	Vitamin C	DPV	10–100	100	—	Commercial multivitamin	Niacin, thiamin, Mg^{2+}, Zn^{2+}, Fe^{2+}, Ca^{2+}	[47]
PdAu/RGO/GCE	Vitamin C	DPV	12.5–700	12 500	—	Real samples	—	[115]
Pt–Gr–CNT/GCE	Vitamin C	DPV	200–900	5000	—	Vitamin C tablet, human serum, urine	—	[116]
rGO/AuNP/PE	Vitamin C	DPV	1–6000	1000	—	Amla juice, vitamin C tablets	—	[117]
Fe_3O_4–NH_2@Gr Sheets/GCE	Vitamin C	DPV	5–1600	74	—	Vitamin C injection	—	[118]
rGO/GCE	Vitamin C	FIA-Amperometry	65–2530	4700	—	Multivitamin drink, milk, fermented milk, and chocolate milk	—	[119]

(Continued)

Table 10.1. (*Continued*)

ss	Analyte	Method	Linear range (μM)	LOD (nM)	LOQ (nM)	Samples	Interference	Ref
3D-rGO/SPE	Vitamin C	LSV	$10–1 \times 10^4$	3400	11 000	Fruit juices	Glucose, fructose, sucrose, citric acid, tartaric acid, malic acids	[120]
HSE–FLG/GCE	Vitamin C	LSV	1–5000	1.8	—	Natural citrus fruit juice	Glucose, fructose, galactose, nitrate, H_2O_2, urea	[79]
rGO/AuNPs	Vitamin C	LSV	10–500	3100	—	Milk, juice, urine	Na^+, NH_4^+, K^+, NO_3^-, Cl^-, glucose, dopamine	[121]
AuNCs/AGr/MWCNT/GCE	Vitamin C	SWV	10–150	270	—	Human urine	Uric acid, ascorbic acid, dopamine, folic acid	[74]
G-30	Vitamin C	SWV	5–1000	17 800		Pharmaceuticals and/or biological fluids	Na^+, K^+, Ca^{2+}, Zn^{2+}, glucose, sucrose, acetaminophen	[122]
Mn–GrOPE	Vitamin C	SWV	$0.022–8.97 \times 10^{-4}$	6.97	—	Rosa canina	Riboflavin, biotin, pyridoxal, niacin, pantothenic acid, thiamin, folic acid, caffeic acid, gallic acid, glucose, fructose, Fe, Cu, Al, Zn, Mg, Ni	[123]
Gr/Fe_2O_3/polyvinyl alcohol/PGE	Vitamin C	SWV	40–4100	234	—	—	Dopamine, uric acid, glucose, L-cystine, Na^+, Fe^{3+}, Zn^{2+}, Ni^{2+}, Cu^{2+}, Cr^{3+}	[124]

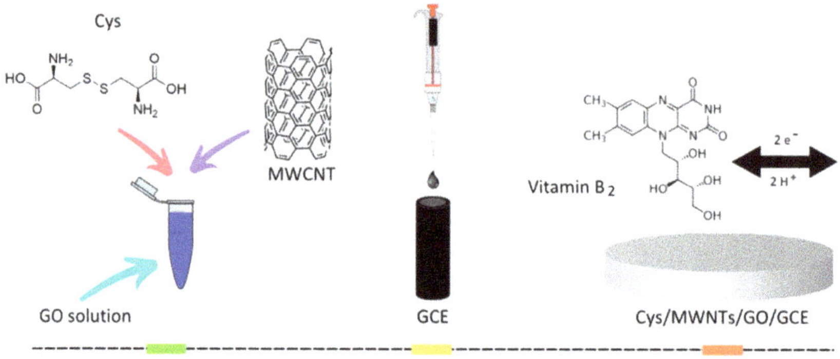

Figure 10.4. Fabrication scheme of Cys/MWCNTs/GO/GCE nanosensor.

carboxylated multi-walled carbon nanotubes (MWCNTs), and GO. The surface morphology of the Cys/MWCNTs/GO/GC electrode was elucidated by scanning electron microscopy (SEM). In addition, the electrochemical properties of the developed composite electrode were investigated in the $K_3[Fe(CN)_6]$ solution. They then found the working ranges for VB_2 and DA under optimum conditions as 2.0 nM–300 μM and 13 nM—250 μM, respectively, on the graphene-based working electrode by CV. Moreover, the limit of detection (LOD, S/N = 3) for VB_2 and DA substances was 1.2 and 1.0 nM, respectively. They also obtained high stability and repeated measurements with relative standard deviation RSD <1.5% at peak currents of both substances. Finally, they performed the analytical application to determine VB_2 and DA in biological samples such as urine and blood with a graphene-supported composite electrode (figure 10.4).

Zeybek and Üğe developed a DPV technique for the determination of vitamins B_2 (VB_2, riboflavin) and B_6 (VB_6, pyridoxine), which are very important for human health, in their study in 2021 [46]. For this, they developed two different electrodes for the analysis of VB_2 and VB_6 substances. The first indicator electrode, ERGO/GCE, was produced by the electrochemical reduction of GO on a GCE. The other working electrode was constructed as f-MWCNTs/GCE by electrochemically depositing functionalized MWCNTs (f-MWCNTs) on the GCE. The surface morphology of both electrodes was characterized by emission scanning electron microscopy (FE-SEM) and Fourier transform infrared (FT-IR) spectroscopy. After that, they determined VB_2 and VB_6 by using the graphene-based modified ERGO/GC electrode exhibiting high electrocatalytic performance. They found the linear operating range for the VB_6 anodic signal to be 20–1000 μM on the ERGO/GCE and also found the LOD value to be 14 nM. The accuracy and precision studies of the sensors developed for the determination of VB_2 and VB_6 were applied to commercially available ampoule and tablet samples. They also found that the graphene-based ERGO/GCE sensor exhibited good selectivity and high stability for the analysis of both vitamins (figure 10.5).

In a study conducted by Szoke *et al* in 2019, nanomaterials containing gold nanoparticles (AuNPs), GO, and Nafion were coated on the GCE surface with the

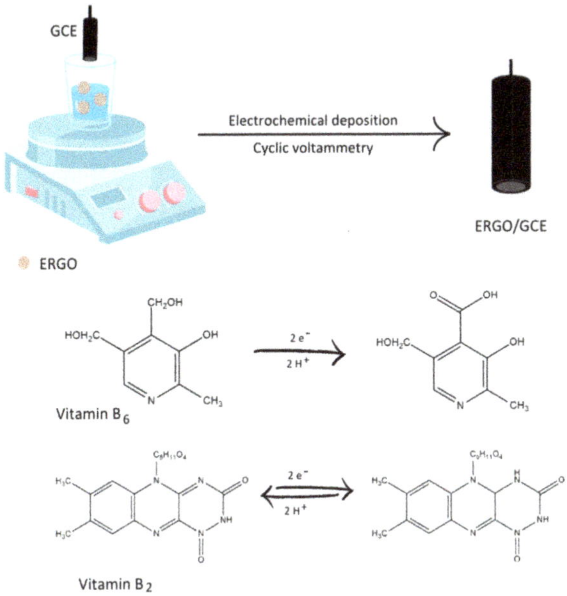

Figure 10.5. Fabrication scheme of ERGO/GCE nanosensor.

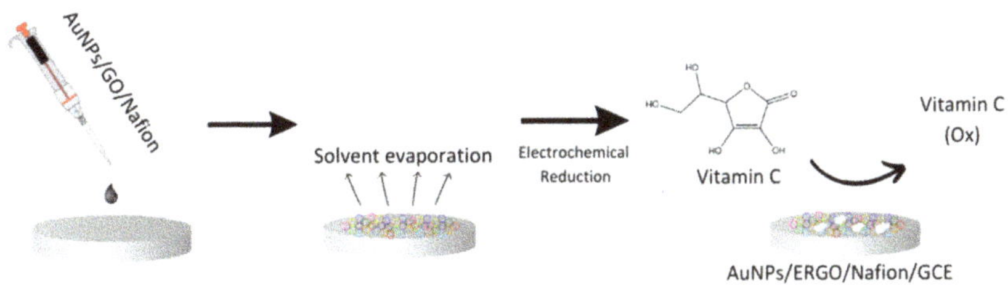

Figure 10.6. Fabrication scheme of Nafion/ERGO/AuNPs/GCE nanosensor.

drop-dry technique [81]. For the composite nanosensor, they used the ERGO form obtained by electrochemical reduction of GO material. The CV, electrochemical impedance spectroscopy (EIS), and amperometry techniques were used to investigate the surface area, catalytic effect, and conductivity properties of newly developed indicator electrode (Nafion/ERGO/AuNPs/GCE). They showed that they were highly selective and sensitive to ascorbic acid (vitamin C) on the nanosensor. For AA determination, they found a linear working range of 5–30 μM and an LOD of 2.76 μM on Nafion/ERGO/AuNPs/GCE by amperometry. They have successfully determined AA with high sensitivity and recovery by standard addition method in pharmaceutical products of different commercial companies containing ascorbic acid on the Nafion/ERGO/AuNPs/GCE (figure 10.6). As a result, they developed a highly sensitive, selective, repetitive, stable new-generation sensor containing graphene for the determination of ascorbic acid (vitamin C), an important vitamin and antioxidant substance (table 10.2).

Table 10.2. Determination with graphene-based electrodes of fat-soluble vitamins.

Modified electrode	Analyte	Method	Linear range (μM)	LOD (nM)	LOQ (nM)	Samples	Interference	Ref
rGO–CuNPs/GCE	Vitamin A	DPV	0.2–3.0	15.3		Plasmatic serum	Uric acid, estriol, ascorbic acid	[125]
Gr/Nafion/SPE	Vitamin A	SWV	0.34–17.4	62.8	—	Dietary supplements	—	[126]
Gr/Nafion/SPE	Vitamin A	AdSWV	0.06–3.49	30.02	99.86	Infant milk, yogurt, parsley		[16]
Gr/Nafion/SPE	Vitamin D	SWV	0.207–12.99	33.79	—	Dietary supplements	—	[126]
Gr/Nafion/SPE	Vitamin D	AdSWV	0.05–2.59	16.37	54.84	Infant milk, yogurt, parsley		[16]
GO/MoS_2/GCE	Vitamin D_3	DPV	0.0002–0.051; 0.051–0.38	0.051	—	Saliva	—	[127]
SiO_2/GO/Ni $(OH)_2$	Vitamin D_3	DPV	0.005–5	3.26	—	Pellet pharmaceuticals	Ascorbic acid, uric acid, dopamine	[128]
GQD–Au/SPE	Vitamin D_3	EIS	0.001–0.5	0.70	2.09	Serum	Oxalic acid, urea, uric acid, vitamin E	[129]
LaNPs–GQDs@ZIF-8@GCE	Vitamin D_3	SWV	0.006 25–1.25	6.10	6.25	Human serum and urine	Vitamin A, vitamin E, vitamin B_2, vitamin B_6, vitamin C, Na^+, Cl^-, Mg^{2+}, glucose cholesterol, uric acid, L-cysteine	[130]
Gr/Nafion/SPE	Vitamin E	SWV	0.18–11.6	27.86	—	Dietary supplements	—	[126]
Gr/Nafion/SPE	Vitamin E	AdSWV	0.046–2.32	17.41	57.81	Infant milk, yogurt, parsley		[16]
rGO/Nafion/GCE	Vitamin E	SWV	0.5–90	60	—	Vegetable oil samples (hazelnut oil, sunflower oil)	Riboflavin, copper, manganese, zinc and selenium, soy oil, starch, glycerin, carrageenan, di-sodium phosphate, ascorbic acid, beta-carotene	[131]
Gr/Nafion/SPE	Vitamin K	SWV	0.04–3.55	8.87	—	Dietary supplements	—	[126]
Gr/Nafion/SPE	Vitamin K	AdSWV	0.04–2.2	15.75	55.24	Infant milk, yogurt, parsley	—	[16]
Gr/SPE	Vitamin K	AdSWV	2.21–33.28	219.65	—	Vegetables	Ca^{2+}, Mg^{2+}, vitamin C, vitamin A, vitamin D, vitamin E	[132]
PEDOT:PSS-CMC-Pd-rGO/GCE	Vitamin K_3	LSV	0.4–90	130	—	Chicken blood, broiler feed and serum of older bovine	Cd2 +, Cu^{2+}, glucose, vitamin B_2, vitamin D_2, methyl parathion, niclosamide, glycine	[133]

10.2 Conclusions

Intensive studies are also carried out on the electrochemical sensor of the graphene material, which has attracted a lot of attention since its discovery and has many applications. The reason for this is that it exhibits excellent conductivity due to its dense π bonds. In addition, due to its single layer, robustness, stability, mechanical, thermal, and optical properties, it attracts great attention in the world of electrochemical sensors. Moreover, GQDs are of great advantage in being obtained from one or several graphene layers and exhibiting outstanding GODs. They are not only chemically but also physically stable and have a fairly large surface-to-mass ratio. It is apparent that since the discovery of graphene, various graphene derivatives (Gr, GQDs, GO, rGO, etc) are widely used in electrochemical analysis. The most important of these are vitamin determinations, which have an important place in terms of human health. Electrochemical determinations of water-soluble vitamins and fat-soluble vitamins with graphene-based nanosensors are discussed in this chapter. For the determination of vitamins in both classes, highly sensitive, selective, repetitive, stable new-generation sensors containing graphene have been developed. Graphene-supported composite electrodes with very wide operating range and low LOD and limits of quantification (LOQ) have been fabricated. As a result, it is seen that the development of graphene-supported nanosensors, using the incredible and most important material of the 21st century, and hybrid electrochemical sensor studies will continue without slowing down.

Abbreviations

ABS	Acrylonitrile-butadiene-styrene graphite composite electrode
AGr	Activated graphene
Au-IDA	Gold interdigitated microelectrodes array
AuNR	Gold nanorod
AuNR/NG	Gold nanorod- decorated N-doped graphene
CMS	Carboxymethyl cellulose sodium
CMC	Carboxymethyl cellulose
FFTSWV	Fast Fourier transform square-wave voltammetry
FLG	Few-layered graphene
G-30	Graphene ink coated glass
GFE	Graphene foam electrode
GNP	Graphene nanoplatelets
GNS	Graphene nanosheets
GONR	Graphene oxide nanoribbon
GQD	Graphene quantum dots
HNONS	Hexagonal nickel oxide nanosheets
HMS	Hollow MnO_2 spheres
LP-HSE	Liquid phase high shear exfoliation
MB	Methyl blue
NG	N-doped graphene
N-Gr	Nitrogen-doped graphene
PDDA	Poly dimethyl diallyl ammonium chloride
PEI	Polyethyleneimine

PGE	Pencil graphite electrode
PGBHA–afGQDs–MnO_2/GCE	Poly(glyoxal-bis(2-hydroxyanil))–amino-functionalized graphene quantum dots–MnO_2 composite on GCE
pNR	Co-polymerizing neutral red
rGO	Reduced graphene oxide
SPE	Screen-printed electrode

References

[1] Semba R D 2012 The discovery of the vitamins *Int. J. Vitam. Nutr. Res.* **82** 310–5

[2] Yang Y, Hu N, Deng J and Yang J 2022 Electrochemical sensing for vitamins *Chemosensors* **10** 494

[3] Aydogdu N, Ozcelikay G and Ozkan S A 2022 Rapid and sensitive electrochemical assay of cefditoren with MWCNT/chitosan NCs/Fe_2O_3 as a nanosensor *Micromachines* **13** 1348

[4] Wilson N R *et al* 2009 Graphene oxide: structural analysis and application as a highly transparent support for electron microscopy *ACS Nano* **3** 2547–56

[5] Stankovich S, Dikin D A, Piner R D, Kohlhaas K A, Kleinhammes A, Jia Y, Wu Y, Nguyen S T and Ruoff R S 2007 Synthesis of graphene-based nanosheets via chemical reduction of exfoliated graphite oxide *Carbon* **45** 1558–65

[6] Paredes J I, Villar-Rodil S, Martínez-Alonso A and Tascón J M D 2008 Graphene oxide dispersions in organic solvents *Langmuir* **24** 10560–4

[7] Lightcap I V, Kosel T H and Kamat P V 2010 Anchoring semiconductor and metal nanoparticles on a two-dimensional catalyst mat. Storing and shuttling electrons with reduced graphene oxide *Nano Lett.* **10** 577–83

[8] Yang D *et al* 2009 Chemical analysis of graphene oxide films after heat and chemical treatments by x-ray photoelectron and micro-Raman spectroscopy *Carbon* **47** 145–52

[9] Khan Q A, Shaur A, Khan T A, Joya Y F and Awan M S 2017 Characterization of reduced graphene oxide produced through a modified Hoffman method *Cogent Chem.* **3** 1298980

[10] Gilje S, Han S, Wang M, Wang K L and Kaner R B 2007 A chemical route to graphene for device applications *Nano Lett.* **7** 3394–8

[11] Campbell E, Hasan M T, Gonzalez-Rodriguez R, Truly T, Lee B H, Green K N, Akkaraju G and Naumov A V 2021 Graphene quantum dot formulation for cancer imaging and redox-based drug delivery *Nanomed. Nanotechnol. Biol. Med.* **37** 102408

[12] Ghosh S, Sachdeva B, Sachdeva P, Chaudhary V, Rani G M and Sinha J K 2022 Graphene quantum dots as a potential diagnostic and therapeutic tool for the management of Alzheimer's disease *Carbon Lett.* **32** 1381–94

[13] Zhang Z, Xu J, Wen Y and Wang T 2018 A highly-sensitive VB2 electrochemical sensor based on one-step co-electrodeposited molecularly imprinted WS2-PEDOT film supported on graphene oxide-SWCNTs nanocomposite *Mater. Sci. Eng.* C **92** 77–87

[14] Cui M, Long L, Wu Y and Gao D 2021 Electrochemical biosensor based on reduced graphene oxide and CMC/silica sol-gel hybrid membranes for the detection of VB6 *Funct. Mater.* **28** 605–11

[15] Huang Z N *et al* 2019 A novel electrochemical sensor based on self-assembled platinum nanochains—multi-walled carbon nanotubes-graphene nanoparticles composite for simultaneous determination of dopamine and ascorbic acid *Ecotoxicol. Environ. Saf.* **172** 167–75

[16] Thangphatthanarungruang J, Yakoh A, Laocharoensuk R, Chotsuwan C, Chailapakul O and Siangproh W 2020 High-efficient of graphene nanocomposite: application to rapidly

simultaneous identification and quantitation of fat-soluble vitamins in different matric samples *J. Electroanal. Chem.* **873** 114361
[17] Zhang J and Zhang N 2018 Preparation of novel cystine-multiwalled carbon nanotubes-graphene oxide-glassy carbon electrode and electrochemical detection of riboflavin and dopamine *Sens. Mater.* **30** 2799
[18] Raril C and Manjunatha J G 2018 Sensitive electrochemical analysis of resorcinol using polymer modified carbon paste electrode: a cyclic voltammetric study *Anal. Bioanal. Electrochem.* **10** 488–98
[19] Amrutha B M, Manjunatha J G, Bhatt A S, Raril C and Pushpanjali P A 2019 Electrochemical sensor for the determination of Alizarin Red-S at non-ionic surfactant modified carbon nanotube paste electrode *Phys. Chem. Res.* **7** 523–33
[20] Manjunatha J G 2019 Electrochemical polymerised graphene paste electrode and application to catechol sensing *Open Chem. Eng. J.* **13** 81–7
[21] Prinith N S, Manjunatha J G and Raril C 2019 Electrocatalytic analysis of dopamine, uric acid and ascorbic acid at poly(adenine) modified carbon nanotube paste electrode: a cyclic voltammetric study *Anal. Bioanal. Electrochem.* **11** 742–56
[22] Prinith N S and Manjunatha J G 2020 Polymethionine modified carbon nanotube sensor for sensitive and selective determination of L-tryptophan *J. Electrochem. Sci. Eng.* **10** 305–15
[23] Pushpanjali P A, Manjunatha J G, Tigari G and Fattepur S 2020 Poly(Niacin) based carbon nanotube sensor for the sensitive and selective voltammetric detection of vanillin with caffeine *Anal. Bioanal. Electrochem.* **12** 553–68
[24] Hareesha N, Manjunatha J G, Amrutha B M, Pushpanjali P A, Charithra M M and Prinith Subbaiah N 2021 Electrochemical analysis of indigo carmine in food and water samples using a poly(glutamic acid) layered multi-walled carbon nanotube paste electrode *J. Electron. Mater.* **50** 1230–8
[25] Hareesha N, Manjunatha J G, Amrutha B M, Sreeharsha N, Basheeruddin Asdaq S M and Anwer M K 2021 A fast and selective electrochemical detection of vanillin in food samples on the surface of poly(glutamic acid) functionalized multiwalled carbon nanotubes and graphite composite paste sensor *Colloids Surf.* A **626** 127042
[26] Garai L 2017 Improving HPLC analysis of vitamin A and E: use of statistical experimental design *Procedia Comput. Sci.* **108** 1500–11
[27] Woollard D C, Bensch A, Indyk H and McMahon A 2016 Determination of vitamin A and vitamin E esters in infant formulae and fortified milk powders by HPLC: use of internal standardisation *Food Chem.* **197** 457–65
[28] Petteys B J and Frank E L 2011 Rapid determination of vitamin B2 (riboflavin) in plasma by HPLC *Clin. Chim. Acta* **412** 38–43
[29] Vidović S, Stojanović B, Veljković J, Pražić-Arsić L, Roglić G and Manojlović D 2008 Simultaneous determination of some water-soluble vitamins and preservatives in multivitamin syrup by validated stability-indicating high-performance liquid chromatography method *J. Chromatogr.* A **1202** 155–62
[30] da Silva D C, Visentainer J V, de Souza N E and Oliveira C C 2013 Micellar electrokinetic chromatography method for determination of the ten water-soluble vitamins in food supplements *Food Anal. Methods* **6** 1592–606
[31] Profumo A, Profumo V and Vidali G 1996 Micellar electrokinetic capillary chromatography of three natural vitamin A derivatives *Electrophoresis* **17** 1617–21

[32] MENG Q 2008 First-order derivative spectrophotometry for the determination of vitamin C in medicament *Chem. Res. Chinese Univ.* **24** 29–31
[33] Arya S, Mahajan M and Jain P 2001 Spectrophotometric determination of vitamin C with iron(II)–4-(2-pyridylazo)resorcinol complex *Anal. Chim. Acta* **427** 245–51
[34] Wang X, Li K, Yao L, Wang C and Van Schepdael A 2018 Recent advances in vitamins analysis by capillary electrophoresis *J. Pharm. Biomed. Anal.* **147** 278–87
[35] Dziomba S, Kowalski P and Bączek T 2012 Field-amplified sample stacking–sweeping of vitamins B determination in capillary electrophoresis *J. Chromatogr.* A **1267** 224–30
[36] Shabangi M and Sutton J A 2005 Separation of thiamin and its phosphate esters by capillary zone electrophoresis and its application to the analysis of water-soluble vitamins *J. Pharm. Biomed. Anal.* **38** 66–71
[37] Amin A 2001 Colorimetric determination of tocopheryl acetate (vitamin E) in pure form and in multi-vitamin capsules *Eur. J. Pharm. Biopharm.* **51** 267–72
[38] Berzas Nevado J J and Gómez Laguna M A 1998 Spectrofluorimetric determination of vitamin K3 *Analyst* **123** 287–90
[39] Gheorghe D-C, van Staden J, (Koos) F, Stefan-van Staden R-I and Sfirloaga P 2022 Gold nanoparticles/nanographene-based 3D sensors integrated in mini-platforms for thiamine detection *Sensors* **23** 344
[40] Pushpanjali P A, Manjunatha J G, Hareesha N, Amrutha B M, Raril C, ALOthman Z A, Alanazi A M and Pandith A 2022 Fabrication of poly(L-Aspartic acid) layer on graphene nanoplatelets paste electrode for riboflavin sensing *Mater. Chem. Phys.* **276** 125392
[41] Qianfen Z, Yong W and Yongnian N 2015 Electrochemical sensor for the detection of riboflavin based on nanocomposite film of polydeoxyadenylic acid/reduced graphene oxide *Chem. J. Chinese Univ.* **36** 1674–80
[42] Derakhshan M, Shamspur T, Molaakbari E, Mostafavi A and Saljooqi A 2020 Fabrication of a novel electrochemical sensor for determination of riboflavin in different drink real samples *Russ. J. Electrochem.* **56** 181–8
[43] Grace A A, Thillaiarasi S and Dharuman V 2021 Binary metal oxide adsorbed graphene modified glassy carbon electrode for detection of riboflavin *Electroanalysis* **33** 993–1006
[44] Li P, Liu Z, Yan Z, Wang X, Akinoglu E M, Jin M, Zhou G and Shui L 2019 An electrochemical sensor for determination of vitamin B 2 and B 6 based on AuNPs@PDA-RGO modified glassy carbon electrode *J. Electrochem. Soc.* **166** B821–9
[45] Lu J, Kou Y, Jiang X, Wang M, Xue Y, Tian B and Tan L 2019 One-step preparation of poly(glyoxal-bis(2-hydroxyanil))-amino-functionalized graphene quantum dots-MnO_2 composite on electrode surface for simultaneous determination of vitamin B2 and dopamine *Colloids Surf.* A **580** 123652
[46] Zeybek B and Üğe A 2021 Development of carbon-based sensors for electrochemical quantification of vitamins B2 and B6 at nanomolar levels *Chem. Pap.* **75** 1323–39
[47] Shadjou N, Hasanzadeh M and Omari A 2017 Electrochemical quantification of some water soluble vitamins in commercial multi-vitamin using poly-amino acid caped by graphene quantum dots nanocomposite as dual signal amplification elements *Anal. Biochem.* **539** 70–80
[48] Asgharian Marzabad M, Jafari B and Norouzi P 2020 Determination of riboflavin by nanocomposite modified carbon paste electrode in biological fluids using fast Fourier transform square wave voltammetry *Int. J. Eng.* **33** 1696–702

[49] Martins E C, Santana E R and Spinelli A 2023 Nitrogen and sulfur co-doped graphene quantum dot-modified electrode for monitoring of multivitamins in energy drinks *Talanta* **252** 123836

[50] Sriramprabha R, Divagar M, Ponpandian N and Viswanathan C 2018 Tin oxide/reduced graphene oxide nanocomposite-modified electrode for selective and sensitive detection of riboflavin *J. Electrochem. Soc.* **165** B498–507

[51] Zhang L, Liu X, Luo L, Hu C, Fu J, Chang X and Gan T 2021 A high-performance voltammetric methodology for the ultra-sensitive detection of riboflavin in food matrices based on graphene oxide-covered hollow MnO_2 spheres *Food Chem.* **352** 129368

[52] Raj M A, Gowthaman N S K and John S A 2016 Highly sensitive interference-free electrochemical determination of pyridoxine at graphene modified electrode: importance in Parkinson and Asthma treatments *J. Colloid Interface Sci.* **474** 171–8

[53] Rison S, Mathew A T, George L, Maiyalagan T, Hegde G and Varghese A 2022 Pt nanospheres decorated graphene-β-CD modified pencil graphite electrode for the electrochemical determination of vitamin B6 *Top. Catal.* (https://doi.org/10.1007/s11244-021-01559-1)

[54] de Oliveira G C, Pereira L C, Silva A L, Semaan F S, Castilho M and Ponzio E A 2018 Acrylonitrile-butadiene-styrene (ABS) composite electrode for the simultaneous determination of vitamins B2 and B6 in pharmaceutical samples *J. Solid State Electrochem.* **22** 1607–19

[55] Cui M, Shao Z, Lu D and Wang Y 2019 Eco-friendly electrochemical biosensor based on sodium carboxymethyl cellulose/reduced graphene oxide composite *Macromol. Res.* **27** 327–33

[56] Liu Y, Li M, Li H, Wang G, Long Y, Li A and Yang B 2019 *In situ* detection of melatonin and pyridoxine in plants using a CuO–poly(l-lysine)/graphene-based electrochemical sensor *ACS Sustain. Chem. Eng.* **7** 19537–45

[57] Beitollahi H and Garkani Nejad F 2018 Voltammetric determination of vitamin B6 (pyridoxine) at a graphite screen-printed electrode modified with graphene oxide/ Fe_3O_4@SiO_2 nanocomposite *Russ. Chem. Bull.* **67** 238–42

[58] Montes C, Soriano M L, Villaseñor M J and Ríos Á 2020 Carbon-based nanodots as effective electrochemical sensing tools toward the simultaneous detection of bioactive compounds in complex matrices *J. Electroanal. Chem.* **878** 114573

[59] Feng-Yun H, Jing Y, Huan L, Xiao-Yan G, Peng-Peng T, Hong Z, Chang-Yun C and Guo-Fei L 2014 Electrochemical behavior and determination of pyridoxine hydrochloride at graphene-carbon nanotubes/poly (nicotinic acid) modified electrode *Chinese J. Anal. Chem.* **42** 1039–43

[60] Vu T V, Nguyen M T T, Do T T, Nguyen H L, Nguyen V and Nguyen D T 2022 Adsorption of copper ions onto poly(1,8-diaminonaphthalene)/graphene film for voltammetric determination of pyridoxine *Electroanalysis* **34** 1478–86

[61] Salimi H and Beitollahi H 2016 Graphene/ZnO nanocomposite for voltammetric sensing of vitamin B6 using modified glassy carbon electrode *Anal. Bioanal. Electrochem.* **8** 732–40

[62] Kıranşan K D and Topçu E 2018 Free-standing and flexible MoS_2/rGO paper electrode for amperometric detection of folic acid *Electroanalysis* **30** 810–8

[63] Mani V, Umamaheswari R, Chen S-M, Govindasamy M, Su C, Sathiyan A, Merlin J P and Keerthi M 2017 Highly sensitive determination of folic acid using graphene oxide nanoribbon film modified screen printed carbon electrode *Int. J. Electrochem. Sci.* **12** 475–84

[64] Mani V, Govindasamy M, Chen S-M, Subramani B, Sathiyan A and Merlin J P 2017 Determination of folic acid using graphene/molybdenum disulfide nanosheets/gold nanoparticles ternary composite *Int. J. Electrochem. Sci.* **12** 258–67

[65] Chekin F, Mishyn V, Barras A, Lyskawa J, Ye R, Melinte S, Woisel P, Boukherroub R and Szunerits S 2019 Dopamine-functionalized cyclodextrins: modification of reduced graphene oxide based electrodes and sensing of folic acid in human serum *Anal. Bioanal. Chem.* **411** 5149–57

[66] Xu H, Bai Z, Wang G, O'Halloran K P, Tan L, Pang H and Ma H 2017 Voltammetric determination of folic acid at physiological pH values by using a glassy carbon electrode modified with a multilayer composite consisting of polyoxometalate ($H_8P_2Mo_{16}V_2O_{62}$) and reduced graphene oxide and prepared via layer-by-layer self-assembly and in-situ photoreduction *Microchim. Acta* **184** 4295–303

[67] Gao X, Yue H, Huang S, Lin X, Gao X P A, Wang B, Yao L, Wang W and Guo E 2018 Synthesis of graphene/ZnO nanowire arrays/graphene foam and its application for determination of folic acid *J. Electroanal. Chem.* **808** 189–94

[68] M, Safaei, Beitollahi, H and Shishehbore M R 2020 Amplified electrochemical sensor employing Fe_3O_4@SiO_2/graphene nanocomposite for selective determination of folic acid *J. Anal. Chem.* **75** 95–100

[69] Chekin F, Teodorescu F, Coffinier Y, Pan G-H, Barras A, Boukherroub R and Szunerits S 2016 MoS_2/reduced graphene oxide as active hybrid material for the electrochemical detection of folic acid in human serum *Biosens. Bioelectron* **85** 807–13

[70] Zhang D, Ouyang X, Ma W, Li L and Zhang Y 2016 Voltammetric determination of folic acid using adsorption of methylene blue onto electrodeposited of reduced graphene oxide film modified glassy carbon electrode *Electroanalysis* **28** 312–9

[71] Wang Z, Han Q, Xia J, Xia L, Bi S, Shi G, Zhang F, Xia Y, Li Y and Xia L 2014 A novel phosphomolybdic acid–polypyrrole/graphene composite modified electrode for sensitive determination of folic acid *J. Electroanal. Chem.* **726** 107–11

[72] xi Ma J, Yang L, Wang L, Wu S Q and Liu Y 2021 Determination of folic acid in food by differential pulse voltammetry with ZnO@GO nanocomposites modified glassy carbon electrode *Int. J. Electrochem. Sci.* **16** 150922

[73] Peng Y and Zhang X 2021 Selective and sensitive determination of folic acid based on molecularly imprinted poly (o-aminophenol) and reduced graphene oxide decorated with Au nanoparticles *Curr. Anal. Chem.* **17** 1201–10

[74] Abdelwahab A A and Shim Y-B 2015 Simultaneous determination of ascorbic acid, dopamine, uric acid and folic acid based on activated graphene/MWCNT nanocomposite loaded Au nanoclusters *Sensors Actuators* B **221** 659–65

[75] Singh R, Jaiswal S, Singh K, Fatma S and Prasad B B 2018 Biomimetic polymer-based electrochemical sensor using methyl blue-adsorbed reduced graphene oxide and functionalized multiwalled carbon nanotubes for trace sensing of cyanocobalamin *ACS Appl. Nano Mater.* **1** 4652–60

[76] Anil Subash S, Manjunatha C, Sudeep M, Rastogi C K, Chaudhary V, Girish Kumar S and Sekhar P 2023 Development of a non-enzymatic vitamin-C electrochemical sensor based on rGO/$Ce_2(SO_4)_3$ hierarchical nanocomposite *J. Electrochem. Soc.* **170** 037504

[77] Zhang W, Du D, Gunaratne D, Colby R, Lin Y and Laskin J 2014 Polyoxometalate-graphene nanocomposite modified electrode for electrocatalytic detection of ascorbic acid *Electroanalysis* **26** 178–83

[78] Banavath R, Abhinav A, Srivastava R and Bhargava P 2022 Highly sensitive ascorbic acid sensors from EDTA chelation derived nickel hexacyanoferrate/graphene nanocomposites *Electrochim. Acta* **419** 140335

[79] Banavath R, Abhinav A, Nemala S S, Srivastava R and Bhargava P 2022 Liquid phase high shear exfoliated few-layered graphene for highly sensitive ascorbic acid electrochemical sensors *Mater. Adv.* **3** 9019–29

[80] Harraz F A, Faisal M, Ismail A A, Al-Sayari S A, Al-Salami A E, Al-Hajry A and Al-Assiri M S 2019 TiO_2/reduced graphene oxide nanocomposite as efficient ascorbic acid amperometric sensor *J. Electroanal. Chem.* **832** 225–32

[81] Szoke A, Zsebe Z, Turdean G L and Muresan L M 2019 Composite electrode material based on electrochemically reduced graphene oxide and gold nanoparticles for electrocatalytic detection of ascorbic acid *Electrocatalysis* **10** 573–83

[82] Sharma V, Sundaramurthy A, Tiwari A and Sundramoorthy A K 2018 Graphene nanoplatelets-silver nanorods-polymer based *in situ* hybrid electrode for electroanalysis of dopamine and ascorbic acid in biological samples *Appl. Surf. Sci.* **449** 558–66

[83] Abu Zahed M, Barman S C, Sharifuzzaman M, Xuan X, Nah J S and Park J Y 2018 *Ex situ* synthesis of hexagonal NiO nanosheets and carboxyl-terminated reduced graphene oxide nanocomposite for non-enzymatic electrochemical detection of H_2O_2 and ascorbic acid *J. Electrochem. Soc.* **165** B840–7

[84] Chen X, Yang Z, Tuo X, Huang H, Huang J, Li L and Yu X 2022 Sea-urchin-structured $NiCo_2O_4$ decorated nitrogen-doped graphene for enhanced electrochemical detection of ascorbic acid *Solid State Sci.* **133** 107000

[85] Liu L, Zhai J, Zhu C, Han L, Ren W and Dong S 2017 One-step synthesis of functional pNR/rGO composite as a building block for enhanced ascorbic acid biosensing *Anal. Chim. Acta* **981** 34–40

[86] Abellán-Llobregat A, Vidal L, Rodríguez-Amaro R, Berenguer-Murcia Á, Canals A and Morallón E 2017 Au-IDA microelectrodes modified with Au-doped graphene oxide for the simultaneous determination of uric acid and ascorbic acid in urine samples *Electrochim. Acta* **227** 275–84

[87] Fu L, Zheng Y-H and Fu Z-X 2015 Ascorbic acid amperometric sensor using a graphene-wrapped hierarchical TiO_2 nanocomposite *Chem. Pap* **69** 655–61

[88] Fernandes Loguercio L, Thesing A, da Silveira Noremberg B, Vasconcellos Lopes B, Kurz Maron G, Machado G, Pope M A and Lenin Villarreal Carreno N 2022 Direct laser writing of poly(furfuryl alcohol)/graphene oxide electrodes for electrochemical determination of ascorbic acid *ChemElectroChem.* **9** e202200334

[89] Anon 2020 Simultaneous detection of dopamine, tyrosine and ascorbic acid using NiO/graphene modified graphite electrode *Biointerface Res. Appl. Chem.* **10** 5599–609

[90] Hasan M M, Li Z and Nagao Y 2023 Palladium nanoparticles combined with reduced graphene oxide and multiwall carbon nanotubes for alkaline ascorbic acid oxidation *Jpn. J. Appl. Phys.* **62** 027003

[91] Chu G 2019 A novel non-enzyme sensor based on a PDDA-RGO/Pt NPs nanocomposite for electrochemical detection of ascorbic acid *Int. J. Electrochem. Sci.* **14** 11406–18

[92] Ji D, Liu Z, Liu L, Low S S, Lu Y, Yu X, Zhu L, Li C and Liu Q 2018 Smartphone-based integrated voltammetry system for simultaneous detection of ascorbic acid, dopamine, and uric acid with graphene and gold nanoparticles modified screen-printed electrodes *Biosens. Bioelectron.* **119** 55–62

[93] Zhao Y, Qin J, Xu H, Gao S, Jiang T, Zhang S and Jin J 2019 Gold nanorods decorated with graphene oxide and multi-walled carbon nanotubes for trace level voltammetric determination of ascorbic acid *Microchim. Acta* **186** 17

[94] Hashemi S A, Mousavi S M, Bahrani S, Ramakrishna S, Babapoor A and Chiang W-H 2020 Coupled graphene oxide with hybrid metallic nanoparticles as potential electrochemical biosensors for precise detection of ascorbic acid within blood *Anal. Chim. Acta* **1107** 183–92

[95] Yang S, Cheng Y, Cheng D, Wang Y, Xu H, Li M, Jiang T and Wang H 2023 N-doped graphene-supported gold nanorods for electrochemical sensing of ascorbic acid with superior sensitivity *J. Electron. Mater.* **52** 2336–46

[96] Ghanbari M H, Mashhadizadeh M H and Norouzi Z 2021 Introducing a novel nanocomposite consisting of TiO_2 nanoparticles@copper oxide/reduced graphene oxide for the electrocatalytic sensing of ascorbic acid *J. Iran. Chem. Soc.* **18** 1329–41

[97] Zhang L, Khungwa J, Liu Y, Li L, Wang X and Wang S 2019 Fabrication of electro-active Pt/IMo_6O_{24}/graphene oxide nanohybrid modified electrode for the simultaneous determination of ascorbic acid, dopamine and uric acid *J. Electrochem. Soc.* **166** H351–8

[98] Faridan A, Bahmaei M and Sharif A M 2022 Simultaneous determination of trace amounts ascorbic acid, melatonin and tryptophan using modified glassy carbon electrode based on Cuo-CeO2-rGO-MWCNTS nanocomposites *Anal. Bioanal. Electrochem.* **14** 201–15

[99] Yang Y J 2015 One-pot synthesis of reduced graphene oxide/zinc sulfide nanocomposite at room temperature for simultaneous determination of ascorbic acid, dopamine and uric acid *Sensors Actuators* B **221** 750–9

[100] Nancy T E M and Kumary V A 2014 Synergistic electrocatalytic effect of graphene/nickel hydroxide composite for the simultaneous electrochemical determination of ascorbic acid, dopamine and uric acid *Electrochim. Acta* **133** 233–40

[101] Li H, Wang Y, Ye D, Luo J, Su B, Zhang S and Kong J 2014 An electrochemical sensor for simultaneous determination of ascorbic acid, dopamine, uric acid and tryptophan based on MWNTs bridged mesocellular graphene foam nanocomposite *Talanta* **127** 255–61

[102] Zou H L, Li B L, Luo H Q and Li N B 2015 A novel electrochemical biosensor based on hemin functionalized graphene oxide sheets for simultaneous determination of ascorbic acid, dopamine and uric acid *Sensors Actuators* B **207** 535–41

[103] Du J, Yue R, Ren F, Yao Z, Jiang F, Yang P and Du Y 2014 Novel graphene flowers modified carbon fibers for simultaneous determination of ascorbic acid, dopamine and uric acid *Biosens. Bioelectron* **53** 220–4

[104] Zhao L, Li H, Gao S, Li M, Xu S, Li C, Guo W, Qu C and Yang B 2015 MgO nanobelt-modified graphene-tantalum wire electrode for the simultaneous determination of ascorbic acid, dopamine and uric acid *Electrochim. Acta* **168** 191–8

[105] Yang B, Wang H, Du J, Fu Y, Yang P and Du Y 2014 Direct electrodeposition of reduced graphene oxide on carbon fiber electrode for simultaneous determination of ascorbic acid, dopamine and uric acid *Colloids Surf.* A **456** 146–52

[106] Yang Y J and Li W 2014 CTAB functionalized graphene oxide/multiwalled carbon nanotube composite modified electrode for the simultaneous determination of ascorbic acid, dopamine, uric acid and nitrite *Biosens. Bioelectron.* **56** 300–6

[107] Huang H, Yue Y, Chen Z, Chen Y, Wu S, Liao J, Liu S and Wen H 2019 Electrochemical sensor based on a nanocomposite prepared from $TmPO_4$ and graphene oxide for simultaneous voltammetric detection of ascorbic acid, dopamine and uric acid *Microchim. Acta* **186** 189

[108] Wang C, Du J, Wang H, Zou C, Jiang F, Yang P and Du Y 2014 A facile electrochemical sensor based on reduced graphene oxide and Au nanoplates modified glassy carbon electrode for simultaneous detection of ascorbic acid, dopamine and uric acid *Sensors Actuators* B **204** 302–9
[109] Luo X, Chen L, Yang J, Li S, Li M, Mo Q, Li Y and Li X 2021 Electrochemically simultaneous detection of ascorbic acid, sulfite and oxalic acid on Pt-Pd nanoparticles/chitosan/nitrogen doped graphene modified glassy carbon electrode: a method for drug quality control *Microchem. J.* **169** 106623
[110] Edris N M M A, Abdullah J, Kamaruzaman S and Sulaiman Y 2019 Ultrasensitive reduced graphene oxide-poly(procion)/gold nanoparticles modified glassy carbon electrode for selective and simultaneous determination of ascorbic acid, dopamine, and uric acid *J. Electrochem. Soc.* **166** B664–72
[111] Prongmanee W and Asanithi P 2019 L-Cysteine/graphene oxide for electrochemical determination of ascorbic acid in the presence of dopamine and uric acid *J. Phys. Conf. Ser.* **1380** 012075
[112] Sha R and Badhulika S 2018 Facile green synthesis of reduced graphene oxide/tin oxide composite for highly selective and ultra-sensitive detection of ascorbic acid *J. Electroanal. Chem.* **816** 30–7
[113] Mohammed Modawe Alshik Edris N, Abdullah J, Kamaruzaman S, Saiman M I and Sulaiman Y 2018 Electrochemical reduced graphene oxide-poly(eriochrome black T)/gold nanoparticles modified glassy carbon electrode for simultaneous determination of ascorbic acid, dopamine and uric acid *Arab. J. Chem.* **11** 1301–12
[114] Pichaimuthu K, Keerthi M, Chen S-M, Chen T-W and Su C 2018 Silver nanoparticles decorated on graphene oxide sheets for electrochemical detection of ascorbic acid (AA) in human urine sample *Int. J. Electrochem. Sci.* **13** 7859–69
[115] Zou C, Zhong J, Li S, Wang H, Wang J, Yan B and Du Y 2017 Fabrication of reduced graphene oxide-bimetallic PdAu nanocomposites for the electrochemical determination of ascorbic acid, dopamine, uric acid and rutin *J. Electroanal. Chem.* **805** 110–9
[116] Ramakrishnan S, Pradeep K R, Raghul A, Senthilkumar R, Rangarajan M and Kothurkar N K 2015 One-step synthesis of Pt-decorated graphene–carbon nanotubes for the electrochemical sensing of dopamine, uric acid and ascorbic acid *Anal. Methods* **7** 779–86
[117] Imran H, Manikandan P N and Dharuman V 2015 Facile and green synthesis of graphene oxide by electrical exfoliation of pencil graphite and gold nanoparticle for non-enzymatic simultaneous sensing of ascorbic acid, dopamine and uric acid *RSC Adv.* **5** 63513–20
[118] Wu D, Li Y, Zhang Y, Wang P, Wei Q and Du B 2014 Sensitive electrochemical sensor for simultaneous determination of dopamine, ascorbic acid, and uric acid enhanced by amino-group functionalized mesoporous Fe_3O_4@graphene sheets *Electrochim. Acta* **116** 244–9
[119] de Faria L V, Lisboa T P, de Farias D M, Araujo F M, Machado M M, de Sousa R A, Matos M A C, Muñoz R A A and Matos R C 2020 Direct analysis of ascorbic acid in food beverage samples by flow injection analysis using reduced graphene oxide sensor *Food Chem.* **319** 126509
[120] Buffon E and Stradiotto N R 2023 Disposable three-dimensional graphene oxide electrode with sandwich-like architecture for the determination of ascorbic acid in fruit juices *Mater. Today Commun.* **35** 105535

[121] Mazzara F, Patella B, Aiello G, O'Riordan A, Torino C, Vilasi A and Inguanta R 2021 Electrochemical detection of uric acid and ascorbic acid using r-GO/NPs based sensors *Electrochim. Acta* **388** 138652

[122] Fu L, Wang A, Lai G, Su W, Malherbe F, Yu J, Lin C-T and Yu A 2018 Defects regulating of graphene ink for electrochemical determination of ascorbic acid, dopamine and uric acid *Talanta* **180** 248–53

[123] Karastogianni S, Diamantidou D and Girousi S 2021 Selective voltammetric detection of ascorbic acid from Rosa canina on a modified graphene oxide paste electrode by a manganese(II) complex *Biosensors* **11** 294

[124] Das T R, Jena S K, Madhuri R and Sharma P K 2018 Polymeric iron oxide-graphene nanocomposite as a trace level sensor of vitamin C *Appl. Surf. Sci.* **449** 304–13

[125] Fernandes V Q, Silva M K L and Cesarino I 2020 Determination of isotretinoin (13-cis-retinoic acid) using a sensor based on reduced graphene oxide modified with copper nanoparticles *J. Electroanal. Chem.* **856** 113692

[126] Thangphatthanarungruang J, Ngamaroonchote A, Laocharoensuk R, Chotsuwan C and Siangproh W 2018 A Novel electrochemical sensor for the simultaneous determination of fat-soluble vitamins using a screen-printed graphene/Nafion electrode *Key Eng. Mater.* **777** 597–601

[127] Park J, Kim M, Kim W, Jo S, Kim W, Kim C, Park H, Lee W and Park J 2022 Ultrasensitive detection of 25-hydroxy vitamin D3 in real saliva using sandwich-type electrochemical aptasensor *Sensors Actuators* B **355** 131239

[128] Canevari T C, Cincotto F H, Landers R and Machado S A S 2014 Synthesis and characterization of α-nickel (II) hydroxide particles on organic-inorganic matrix and its application in a sensitive electrochemical sensor for vitamin D determination *Electrochim. Acta* **147** 688–95

[129] Wadhwa S, John A T, Nagabooshanam S, Mathur A and Narang J 2020 Graphene quantum dot-gold hybrid nanoparticles integrated aptasensor for ultra-sensitive detection of vitamin D3 towards point-of-care application *Appl. Surf. Sci.* **521** 146427

[130] Anusha T, Bhavani K S, Shanmukha Kumar J V and Brahman P K 2021 Synthesis and characterization of novel lanthanum nanoparticles-graphene quantum dots coupled with zeolitic imidazolate framework and its electrochemical sensing application towards vitamin D3 deficiency *Colloids Surf.* A **611** 125854

[131] Filik H, Avan A A, Aydar S and Çakar Ş 2016 Determination of tocopherol using reduced graphene oxide-Nafion hybrid-modified electrode in pharmaceutical capsules and vegetable oil samples *Food Anal. Methods* **9** 1745–53

[132] Jesadabundit W, Chaiyo S, Siangproh W and Chailapakul O 2020 Simple and cost-effective electrochemical approach for monitoring of vitamin K in green vegetables *Chem. Electro. Chem.* **7** 155–62

[133] Zhang Z, Zhang J, Zhang H, Xu J, Wen Y and Ding W 2016 Characterization of PEDOT: PSS-reduced graphene oxide@Pd composite electrode and its application in voltammetric determination of vitamin K3 *J. Electroanal. Chem.* **775** 258–66

IOP Publishing

Graphene-Based Sensors

Jamballi G Manjunatha

Chapter 11

Graphene-based electrochemical platform for well-known phenolic compounds from natural sources

Mohamed Brahmi, Nurgul K Bakirhan, Merve Yuzbasioglu Baran, Sıla Ozlem Sener, Duygu Sevim and Abdesselam Tahani

Graphene-based nanomaterials are among the materials that are increasingly used with their many advantages. In recent years, the mechanical, electrical, thermal-optical properties of graphene and its potential applications in nanoelectronics, catalysis, sensors, supercapacitors, batteries and transistors have attracted attention.

The most active area of phenolic compounds research is its contributions to human health. Different research groups have determined that quinones, glycosides and flavonoids show anti-inflammatory, antioxidant, antimicrobial, antibacterial and anti-carcinogenic effects.

Electrochemical studies are of great interest due to their excellent features in the analytical chemistry field and various applications. The sensitive methods in electrochemical analysis are preferred for important compound analysis of different plants, environmental samples, natural samples or biological and pharmaceutical samples. In this chapter, detailed knowledge has been presented for the readers about the analysis of important phenolic compounds through the electrochemical methods.

11.1 Introduction

Carbon is one of the indispensable elements of human life. Throughout the Earth, in the atmosphere and Nature, gas, liquid and solid fuels, and petroleum products all contain carbon. It is exposed on the Earth, in the atmosphere and in the oceans by volcanic processes [1]. In addition to many foods and minerals, the human body also contains the element carbon. Many biomaterials containing carbon due to their biocompatibility are of vital importance for humans. In addition to their mineral

doi:10.1088/978-0-7503-5578-0ch11

needs, plants in Nature need carbon dioxide, which contains carbon, in order to perform photosynthesis [2]. All of the polymer and plastic materials that have an indispensable place in our lives, textile products, dyestuffs, medicines, and health materials contain carbon.

Graphene is structurally a two-dimensional (2D) sheet of three-dimensional (3D) graphite in layers, with Å as thickness unit and μm as length unit. Graphene has a surface area of approximately 2630 $m^2\ g^{-1}$, a thermal conductivity of 5000 $W\ m^{-1}\ K^{-1}$ and an optical transmittance of around 97.7% in the form of a honeycomb [3]. There have been continuous studies recently on graphene and its structure. Graphene and its electronic optical properties have been used in many applications. The demonstration of graphene to be one atom thick in 2004 by Geim and Novoselov led to an increased interest in graphene [4]. It is 2D and stable under high crystallographic quality and favorable environmental conditions, making graphene a promising candidate for many technological applications [5].

Graphene, a carbon compound, is used for various purposes. Although it is harder than diamond, it is also flexible. It has a light structure. Like the carbon compound, the graphene oxide (GO) structure is also biocompatible [6]. Although graphene has *sp*2 hybrid structure, it is known that it is not a polyaromatic hydrocarbon [7]. Novoselov *et al* synthesized graphene as a monolayer in 2004 and discovered the existence of graphene's physical and electronic properties [4]. Andre Geim and Konstantin Novoselov received the Nobel Prize in Physics for their work on graphene in 2010 for their six years of work, for their recognition, isolation, production and characterization of graphene [4].

Graphene has been used in many potential applications such as electronic devices, solar cells, sensors, nanomaterials, batteries, supercapacitors, hydrogen storage since 2004 [8]. Graphene is used as a good support material for catalysts with its unique internal structure, high chemical stability, large surface area and excellent electron transport properties [4]. Despite its large surface area, graphene is not a suitable method for dye removal from wastewater due to its low solubility in water. Therefore, GO is more suitable for such applications. Hydroxyl, epoxy, carboxylic acid groups on GO increase the solubility in water. Therefore, GO is suitable for dye removal from wastewater [9].

Electrochemistry is a branch of chemistry that deals with the relationship between electrical effects. In electrochemical methods, an electrical effect is applied to the electrode-solution assembly and the response of the apparatus to us is measured. This answer gives information about the features of the mechanism. Electrochemical methods are preferred to other analytical methods because of their easy usage, low cost, low solvent consumption, sensitivity and rapid response time [10–12]. In electrochemistry, cyclic voltammetry (CV), chronoamperometry (CA) and chronocoulometry (CC) techniques are used, as well as differential pulse voltammetry (DPV) and square-wave voltammetry (SWV) techniques for quantitative determination. Electrochemical impedance spectroscopy (EIS, impedimetry) technique, which is a very sensitive analysis and characterization technique, has been widely used [13].

Phenolic compounds are a large group of plant secondary metabolites with important roles in plant physiology, commonly found in various higher plant organs such as vegetables, fruit, spices, grains, legumes, and nuts. In recent years, biological activity studies on phenolic compounds have shown that these compounds have antioxidant, antimicrobial, anti-carcinogenic, and anti-inflammatory activity and are effective in preventing cardiovascular diseases, cancers, diabetes, and oxidative stress-related diseases. Both quinone and flavonoid derivatives are enormous groups of the phenolic compounds [20].

Quinone compounds with a ketone structure are widely used compounds with significant biological activity [14]. These compounds are also of great importance due to their use in the treatment of some diseases, such as in the protection and treatment of organs such as kidney, lung, brain and heart against diseases, and in medicine as bactericide, fungicide, insecticide, anti-tumor antibiotics. Some derivatives are also used in drug synthesis [14]. Flavonoids are natural compounds found in the seeds, leaves, fruit and flowers of most plants. It has been determined by various studies that flavonoids capture hydroxyl radicals, superoxide anions and lipid peroxy radicals, makng them a very good antioxidant [15]. These compounds, consisting of two phenyls and a heterocycle, are divided into various subclasses such as flavones, flavonols, flavanones, chalcones, dihydrochalcones, and anthocyanidins according to the different oxidation degrees of the hetero ring [16]. Similarly, flavonoid glycosides are compounds that are widely found in the plant kingdom and have very different biological effects such as antioxidant, hepatoprotective, UV-light protector, antibacterial and anti-carcinogenic. Although they are widely found in plants, their low amount and important pharmacological activities have led researchers to the isolation and synthesis of these compounds [17]. While the structure of only 360 of the flavone and flavonol glycosides was known in 1975, this number doubled to 720 structures over the following five years. Between 1981 and 1985, 90 more new flavone glycoside compounds were discovered [18]. It is known that more than four thousand flavonoids and flavonoid glycoside derivatives have been isolated from plants to date [19].

In this chapter, graphene-based electrochemical sensors are discussed for the sensitive detection of quinones and flavonoid derivatives.

11.2 Classification of some popular phenolic compounds in natural resources

Phenolic compounds, which have a common chemical structure containing an aromatic ring with one or more hydroxyl substituents, are divided into several classes [20, 21].

Simple phenolic compounds, tannins, coumarins, and their glycosides, quinones and quinone derıvatives, chromones and xanthones, flavone and flavonoid glycosides (flavones, flavonols, flavanones, isoflavones, biflavones), anthocyanidins and glycosides, stilbenes, lignans, and lignins can be given as pharmaceutically important phenolic classes [22].

In this section, detailed information will be given about the classes and compounds that have studies on quinone and flavonoid derivatives which are sensitive graphene-based electrochemical sensors.

11.2.1 Quinones and quinone derivatives

Quinones are secondary metabolites that are found in animals, plants, fungi, and microorganisms. They are common in Liliaceae (*Aloe*, etc), Hypericaceae (*Hypericum*, etc), Polygonaceae (*Rheum*, *Rumex*, *Polygonum*, etc), Rhamnaceae (*Rhamnus*), Rubiaceae (*Rubia*, *Galium*, etc), Caesalpiniaceae (*Cassia*), Boraginaceae (*Alkanna*, *Arnebia*) and Juglandaceae (*Juglans*, etc) families [23].

Quinones can undergo oxidation and reduction reactions. They are easily reduced to semiquinones and/or hydroquinones that are autoxidized by molecular oxygen, turning into superoxide anion radicals. This process is called the quinone redox cycle and ends up with the formation of reactive oxygen species (ROS) such as hydrogen peroxide and hydroxyl radicals. ROS ruin the oxidant–antioxidant balance and cause oxidative stress. Furthermore, ROS can interact with molecular lipids, proteins, DNA, and RNA causing irreversible damage to these molecules [24, 25].

Quinone derivatives are classified according to their ring structure as an *ortho* or a *para*-substituted dione to an aromatic ring: 1,4-benzoquinone (cyclohexadienedione), 1,2-benzoquinone (*ortho*-quinone) and condensed polycyclic aromatic rings: naphthoquinones and anthranoids [26].

Benzoquinones are the simplest quinones found in plants, animals, and bacteria. These compounds play important roles in bioenergy transport, oxidative phosphorylation, and electron transfer processes and show various activities such as anticancer, anti-inflammatory, antimicrobial, and antiviral [27].

11.2.1.1 Naphthoquinones

Naphthoquinones are found in some actinomycetes, fungi, lichens, algae, and plants. The chemical structure consists of naphthalene skeletons substituted in the C-1 and C-4 (1,4-naphthoquinones) or C-1 and C-2 (1,2-naphthoquinones) positions [28].

The 1,4-naphthoquinones with the highest cytotoxic activity among the naphthoquinones contain two quinone groups that can accept one or two electrons to form radical anion or di-anion species. They show a strong cytotoxic effect, possibly due to the quinone redox cycle that causes the formation of ROS and the arylation reactions responsible for the biological effect in quinones [29].

The enantiomeric naphthoquinones, *S*-Alkannin (figure 11.1(A)) and *R*-Shikonin (5,8-dihydroxy-1,4-naphthoquinone) (figure 11.1(B)) are lipophilic red pigments that are yielded from the roots of many Boraginaceae family plants, such as *Alkanna tinctoria*, *Lithospermum erythrorhizon* and *Arnebia euchroma*. *S*-Alkannin and *R*-Shikonin are effective pharmaceutical substances with wound healing, antibacterial, anti-inflammatory, antitumoral, and antithrombotic activities. In Asia, Shikonin is currently highly valued for both its therapeutic and coloring properties. In Europe, Alkannin is an active ingredient that is frequently used in the formula of ointments used for the treatment of burns and wounds [30, 31].

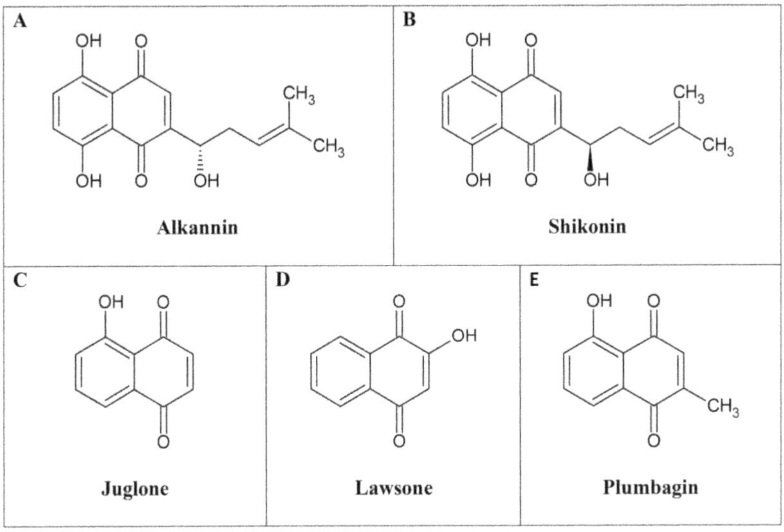

Figure 11.1. Some natural compounds of the naphthoquinones class.

Juglone (5-hydroxy-1,4-naphthoquinone) (figure 11.1(C)) is yielded from the leaves, roots, bark, and nuts of different walnut species (*Juglans nigra*, *J. regia* and *J. cinerea*). It has antimicrobial, cytotoxic, and wound healing properties. It is a safe dye generally used in nutrition and cosmetic production, with herbicide activity [32–34].

Lawsone (2-hydroxy-1,4-naphthoquinone) (figure 11.1(D)) is a natural dye extractable from the leaves of *Lawsonia inermis* which has been traditionally used for thousands of years as a skin and hair dye (commonly known as henna plant). Besides its dyeing properties, lawsone has antifungal, antimicrobial, antimalarial, cytotoxic, molluscicidal, and antioxidant activity [33, 34].

Plumbagin (5-hydroxy-2-methyl-1,4-naphthoquinone) (figure 11.1(E)) is also a natural dye, it is mainly isolated from the roots of the *Plumbago* spec. That is used for the treatment of tuberculosis and leprosy [33, 34].

11.2.1.2 Anthranoids

Anthranoids are characterized by carrying an anthracene ring in their structure, there is a quinone structure in the ninth position of the anthracene ring, they are grouped as anthrone, anthraquinone, and diantron according to the group in the 10th position, and they can transform into each other by oxidation–reduction reactions [35].

Anthraquinones are commonly found in taxa belonging to Fabaceae, Liliaceae, Polygonaceae, Rhamnaceae, and Rubiaceae families. While 1,2-dihydroxyanthraquinones are dyeing materials, 1,8-dihydroxyanthraquinones are generally used for their laxative effects [36, 37].

Alizarin (1,2-dihydroxy anthraquinone) (figure 11.2(A)) is one of the common and safe natural colorants which has different colors under varying pH. It has also

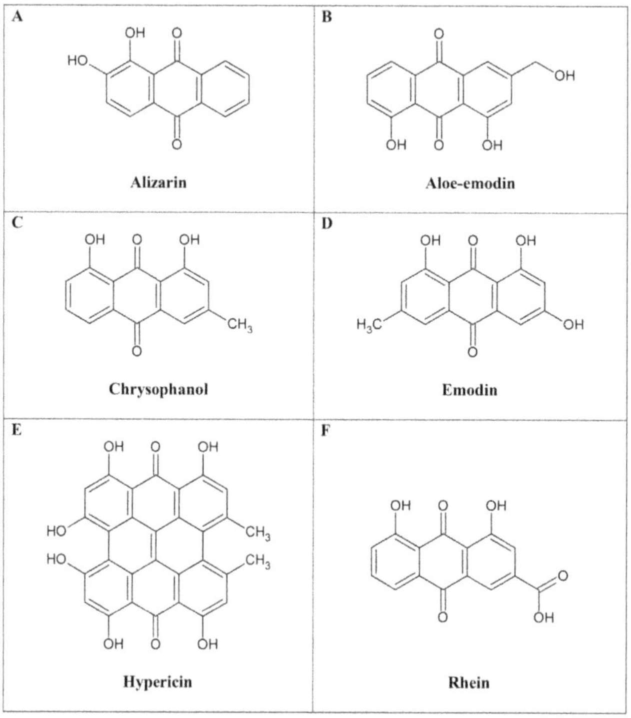

Figure 11.2. Some natural compounds of anthranoids class.

important biological activities such as antioxidant, antibacterial, and antitumoral [38, 39].

Aloe-emodin (figure 11.2(B)) is a 1,8-dihydroxy-3-hydroxymethyl-anthraquinone that occurs in *Cassia* spec., *Aloe vera* and *Rheum palmatum.* Aloe-emodin exhibits antiproliferative, antiviral, anti-inflammatory, antiparasitic, neuroprotective, and hepatoprotective activities [40, 41].

Chrysophanol (1,8-dihydroxy-3-methylanthraquinone) (figure 11.2(C)) is one of the significant anthraquinone mostly isolated from *Rhei rhizoma* which has anti-inflammatory, anti-depressant, anti-cancer and neuroprotective effects. It is included in the composition of some cosmetic products and drugs [40].

Emodin (3,8-trihydroxy-6-methylanthraquinone) (figure 11.2(D)) is an anthraquinone found in several traditional Chinese medicine plants like *Polygonum cuspidatum, Polygonum multiflorum* and *R. palmatum.* Emodin has been used for more than 2000 years in traditional Chinese medicine. Emodin is still present in numerous herbal medicines. Literature data has indicated that emodin has a broad range of pharmacological effects, including anti-cancer, hepatoprotective, anti-inflammatory, antioxidant, and antibacterial capabilities. However, emodin can cause hepatotoxicity, renal toxicity, and reproductive toxicity at high dosages and long-term usage. Because of its high glucuronidation, emodin has low oral bioavailability in rats, according to pharmacokinetic studies [41, 42].

Hypericin (figure 11.2(E)) is a naphthodianthrone, an anthraquinone derivative commonly found in St. John's Wort (*Hypericum* spec.). It has been used for centuries in traditional medicine as one of the main secondary metabolites of *Hypericum perforatum*. It has anti-depressant, anti-cancer, and antiviral effects. It has a large chromophore group in the molecule structure, so it causes phototoxicity. Due to its photodynamic activation, Hypericin is utilized as a diagnostic and therapeutic agent for cancer patients [43, 44].

A lipophilic anthraquinone, rhein (4,5-dihydroxy anthraquinone-2-carboxylic acid) (figure 11.2(F)) is abundantly found in Chinese medicine plants such as *R. palmatum*, *Cassia tora*, *P. multiflorum*, and *Aloe barbadensis*. Rhein is a well-studied biologically active substance. It is shown that rhein has hepatoprotective, kidney protective, anti-inflammatory, antioxidant, antitumoral, and antimicrobial activities [41, 45].

11.2.2 Flavone and related flavonoid glycosides

Flavonoids are important plant secondary metabolites having a polyphenolic structure, and are the largest group of naturally occurring phenols [19]. These are the most diverse phenolic compounds that are closely related to the flower color formation of plants and play a variety of biological activities in plants. In recent years, flavonoids have been important components in various nutraceutical, pharmaceutical, medical, and cosmetic applications with their antioxidant, anti-inflammatory, anti-mutagenic, and anti-carcinogenic properties [19, 20].

Flavonoids are a series of compounds with diphenylpropanes as the basic skeleton and 2 aromatic rings (A- and B-rings) connected to each other through the central 3-carbon bridge (C-ring) [20].

Flavonoids can be divided into different subgroups according to the carbon, the degree of unsaturation and oxidation of the central 3-carbon bridge (C-ring), and where the B-ring attached to the C-ring (figure 11.3) [19]. These subgroups are given below:

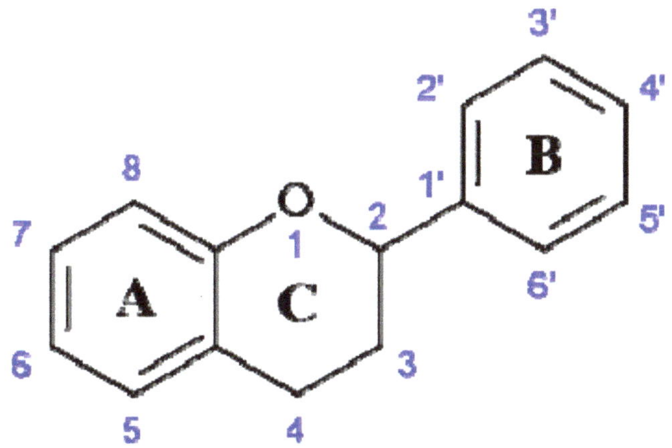

Figure 11.3. Basic flavonoid structure.

11.2.2.1 Flavone

Flavones, are an important subgroup of flavonoids, and are widely present in leaves, flowers, and fruits as glucosides. The C ring has a double bond between the second and third positions and a ketone at the fourth position [19].

The naturally occurring flavonoid compound apigenin (figure 11.4(A)), also known as 4′,5,7-trihydroxyflavone, is found in a variety of medicinal plants, fruits, functional foods, and veggies. Apigenin possesses various pharmacological properties including anti-cancer, anti-angiogenic, anti-inflammatory, antioxidant, hepatoprotective effect, etc. In addition, apigenin has been proven to be effective in the treatment of many diseases such as auto-immune diseases, Alzheimer's disease, arthritis, cardiovascular diseases, HIV, ischemia–reperfusion injury, lung fibrosis, multiple sclerosis, Parkinson's disease, nephrotoxicity [21, 46–48].

Chrysin (figure 11.4(B)), identified as 5,7-dihydroxyflavone, is a natural compound found in propolis, honey, mushrooms, and various plant sources. Scientific studies revealed that Chrysin has lots of therapeutic effects like anti-cancer, anti-diabetic, antioxidant, anti-inflammatory, anti-allergic, hepatoprotective, and neuroprotective. Furthermore, Chrysin is one of the important metabolites for reproductive and cardiovascular health [49–51].

Luteolin (3,4,5,7-tetrahydroxyflavone) (figure 11.4(C)) occurs naturally in various vegetables, fruits, and medicinal herbs. Pharmacological effects including anti-allergic, anti-cancer, anti-inflammatory, antioxidant, antiaging, and neuroprotective are attributed to luteolin [52–55].

11.2.2.2 Flavonol

Flavonols are flavonoids with a ketone group. Compared with flavones, flavonols have a hydroxyl group in position 3 of the C ring, which may also be glycosylated [19].

A natural phenolic compound, fisetin (figure 11.5(A)) identified as 3,3′,4′,7-tetrahydroxyflavon is present in fruit, vegetables, and various medicinal plants. A wide variety of pharmacological effects, including anti-angiogenic, anti-carcinogenic, anti-diabetic, anti-inflammatory, antimicrobial, anti-osteoporotic, anti-tumorigenic, cardioprotective, and neuroprotective effects have been demonstrated for fisetin in numerous preclinical studies [56–58].

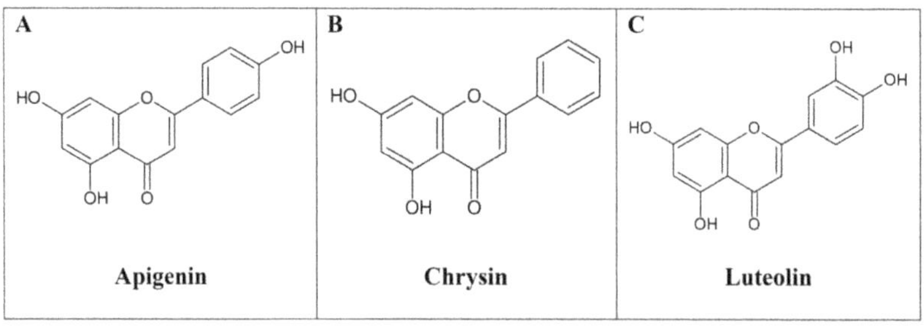

Figure 11.4. Some natural compounds of flavone class.

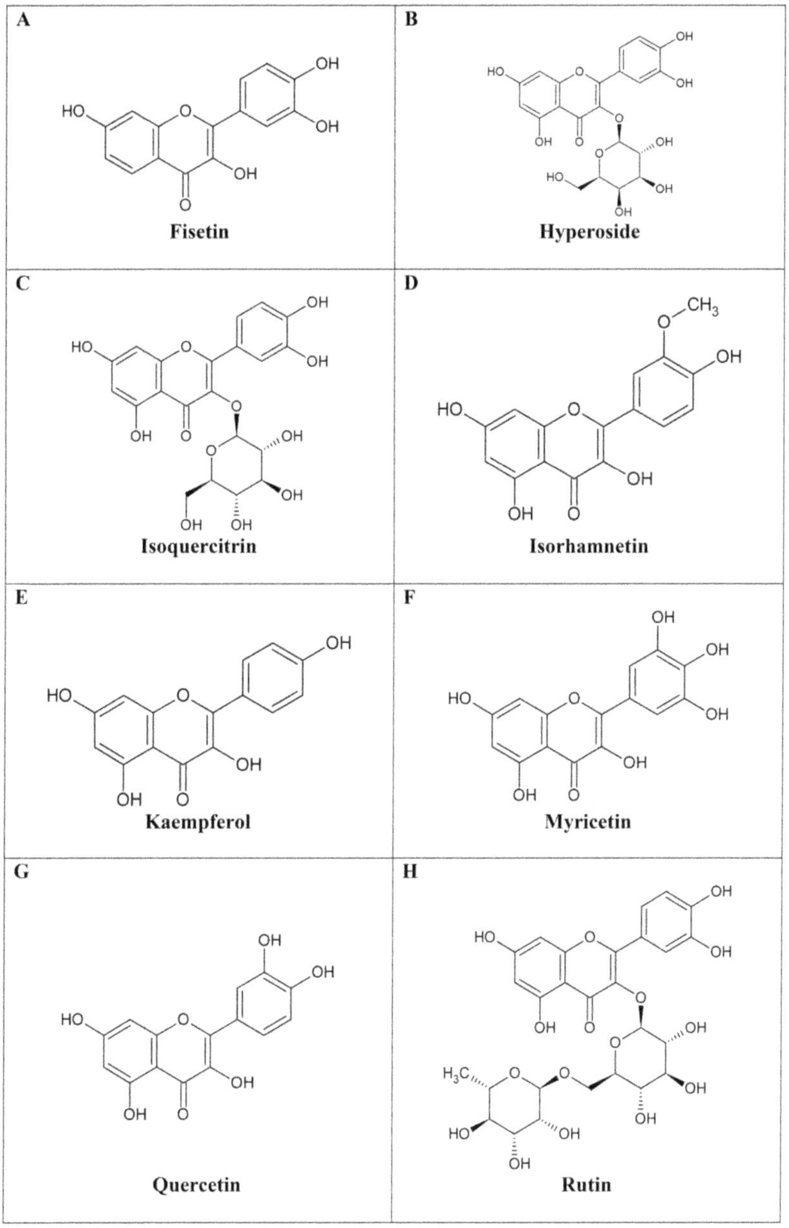

Figure 11.5. Some natural compounds of flavonol class.

Hyperoside (figure 11.5(B)) identified as quercetin-3-*O*-β-D-galactopyranoside is a natural flavonol glycoside which is included in different plants. According to scientific studies, hyperoside has a wide range of biological impacts, including antibacterial, anti-cancer, anti-depressant, anti-inflammatory, antioxidant, antiviral, and protective properties. Hyperoside has therapeutic efficiency for the treatment of arthritis, cancer, colitis, diabetic nephropathy, myocardial ischemia–reperfusion, and pulmonary fibrosis due to these effects [24, 59, 60].

The flavonol glycoside, isoquercitrin (quercetin-3-*O*-β-D-glucopyranoside) (figure 11.5(C)) is frequently present in foods and beverages obtained from plants. Isoquercitrin has become popular due to its anti-allergic, anti-angiogenic, anti-inflammatory, anti-diabetic, antioxidant, and cardioprotective effects [61, 62].

Isorhamnetin (figure 11.5(D)), a flavonoid compound is available in some medicinal plants including *Hippophae rhamnoides* L. and *Ginkgo biloba* L. The pharmacological effects of isorhamnetin have been the subject of numerous investigations. The effects of Isorhamnetin such as antiapoptotic, anti-inflammatory, anti-diabetic, antioxidant, anti-obesity, anti-adipogenic, antioxidant, anti-tumor, and cardioprotective effects have been proven by current studies [63–66].

Kaempferol (figure 11.5(E)) known as 3,5,7,4′-tetrahydroxyflavone is found in various edible and medicinal plants. Kaempferol possesses a wide variety of pharmacological properties, containing analgesic, anti-allergic, anxiolytic, antioxidant, anti-cancer, anti-inflammatory, antimicrobial, anti-osteoporotic, cardioprotective, and neuroprotective effects as demonstrated by preclinical scientific studies [57, 67, 68].

Myricetin (figure 11.5(F)) identified as 3,5,7,3′,4′,5′-hexahydroxyflavone is a flavonoid component that is available in various natural plants. Myricetin has been indicated in previous scientific studies to have therapeutic importance because of its protective effect against cardiovascular, neurological damage, and liver injury. Myricetin also has been proven to possess a wide range of pharmacological effects like anti-cancer, antibacterial, anti-diabetic, anti-inflammatory, anti-obesity, anti-tumor, and antiviral [57, 69, 70].

Quercetin (3,3′,4′,5,7-pentahydroxyl-flavone) (figure 11.5(G)) is a natural flavonol that has long been a part of human nutrition. Quercetin has become an attractive secondary metabolite because of its health benefits like anti-cancer, anti-inflammatory, antimicrobial, antioxidant, antiviral, and hepatoprotective activities as well as healer effect for neurological and cardiovascular diseases [57, 71].

Rutin (figure 11.5(H)) identified as 3,3′,4′,5,7-pentahydroxyflavone-3-rhamnoglucoside is a flavonol, widely presented in medicinal plants and plant-based foods. Recent studies demonstrated that many pharmacological activities like anti-allergic, anti-carcinogenic, anti-diabetic, antifungal, antimicrobial, antioxidant, antihypertensive, cytoprotective, cardioprotective, vasoprotective, and neuroprotective activities [72].

11.2.2.3 Flavanone

Flavanones are another important class that is generally present in all citrus fruit such as oranges, lemons, and grapes. In flavanones, also called dihydroflavones, unlike flavones, the double bond between the second and third positions in the C ring is saturated [19].

Due to their free radical scavenging properties, flavanones are associated with several health benefits. Citrus flavonoids (such as hesperetin, hesperidin, naringenin, naringin, eriodictyol, etc) show various biological effects as antioxidant, anti-inflammatory, blood lipid-lowering and cholesterol-lowering agents [19].

Hesperetin (figure 11.6(A)) is an aglycone metabolite of hesperidin (figure 11.6(B)). Two of them have been documented to possess a wide range of pharmacological properties including cardiovascular, anti-inflammatory, anti-obesity, hypoglycemic, lipid-lowering and beneficial effects on bone and Alzheimer's disease [73, 74]. And recently hesperidin and hesperetin attracted attention for their anti-cancer, antiviral, and anti-COVID-19 (coronavirus disease-2019) properties [73, 75–77].

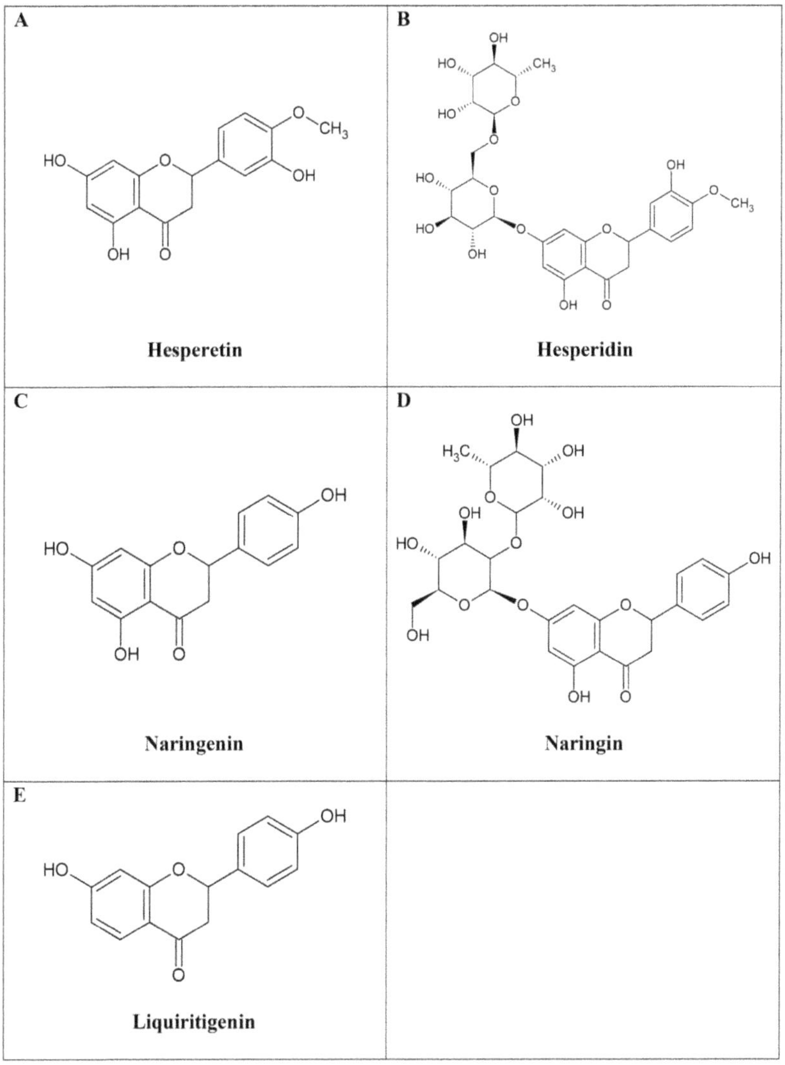

Figure 11.6. Some natural compounds of flavanone class.

Several biological activities are expressed for naringenin (4′,5,7-trihydroxyflavanone) (figure 11.6(C)), such as antioxidant, anti-tumor, antiviral, antibacterial, anti-inflammatory, anti-adipogenic, and cardioprotective effects [74, 78].

Naringin (4′,5,7-trihydroxyflavanone-7-rhamnoglucoside) (figure 11.6(D)) is the glycoside form of naringenin. It possesses anti-cancer activity like naringenin [78]. Also, it can be used to treat or prevent various diseases, such as heart disease, diabetes, obesity, and metabolic syndrome [79].

Liquiritigenin (7,4′-dihydroxyflavanone) (figure 11.6(E)) is a flavonoid obtained from various kinds of licorice (*Glycyrrhiza* species). It possesses hepatoprotective activities against cadmium-induced toxicity in rat-derived hepatocyte cell lines [80, 81]. In addition, liquiritigenin has been shown to reduce the production of iNOS and proinflammatory cytokines by showing anti-inflammatory effects resulting from the inhibition of NF-κB activation in macrophages [82, 83].

11.2.2.4 Flavononol

Also called dihydroflavonols or catechins, flavononols are the 3-hydroxy derivatives of flavanones. Since the hydroxyl group is always attached to the third position of the C ring, they are also referred to as flavon-3-ols. There is no double bond between second and third positions, unlike many flavonoids [19].

Taxifolin (3,5,7,3,4-pentahydroxy flavanone or dihydroquercetin) (figure 11.7) is commonly found in milk thistle, onions, Douglas fir bark, and French maritime pine bark. It is an important component of dietary supplements, having rich antioxidant properties. Taxifolin has various pharmacological effects such as anti-inflammatory, antimicrobial, hepatoprotective, cardiovascular, anti-angiogenic, anti-cancer, anti-Alzheimer's, etc [84, 85].

11.2.2.5 Isoflavone

Isoflavonoids are a different subgroup from the other groups of flavonoids. Isoflavonoids, which have a very restricted distribution in the plant kingdom, are mostly found in soybeans and other leguminous plants. Isoflavones such as genistein and daidzein are often referred to as phytoestrogens because of their estrogenic activity [19, 86]. In addition, studies have shown that they can be used to prevent some momentous diseases, such as hypertension and cancer, and ameliorate intestinal health [20, 87–89].

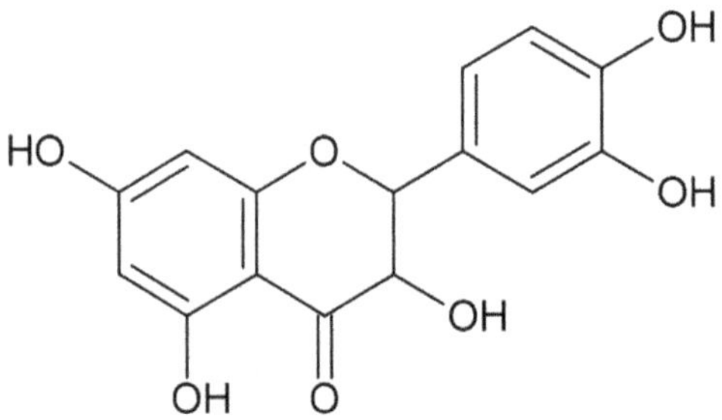

Figure 11.7. Molecular structure of taxifolin.

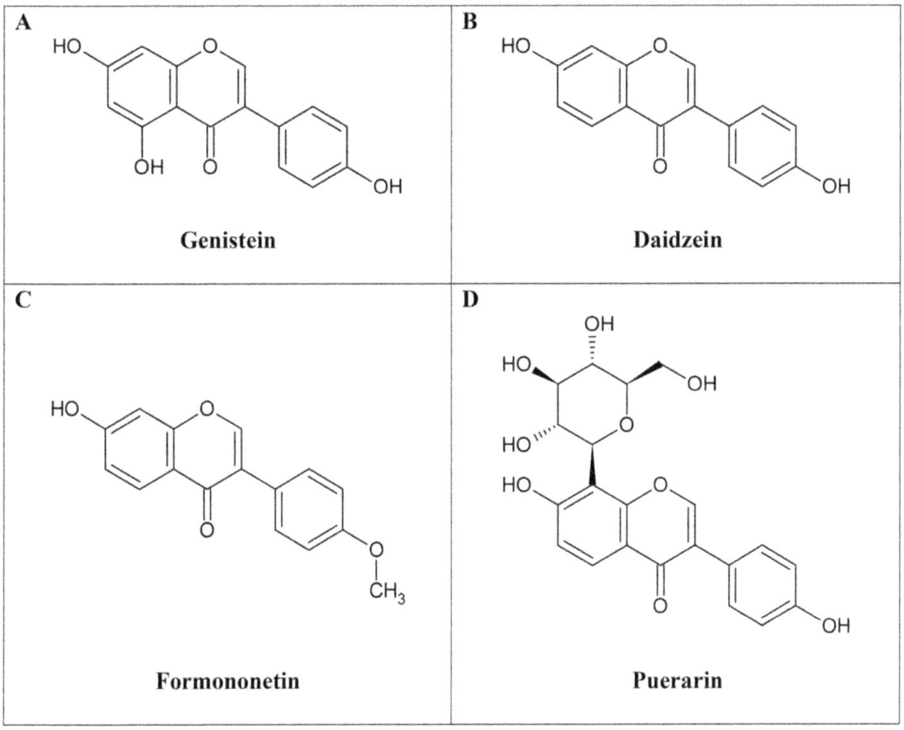

Figure 11.8. Some natural compounds of the isoflavone class.

Genistein (4,5,7-trihydroxyisoflavone) (figure 11.8(A)), most found in soybeans, is a compound with numerous health benefits. It is found to be a potent agent in both prophylaxis and treatment of cancer and various other chronic diseases because of its antioxidant activity. Genistein possesses suppressive effects on both the cell-mediated and humoral components of the adaptive immune system. Also, this compound, which can reduce atherogenic condition and hypercholesterolemia, has a significant cardioprotection activity. In addition, Genistein may inhibit cancer progression by inducing apoptosis or inhibiting proliferation, which has a prominent role in preventing DNA damage [90].

Daidzein (4,7-dihydroxyisoflavone) (figure 11.8(B)), is a naturally occurring phytoestrogen with various biological activities, such as antihemolytic, antioxidant, and anti-inflammatory activities [91, 92]. Hence to its estrogenic activity, daidzein is thought to be a therapeutic strategy for estrogen-dependent health conditions, such as breast and prostate cancer, diabetes, osteoporosis, and cardiovascular diseases. However, daidzein also has the ability to reduce oxidative damage, regulate the immune reaction, and induce apoptosis, directly linked to their anti-cancer effects [92, 93].

It has been reported that formononetin (7-hydroxy-4′-methoxyisoflavone) (figure 11.8(C)) has been isolated from different plants of the Fabaceae family, especially from *Trifolium* genus, known as a rich source of formononetin.

The biological and pharmacological activities of formononetin are well-documented as anti-tumor, antiproliferative, growth inhibitory, vasodilation, antiapoptotic, cardioprotective, mammary gland proliferative, antioxidant, antimicrobial, anti-inflammatory activity, and neuroprotective effect [94].

Puerarin (7,4′-dihydroxy-8-C-glucosylisoflavone) (figure 11.8(D)), an isoflavone glycoside derived from *Pueraria lobata* (Willd.) Ohwi has been identified as a biologically active component with diverse various pharmacological activities, including cardioprotection, vasodilation, anti-inflammation, antioxidant, attenuating insulin resistance, neuroprotection, alleviating pain, promoting bone formation, inhibiting alcohol intake, and anti-cancer. Experimental and clinical investigations have reported that puerarin is widely used in the treatment of cardiovascular diseases (such as atherosclerosis, cardiac hypertrophy, heart failure, myocardial infarction, stroke, and hypertension), diabetes, and its complications, Alzheimer's disease, Parkinson's disease, osteonecrosis, endometriosis, and cancer [95, 96].

11.3 Nanomaterials for improved electrochemical sensing

Nanomaterials are a rapidly expanding field of research and a promising technological sector owing to their exceptional characteristics and prospective usages. Their physicochemical characteristics vary considerably from their bulk counterparts, which make the nanomaterials a desirable choice for various applications that necessitate nanoscale structures or substantial surface area. The term 'nanometer' (nm) refers to a unit of measurement that represents 10^{-9} meters in length, so nanomaterials are typically recognized as materials with one dimension at least in the range of 1–1000 nm [97]. However, the FDA generally categorizes the nanomaterials under the characteristics of one dimension at the minimum and sized between 1 and 100 nm [98].

Nanomaterials are often referred to in the literature by different names such as nanofibers, nanoplates, nanowires, and quantum dots, but they are all characterized by their nanoscale size, shape, and surface area [99]. In electrochemistry, nanomaterials provide a remarkably high number of active sites, which basically improve the sensitivity and selectivity of electrochemical sensing applications [100]. This has encouraged several researches to the development of ultra sensors with improved detection limits, faster response times, and high accuracy. Such sensors have shown promising results in applications such as the detection of numerous pollutants as environmental monitoring [101], as well as for the recognition of biological molecules in the biomedical field [102].

Over the past few years, the literature has been rich with various types of naturally occurring or synthesized nanomaterials, as shown in figure 11.9. In the following discussion, we will explore different types of nanoparticles along with their corresponding properties and electrochemical applications.

Metal-based nanoparticles, particularly noble metals such as gold [103], platinum [104], and rhodium [105], have been widely employed for electrode modification owing to their environmentally friendly behavior, high conductivity, and extensive surface area [106]. To maximize the efficacy of these nanoparticles, studies have

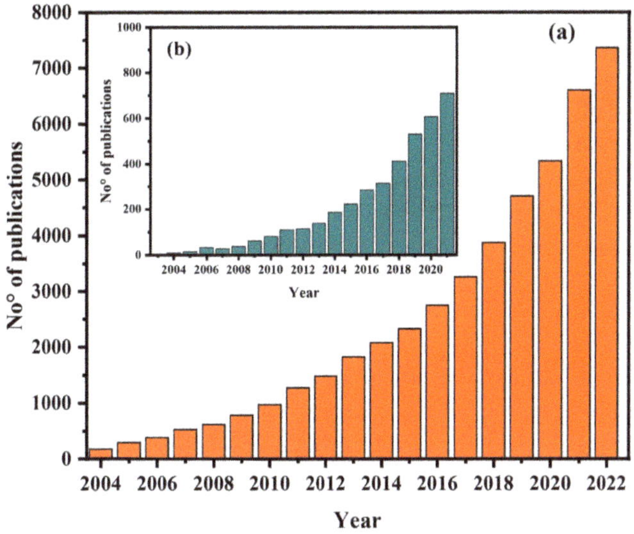

Figure 11.9. Graphical representation of number of publications per year: (A) nanomaterials, (B) nanomaterials for electrochemical applications. Data collected in March 15, 2023 from the Scopus database.

demonstrated the use these metallic nanoparticles in electrode modification through the creation of nanoparticle dispersions on a polymeric or carbonaceous network [107], which provide to the formed composite a reversible oxidation/reduction behavior and consequently enhance the electrocatalytic activity of this nanomaterials [108]. Interest in metallic nanoparticles has also gained momentum in the last decade for their incorporation into metal–organic frameworks (MOFs), which is basically related to the porous and stable structure of MOFs, which facilitate their modification with metal nanoparticles [109]. These intricate new structures offer fascinating properties that complement the characteristics of nanoparticles and open a wide range of electrochemical areas. For instance, MOFs have been utilized to elaborate cancer biomarkers that possess exceptional selectivity and ultra-sensitivity [110], as well as in the environmental field where Yadav *et al* have developed a highly sensitive electrode for nitrobenzene detection using zinc-based MOF incorporated with Au nanoparticles [109].

On the other hand, metal oxide nanoparticles have demonstrated excellent performance in electroanalytical detection of bioactive molecules, attributed to their flexible structure, ease of functionalization, high conductivity, and biocompatibility [111]. Iron oxide nanoparticles, featuring both Fe^{2+} and Fe^{3+} oxidation states, exhibit superior electrical conductivity and have found utility in biosensing applications such as heavy metal detection [112], bisphenol [113] also drug detection [114, 115]. However, in some cases, their sensitivity has been enhanced using other metals including nickel, silver, or gold as doping agents [116–118]. Additionally, the oxide forms of metals have been also reported to be effective as reinforcing nanoparticles for carbonaceous and polymer systems in electroanalysis such as manganese and copper oxides [106]. Overall, the metal or metal oxide nanoparticles offer promising solutions for a wide range of applications in nanoelectronics, biomedicine, and environmental sensing.

Ceramics nanoparticles such as titanium dioxide (TiO_2) have emerged as an additional class of materials with significant potential for use in electrochemical sensing applications [119, 120]. Due to their surface area-to-volume ratio, TiO_2 nanoparticles display a significant potential active surface area for interacting with analytes. This, coupled with their biocompatibility, low cost, and excellent electrical properties, renders them attractive candidates for use in biosensors [106]. However, the use of TiO_2 in electrochemical sensing typically requires modification with enzymes or noble metals to overcome the low stability of TiO_2 films on the electrode surface [119, 121].

Recently, clay minerals have attracted considerable interest due to their distinct properties, including hydrophilicity, nanoparticle size, porous structure, high surface area, and ease of modification [122]. In the field of electrochemistry, several studies have explored the use of clay minerals, including bentonite and nanotubes, for various applications [123]. Clay materials have displayed potential in various areas such as biosensor development and energy storage due to their favorable properties. For example, the clay minerals with high surface area allow for enhanced interactions with analytes, while their porous structure can facilitate mass transfer processes. Additionally, the ability to modify these materials further enhances their suitability for a range of electrochemical applications [122, 124].

In summary, the field of electrochemical sensing has benefited greatly from the properties of nanomaterials. Although numerous nanomaterials have been studied for electrochemical applications, carbon-based nanomaterials (CBNs) hold particular potential owing to their distinct features and possible applications. The upcoming section will explore in-depth the properties and prospective applications of CBNs.

11.3.1 Carbon nanomaterials

CBNs have their unique structures and properties that make them highly attractive for a range of applications. As a result, they have garnered significant attention in various fields, such as electrochemical sensing. These nanomaterials are primarily composed of carbon, which can exist in a range of physical forms in Nature, from hard substances like diamonds to soft materials like graphite [125]. In addition to these well-known allotropes of carbon, a diverse range of low-dimensional CBN forms has emerged, including amorphous carbon, carbon nanotubes (CNTs), fullerene, GO, and graphene quantum dots (GQDs) which offer an outstanding range of new application areas [126].

From a structural standpoint, fullerenes [127] are molecular forms of carbon, characterized by spherical structures formed by a pentagons and hexagons arrangement of carbon atoms. In contrast, CNTs have a cylindrical structure with diameters of a few nanometers, consisting of rolled graphene sheets. CBNs can be classified into two main types based on their structure: single-walled nanotubes and multiwalled nanotubes [125]. Graphene, on the other hand, is an additional form of carbon that has two dimensions arranged in a hexagonal crystal lattice. In the structure of graphene, the carbon atoms undergo sp^2-hybridization and are linked together in the crystal lattice through σ and π bonds. Graphene has high mechanical rigidity, thermal stability, and unique electrical properties, making it a promising material for various applications [128]. While fullerenes and CNTs have unique mechanical and electrical properties, graphene stands out with its 2D structure and high surface area.

11.3.2 Carbon nanomaterials for electrochemical sensing: properties and potential

Carbon nanomaterials (CNMs) have been widely studied and utilized in the electrochemical field due to their exceptional properties. Recently, graphene, CNTs, diamond, and diamond-like carbon have gained popularity as electrode materials for electrochemical sensing applications [126]. The remarkable charge storage and electron transfer properties of these materials can be can be tailored for electrochemical applications. Unlike bulk materials, CBNs exhibit high sensitivity and selectivity even in harsh environments and across a broad temperature and dynamic range [129, 130].

When compared to traditional materials, CNMs have demonstrated superior performance on the electrode modification for enhancing electrochemical detection due to the available features including high electron mobility, conductivity and surface area [126]. Additionally, CNM performance could be easily enhanced through surface functionalization, either by organic compounds, polymers, DNA, enzymes or through inorganic modification all depend on the desired applications [131–134]. The CNMs have been reported to have varying densities of states, which impact their potential applications. For this, several studies have been conducted on the adjusting of the atomic bonding structures of CNMs. For instance, CNT can be selectively peeled off in a controlled manner or through MWCNTS oxidation. The planar structure of graphene and the tubular structure of nanotubes provide the opportunity to expose a large number of surface atoms for chemical bonding with target molecules in biosensing applications [135].

11.3.2.1 Synthesis methods of carbon nanomaterials

11.3.2.1.1 Carbon nanotubes synthesis

Various methods can be used for the synthesis of CNTs, but the most commonly utilized and important ones are chemical vapor deposition (CVD) [136], followed by electric arc discharge methods and laser ablation [137]. In the CVD method, hydrocarbon gases were used as carbon source including methane acetylene and ethylene, and substrates like silica, zeolite, and iron-coated silicon plates are commonly employed. Other papers reported the use of metal catalyst including Fe, Ni, and Co to produce single-walled CNTs [136]. CVD can be thermally activated, photo-assisted, or plasma-assisted, and the conditions maintained inside the furnace include a temperature of 500 °C–900 °C and argon gaze to maintain an inert atmosphere [138]. One of the major disadvantages of CVD method for the synthesis of CNTs is the environmental concerns associated with the use of hydrocarbon gases. Additionally, CVD method can result in discontinuous CNTs [137]. However, researchers have recently developed the floating catalyst CVD (FC-CVD) technique as a modification of the CVD method [139]. This approach addresses both the environmental concerns and the discontinuity issue of CNTs. For example, Sehrawat *et al* have presented a new green approach to synthesizing highly conductive CNT aerogels using FC-CVD [140], while other studies have also reported the effectiveness of FC-CVD in addressing the discontinuity of CNTs [141]. The electric arc discharge method [142] involves using pure graphite rods as electrodes, while helium gas is often used as an inert gas to cool and condense atoms into CNTs. The temperature is maintained at 3000 °C–3500 °C, and the obtained CNT can be either single-walled or multi-walled, all depending on the incorporating catalyst nanoparticles. The laser ablation method involves using solid graphite as a target source, and irradiates it with a laser source such as CO_2 or Nd: YAG to vaporize the target material into vapor carbon atoms. These atoms are then deposited onto a substrate through physical vapor deposition (PVD) [143].

11.3.2.1.2 Fullerenes synthesis

Since the discovery of fullerenes, they have been extensively studied. The first method proposed for their synthesis was through the evaporation of graphite electrodes in a helium atmosphere [144], it was found to have low yields for potential industrial applications. As such, other methods were developed to synthesize fullerene efficiently. One such method is laser ablation of graphite [137], which allows for the generation of significant quantities of fullerene under high-temperature conditions. An alternative approach for fullerenes synthesis is the arc-discharge, where the evaporation of graphite is achieved using an electric arc generated between two electrodes. However, this method has the limitation of high temperature conditions, which prevents the detection of occurring types. Fullerenes possess significant potential for utilization in materials science and biological applications, but their hydrophobicity and tendency to agglomerate limit their solubility in polar solvents, which can be overcome through chemical functionalization [137]. The functionalization of fullerenes consists of their chemical modification of the surface, by adding extensions and changing their solubility and electrochemical properties, resulting in increased application areas and diversity [145]. Reactions such as nucleophilic cyclopropanation, Diels Alder reaction, Prato reaction, and cycloaddition can be used to modify the surface of fullerenes [146].

11.3.2.1.3 Synthesis of graphene

Graphene can be synthesized through various processes depending on the desired size, purity, and particular product properties. In the late 1970s, thin graphitic films were produced on transition metal surfaces, followed by the synthesis of few-layer graphite via chemical decomposition methods in 1975 [147–149]. However, isolating and transferring these films to insulating substrates for electrical property characterization was challenging. In the late 1990s, researchers successfully isolated thin layers of graphene by mechanically exfoliating graphite onto SiO_2 substrates, but electrical property characterization was not reported until 2005 [150]. The real breakthrough in graphene research came in 2004 when Geim and co-workers [151] first isolated graphene onto a SiO_2 substrate and measured its electrical properties. Since then, numerous papers have been published on developing other methods for graphene synthesis, including mechanical and chemical exfoliation, thermal or chemical synthesis, and CVD synthesis [149].

Graphene 2D synthesis

Top-down approach

The top-down approach is an effective technique for synthesizing 2D materials that have a layered structure consisting of large crystals. This method involves the application of an external shearing force to the material to separate the 2D crystals from bulk or powdered materials. There are various methods of implementing this approach, including mechanical, chemical or electrochemical exfoliation [152].

Mechanical exfoliation

Mechanical exfoliation is a top-down method of producing graphene, whereby a layered graphite surface is subjected to stress using either a simple scotch tape or an AFM tip. What exists between the layers of graphene in graphite enable the separation of single-layer graphene or few-layer graphene from graphite. This method was first established as a means of extracting graphene flakes on a substrate [153]. The external force required to overcome the weak van der Waals forces is approximately 300 nN μm^{-2} enabling the separation of a single graphene layer [149]. Producing large-scale, high-quality, and defect-free graphene using the exfoliation process is challenging in terms of scalability and reproducibility, which highlights the need for further improvement in this technique for its use in nanotechnology [154]. For this, new chemical procedures were elaborated upon to overcome this limitation of the mechanical methods.

Chemical exfoliation

Chemical exfoliation is reported to be an effective alternative to the mechanical approach [149]. Generally, it consists of a two-step process to achieve a monolayer graphene. In the first step, interlayer van der Waals forces are weakened, and interlayer spacing is increased by adding alkali metals to a graphite solution, leading to the formation of graphene-intercalated compounds (GICs) [138]. The second step involves exfoliating graphene with single to few layers by rapid heating or sonication. The choice of alkali metals is based on their ability to react with graphite, as well as their atomic ratio, which enables their intercalation into the graphene interlayer. Unlike the previous mechanical method, this chemical method has been reported to produce a significant quantity of graphene at a relatively low temperature [149].

Electrochemical exfoliation

Electrochemical exfoliation is a promising green approach for synthesizing high-quality graphene with few layers, large lateral size, and low oxygen content [155]. Graphene obtained through traditional exfoliation methods often contains a high level of oxygen and has a small lateral size, likely due to the introduction of impurities and fragmentation of the graphene sheets. However, electrochemical methods for graphene exfoliation have demonstrated a degree of control over the structural properties [154]. The electrochemical process for graphene exfoliation entails immersing electrodes in an electrolyte solution and applying electrical current between them. This process can be carried out using either two- or three-electrode systems in both aqueous and non-aqueous electrolytes. One of these two electrodes serves as a graphene source, and it is usually made of graphite. The process of exfoliating graphene occurs in two distinct steps, intercalation and expansion [149]. With the application of electrical current, electrolyte ions intercalate between the graphene sheets in graphite, leading to the expansion of the sheets and the eventual exfoliation of the graphene layers. The exfoliation can be either cathodic or anodic, depending on the charge at the working electrode. The application of appropriate reduction potentials is necessary to allow electrolyte anions to produce gas molecules under the intercalation and expansion processes, which, in turn, facilitates the exfoliation process by generating gas bubbles. Several studies have reported the use of various electrode potentials to facilitate intercalation and expansion processes, and splitting the overall process into two steps has enabled the production of high-quality graphene [154–156]. This electrochemical process has also been conducted in non-aqueous media and ionic liquid electrolytes [157], where different mechanisms were proposed. Overall, comprehending the mechanism behind graphene exfoliation is essential to enhance the process's efficiency by determining the optimal conditions for intercalation and expansion steps.

Bottom-up approach

The bottom-up approach for synthesizing 2D materials involves creating products from atomic-level materials, while the top-down approach involves obtaining 2D products by exfoliating bulk materials [152]. When it comes to 2D graphene, common methods include CVD, epitaxial growth, laser-related methods, and electric discharge. In the following discussion, we will delve deeper into the CVD and epitaxial growth methods for synthesizing graphene.

Chemical vapor deposition

The synthesis of graphene using CVD involves introducing a particular mixture of gases, including a carbon precursor (such as methane), gas carriers (like nickel or argon), and sometimes hydrogen as an activation gas, into a chamber. The carbon source is then decomposed due to the temperature effect, and it undergoes a reaction on the substrate, commonly on copper foils or films, forming graphene on the surface [149]. As the resulting graphene is generated at the surface of a catalyst, it is necessary to transfer and remove the metal and depositing graphene on the desired surfaces for various applications. However, this transfer process can be aggressive and lead to surface deterioration and contamination. Various approaches have been proposed and tested for synthesizing graphene using CVD, including the use of a range of precursors such as highly purified gases or reactants, as well as solids, liquids, and even waste materials that are rich in carbon [128]. Additionally, various substrate materials, such as nickel, copper, platinum, germanium, and copper–nickel alloys, have been studied, with copper and nickel being the most popular and studied. Finally, the process is completed by cooling down and removing the by-products and non-reacted gases that were generated during the synthesis process [4].

Plasma-enhanced chemical vapor deposition

Plasma-enhanced CVD (PECVD) is a another approach to synthesizing graphene that requires chemical reactions of reacting gases and plasma in a vacuum chamber [158]. This produces thin films deposed on the substrate. The process can be performed at lower temperatures and reduced deposition duration compared to CVD, making it more appropriate for industrial-scale production. The growth of uncatalyzed graphene can also be produced by controlling process parameters, and various substrates have been reported for PECVD [149].

PECVD results in lower energy consumption and avoids the creation of undesired by-products or less crystalline carbon compared to the CVD method. Although PECVD has been demonstrated to generate high-purity graphene with a high degree of crystallinity, research is still ongoing to produce single-layer, large-scale, and uniformly structured graphene using this method [154].

Epitaxial growth

An effective approach for producing high-quality graphene films is epitaxial growth on single crystalline silicon (SiC) substrates. By using this method, it becomes possible to produce single-crystal graphene with controlled thickness and uniformity over a large area, making it ideal for industrial applications [159]. However, the process has some limitations, including the high growth temperature required and the challenge of transferring the graphene films to other substrates. The growth of graphene on SiC substrates occurs on both the Si-terminated and C-terminated faces, with the C-terminated face being preferred due to the absence of self-limiting growth. Ultrathin graphene films, which are only a few monolayers thick, have been successfully synthesized by several research groups using different growth methodologies, including *ex situ* graphitization, encapsulating SiC crystals in graphite enclosures, and controlling Si vapor pressure using disilane gas. Additionally, the thickness of the graphene films can be controlled by adjusting the growth temperature, with higher temperatures resulting in thicker films [149, 160]. Overall, epitaxial growth of graphene has demonstrated significant potential for producing high-quality graphene films with controlled thickness and uniformity, thus offering potential for industrial-scale fabrication of graphene-based devices.

Graphene 3D synthesis

From 0D dimension carbon materials such as fullerenes, to 1D and 2D graphene materials, all have exhibited significant impact and interesting properties that serve many important applications. However, due to the high demand for these carbon materials, the development of 3D graphene-based materials has become crucial in filling the gap in energy storage, environmental, and electrochemical applications [161]. The synthesis of 3D graphene is generally categorized into three major types:

materials derived from GO (3D rGO), those based on hydrocarbons, and those based on inorganic carbon compounds. In the next section, we will discuss the different synthesis methods reported for the synthesis of 3D graphene-based materials.

Synthesis of 3D reduced graphene oxide

GO is a 2D material that has gained a lot of attention in recent years due to its unique properties, such as high surface area, excellent mechanical properties, and semi-conductivity [162]. However, GO is not suitable for most practical applications due to its hydrophilic nature and poor electrical conductivity. Therefore, reducing GO to obtain rGO has been a focus of research. There are various methods available for GO reduction, including chemical reduction and photoinduced reduction [163].

One of the most promising methods for GO reduction is photoinduced reduction [161], which involves using photon energy to reduce GO without any chemical reagents. This method has several advantages, including high efficiency, little physical contact, and fast processing speed. Photoreduction of GO was first reported in 2008, where ultraviolet light was used along with semiconductor photocatalysts like TiO_2 to reduce GO in solutions [164]. It was later discovered that GO reduction can be achieved directly using ultraviolet light, which leads to the formation of CO, O_2, H_2O, and CO_2. Additionally, xenon lamp-equipped photographic camera lights and laser lights can also thermally deoxygenate GO [165]. The reduction of GO through photoenergy occurs through two mechanisms: photothermal effect and photochemical reduction. Photothermal effect is more effective at temperatures in the range of 200 °C–230 °C. The photochemical process of reducing GO can be induced by a laser emitting long wavelengths, benefiting the strong coupling between electrons and photons as well as the absorption of multiple photons. Both mechanisms cause a powerful excitation of the GO surface, resulting in rapid particle ejection and plasma plume formation. During energy transfer from plume to lattice, oxygen groups detach from GO as the plume expands. This process generates carbon monoxide and carbon dioxide, causing the internal water to evaporate and leading to an increase in interlayer pressure. This, in turn, causes the rGO layer to expand and form a porous structure.

Chemical reduction is another major method for GO reduction. In the early stages of progress, hydrazine was reported as an effective agent for achieving rGO dispersions from low concentration GO suspensions [166]. The literature has reported several reducing agents that can be used in the process of obtaining a 3D network of reduced graphene, including HI, $NaBH_4$, NaH, and NH_2OH, while keeping the reduction temperature within the range of 60 °C–100 °C. The resulting residue and absorbed water are subsequently eliminated via washing and lyophilization processes [167].

Both photoinduced and chemical reduction methods have their own advantages and disadvantages. Photoinduced reduction is a chemical-free process that can produce rGO with high efficiency and processing speed, but it requires specific equipment and careful control of reaction conditions. On the other hand, chemical reduction methods are more flexible and accessible, but they involve the use of chemical reagents that can introduce impurities and reduce the quality of rGO. Overall, the choice of GO reduction method depends on the specific application and desired properties of rGO [163].

The electrochemical reduction of GO is a widely used technique for synthesizing 3D graphene-based electrodes for electrochemical devices [168]. The electrochemical reduction of GO leads to the formation of a 3D graphene network that is tightly assembled onto the electrode surface [169]. This process creates high porosity inside the deposited sheets due to hydrogen evolution under bias voltage. Different materials can be used as electrode bases, including stainless steel mesh, Pt foil, Ni foam, and Au fiber [161]. The resulting 3D rGO structures can have varying hierarchical 3D frameworks, and the electrochemical approach can also be used to reduce GO films that have been precast onto an electrode, improving the control over the stacking of the GO sheets during the casting step. The electrochemical reduction of GO offers a fast and effective way to synthesize 3D graphene-based electrodes with high porosity and excellent performance [169].

Hydrocarbon-based 3D graphene materials

One promising method for large-scale production of graphene is the synthesis of graphene from hydrocarbon-based materials. Renewable sources of hydrocarbons, such as natural botanical precursors and waste products, offer a low-cost and abundant resource for the production of high-quality graphene. One example of a solid natural hydrocarbon precursor used for graphene synthesis is solid camphor ($C_{10}H_{16}O$), which is extracted from the latex of the *Cinnamomum camphora* tree belonging to the Lauraceae family. Camphor has proven to be a popular choice for synthesizing graphene due to its easy sublimation at room temperature and melting point of 180 °C [170]. A recent study by Zheng *et al* demonstrated the synthesis of high-quality 3D graphene using camphor as a carbon source and Ni foam catalysts through CVD.

A viable method for producing 3D graphene materials using organic hydrocarbons is through an organic-based framework. This technique is comparatively easy and efficient in synthesizing carbon-rich materials. It involves heating 3D polymeric structures, including highly cross-linked frameworks, hydrogels, or aerogels, in a process known as thermal annealing or carbonization [171]. This process results in the production of a 3D graphene-based structure. Patil *et al* used this approach to create 3D carbon foam for supercapacitor applications from polymers with high internal phase emulsion [172]. They subjected these polymers to sulphonation and pyrolysis at a temperature of 850 °C, resulting in a 3D configuration of graphene with a monolithic structure that features interconnected open voids and a high specific surface area. In another similar work authors [173], also demonstrated the direct carbonization of melamine foam at the similar temperature range to produce a unique elastic carbon foam with exceptional porosity, specific surface area, and low density. This technique allows the production of 3D porous graphene monolithic structures with exceptional adsorption characteristics.

11.4 Graphene-based nanomaterials for determination of some phenolic compounds

Quinones and derivatives as discussed earlier, have unique characteristics and play several important roles in biological processes [174], such as maintaining the electron transport chain in plants and animals [175], and serving as a source of some vitamins [176, 177]. Quinones also have various clinical and industrial applications [178]. However, quinones can also have toxic effects on living cells and can be a risk factor for respiratory diseases in the environment [174, 179]. For all of these, the detection and determination of quinones in real biological and environmental samples became a need and a focus of many researches, using several techniques [26, 180, 181]. Electrochemical detection of quinones is also an active area of research, as their structures constructed of conjugated cyclic diones could easily undergo a two-electrons redox reaction, which makes them an electroactive component [182]. In 1993, Antonios and co-authors, have used a glassy carbon electrode (GCE) for the determination of total Alkannin by the use of voltammetric technique [183], with a detection limit of 0.55 mg l^{-1}. Since then, several materials modified electrodes have been developed [174, 184]. Among these, graphene-modified electrodes have been used extensively for determination of many components of quinone derivatives. Shikonin, a pigment of the naphthoquinone family, has received much attention due to its various pharmaceutical properties [185], which increase the necessity of developing highly sensitive materials for its detection. Jing *et al* investigated the functionalization of graphene sheets using poly(diallyldimethylammonium chloride) polyelectrolyte for the modification of GCEs [186]. DPV analysis showed a limit of detection of 3.157×10^{-8} mol l^{-1}, and the modification of GCE using functionalized graphene sheets exhibited high stability, selectivity for Shikonin, and good reproducibility. Doxorubicin and daunorubicin are examples of anti-cancer drugs that belong to the anthracycline class. These drugs incorporate anthraquinone groups into their structure, which gives them electroactive properties. Kong *et al* prepared a GCE modified with rGO loaded with gold–silver bimetallic nanoparticles for the detection of daunorubicin. The reduction of quinone groups was used for the determination of the anti-cancer drug, resulting in a highly sensitive electrode with a low detection limit of 3 ng ml^{-1}, in addition to successful recovery from human serum samples [187]. In another work by the same group, daunorubicin was quantified using rGO and single-walled CNTs loaded with platinum nanoparticles as nanomaterials for GCE modification, and the same detection limit was observed [186]. The presence of quinone–hydroquinone in the chemical structure of Doxorubicin provides a redox platform for its electroanalytical determination. Hashemi *et al* used a polyrhodanine/GO/Fe_3O_4 nanocomposite for the electrochemical detection of doxorubicin hydrochloride from human blood plasma samples. They achieved optimal accuracy over 99% [188]. Other works have also reported the use of graphene-based nanomaterials for the detection of doxorubicin. For example, highly conducting polypyrrole polymer was incorporated with the electrical and thermal properties of graphene to develop a highly sensitive electrode for doxorubicin detection [189]. Hydroquinone on the other hand, presents a serious concern in the environment field due to their high toxicity and difficult degrading behavior [190]. Over the past decade, several graphene-based sensors have been developed for the determination of hydroquinone in real samples. Liu and co-authors increased the surface area and conductivity of GCE by modifying it with a graphene sheet-ionic liquid nanocomposite [191]. The limit of detection reached 2.0×10^{-8} M with promising results regarding the simultaneous determination of hydroquinone on real wastewater samples. In another paper, a graphene nanosheet has been used in combination with a carbon nanosphere to modify a GCE, enhancing its electro-analytical sensitivity toward hydroquinone [192]. The modified electrode was also successfully applied for the determination of hydroquinone in a real soil sample, with a recovery percentage ranging from 102.42% to 103.52%.

Flavonoids are a group of polyphenols that are bio-synthesized from the shikimic acid pathway. Their basic structure consists of aromatic phenolic rings combined with a heterocycle containing an oxygen atom. Flavonoids and derivatives have specific functional groups, such as hydroxyl and methoxy bonding into their heterocycle ring [193].

Flavonoids are known for their antioxidant activity and radical scavenging properties, which are largely due to their phenolic groups [194]. These compounds have the ability to donate hydrogen and scavenge radicals through either an H-atom or electron transfer process. The radical electron is then stabilized through delocalization, which ultimately leads to the termination of the chain reaction [193]. Flavonoids have also been reported to possess the ability to inhibit certain redox-active enzymes through chemical interactions, which are further strengthened by electrostatic interactions. The presence of oxygenated groups on the flavonoid structure tends to favor such enzymatic inhibition [195]. Understanding the redox behavior of flavonoids and their analytical determination is crucial due to their numerous biological activities. Electroanalysis provides a useful approach to evaluate the stability, antioxidant activity, pro-oxidant effect, and other chemical structural aspects of flavonoids' redox behavior.

Given the growing interest in the electrochemical analysis of flavonoids, there has been a surge in published literature over the past decade. Many reviews and research papers have reported on various electroanalytical techniques, such as CV [196, 197] and DPV [198–200], as effective methods for determining the redox behavior and chemical characteristics of flavonoids. Various type of electrodes/modified electrodes have been utilized for this purpose, such as metal oxide-based electrodes [201, 202], and carbon-based materials [198]. Among these modifying agents, graphene-based nanomaterials have gained significant interest in the determination of flavonoids and other bioactive compounds, owing to their well-known interesting properties such as high surface area, high conductivity, and promising mechanical and electrical properties, as well as their ease of functionalization. Table 11.1 summarizes the recent published papers concerning the use of graphene-based nanomaterials for detection of various kind of flavonoids. Graphene has been used directly to modify GCEs and used subsequently for electroanalytical detection of quercetin [203]; a linear range of concentration was found between 0.006 and 10 μM with a detection limit of 3.6 nM. The oxidized form of graphene, namely GO, contains a large number of oxygen functional groups, which provide this graphene derivative with a high surface area and excellent electrochemical properties as well as increased hydrophilicity that enhance its dispersity, Arvand *et al*, combined GO with chitosan for electrochemical sensitive determination of rutin with a limit of detection reached 560 nM [204]. The abundance of quercetin and rutin flavonoids in numerous food and drugs has led to extensive research aimed at developing highly sensitive

Table 11.1. Graphene-based electrochemical sensors for determining of flavonoids.

		Electrode	Technique	LOD (nM)	Reference
Flavone	Baicalein	Graphene nanosheet modified with 2,6-dimethyl-β-cyclodextrin (DM-β-CD) modified GCE (DM-β-CD–GNs–GCE)	CV	10	[212]
		Polydopamine (PDA) functionalized reduced graphene oxide (rGO) modified GCE (PDA–rGO–GCE)	CV, DPV	3.1	[213]
		Reduced graphene oxide nanoribbons modified GCE (ErGONRs–GCE)	AdSDPV, SWV	8.5	[214]
		Poly(dimethyldiallylammonium chloride) (PDDA) loaded with ruthenium oxide (RuO_2) functionalized reduced graphene oxide (rGO) modified GCE (RuO_2–PDDA–rGO–GCE)	CV, DPV	0.6	[215]
		Thermally reduced graphene oxide (TRGO) modified GCE (TRGO–GCE)	CV, DPV	6.0	[216]
		molybdenum trioxide (MoO_3) reduced graphene oxide (rGO) modified GCE (MoO_3–rGO–GCE)	CV, DPV	0.381	[208]
		Graphitized carbon–nitride (C_3N_4), single-walled carbon nanotubes (SWCNTs), and reduced graphene oxide (rGO), combined with electrodeposited cyclodextrin–metal–organic framework (CD–MOF) modified GCE (C_3N_4–SWCNTs/rGO/CD–MOF–GCE)	CV, DPV	0.46	[217]
	Luteolin	Chitosan/reduced graphene oxide (CS/rGO) aerogel with dispersed zirconium dioxide (ZrO_2) modified GCE (CS/rGO–ZrO_2)	CV, DPV	1	[218]
		Poly(diallyldimethylammonium chloride) (PDDA)-functionalized graphene sheets-multi-walled carbon nanotubes (G-CNTs) combined with β-cyclodextrin modified GCE (PDDA-G-CNTs/β-CD–GCE)	CV, DPV	20	[219]
		Graphene (GR), tin oxide (SnO_2) composite modified carbon ionic liquid electrode (CILE) (GR–SnO_2–CILE)	DPV	9.8	[42]
		Graphene nanosheets (GNs), hydroxyapatite (HA) composite modified GCE (GNs/HA/GCE)	CV, DPV	10	[220]

		Poly(3,4-ethylenedioxythiophene) (PEDOT) and graphene-like two-dimensional layered molybdenum disulfide (MoS_2) nanocomposite modified GCE (PEDOT–MoS_2–GCE)	DPV	40	[221]
		Nickel (Ni) clusters supported on a composite of graphene oxide (GO) and multi-walled carbon nanotubes (MWCNTs) modified GCE (Ni/GO–MWCNTS–GCE)	CV, DPV	0.34	[207]
		Chitosan (CS)–graphene (GR) modified GCE (CS–GR–GCE)	CV, DPV	0.59	[222]
Flavanol	Fisetin	MXene and amino-functionalized carbon nanotubes (NH_2–CNTs) modified GCE (MXene–NH_2–CNT–GCE)	CV, DPV	1.0	[223]
	Kaempferol	Graphene and poly(ethylene-*co*-vinyl acetate) (EVA) composite modified GCE (GR/EVA–GCE)	HV	220	[224]
		Zinc tungstate ($ZnWO_4$) and porous nanocarbon composite modified GCE	CV, DPV	2.03	[225]
	Hyperoside	Chitosan-functionalized nitrogen-doped graphene (N-GR) and gold nanoparticles (AuNPs) modified GCE (AuNPs/N-GR@CS–GCE)	CV, EIS	64.2	[226]
	Myricetin	Composite of platinum nanoparticles, reduced graphene oxide (rGO), and multi-walled carbon nanotubes (MWCNT) modified GCE (Pt–rGO–MWCNT–GCE)	CV, DPV	10	[206]
	Isoquercetin	SH-β-cyclodextrin functionalized graphene (SH-β-CD-GR) and palladium nanoparticles (PdNP) modified GCE (SH-β-CD-GR/PdNP/GCE)	DPV	0.0016	[227]
		DM-β-cyclodextrin functionalized graphene nanosheets modified GCE (DM-β-CD–GNs/GCE)	CV, DPV	4	[212]
	Quercetin	Three-dimensional reduced graphene oxide aerogel (3D-rGA).\	DPV	65	[228]
		Graphene-modified GCE GR/GCE	DPV	3.6	[203]
		Polypyrrole (PP) film with incorporated graphene oxide (GO)	DPV	48	[229]
		cadmium telluride nanoparticle incorporated on boron, sulfur co-doped reduced graphene oxide composite CdTe/BS–rGO–GCE	CV, DPV	18	[230]
			DPV	5.0	[205]

(*Continued*)

Table 11.1. (*Continued*)

		Electrode	Technique	LOD (nM)	Reference
		Three-dimensional Pd nanoparticles–porous graphene–carbon nanotubes modified GCE (Pd/pGN–CNT/GCE)			
		Graphene quantum dots (GQD) and gold nanoparticles (AuNP) modified GCE (GQD/AuNP/GCE)	DPV	2.0	[210]
		Graphene oxide (GO) functionalized with silver@silica, polyethylene glycol, and gold nanoparticles modified printed circuit board (PCB), GO–Ag@SiO_2–PEG–Au/PCB	DPV	3.57	[231]
		1-Pyrenebutyrate functionalized reduced graphene oxide (PB–rGO) decorated with mercapto-β-cyclodextrin (TCD) and Au nanoparticles (AuNP) modified GCE (PB–rGO/TCD–AuNP–GCE)	CV	1.83	[232]
		Aminated graphene quantum dots (AGQD) functionalized with thiolated β-cyclodextrin (TCD) and gold nanoparticles (AuNP) modified GCE (AGQD–TCD–AuNP–GCE)	CV	0.285	[233]
	Rutin	Composite of platinum nanoparticles, reduced graphene oxide (rGO), and multi-walled carbon nanotubes (MWCNT) modified GCE (Pt–rGO–MWCNT–GCE)	CV, DPV	5	[206]
		N-doped graphene (NG) decorated with Au–Ag nanothorns (NT). Modified GCE (Au–Ag NT/NG–GCE)	CV, DPV	15	[234]
		Tin oxide (SnO_2) and nitrogen-doped graphene (NG) composite modified GCE (SnO_2/NG–GCE)	CV, DPV	0.2	[209]
		Graphene quantum dots (GQD) and poly(3,4-ethylenedioxythiophene) (PEDOT) bilayer film modified GCE (GQD/PEDOT/GCE)	CV	11	[211]
		Graphene oxide (GO)–chitosan (CS) composite modified GCE (GO/CS–GCE)	DPV	560	[204]
	Isorhamnetin	Reduced graphene oxide (rGO) modified GCE (ERGO/GCE)	DPV	3.2	[235]
Flavanone	Hesperidin	Nano-graphene platelet and brilliant green composite modified carbon paste electrode (CPE) (nGp–Bg)/MCPE	DPV	50	[236]
			CV	8.2	[237]

		gold nanoparticles (AuNP) and reduced graphene oxide (rGO) composite modified GCE (AuNPs/rGO–GCE)			
	Hesperetin	Reduced graphene oxide/single-walled carbon nanotube (SWCNT) modified GCE (rGO–SWCNT–GCE)	CV	20	[76]
		Laser-induced graphene chips	CV	15	[238]
	Naringenin	SnS_2 decorated reduced graphene oxide nanoribbons modified GCE (SnS_2–rGONR–GCE)	DPV	313	[239]
	Naringin	Laser-induced graphene chips	CV	11	[238]
Flavonol	Taxifolin	Reduced graphene oxide modified GCE (ERGO–GCE)	SWASV	2	[240]
		Co_3S_4@MoS_2 loaded on reduced graphene oxide (rGO) modified GCE (rGO–Co_3S_4@MoS_2/GCE)	CV	1.67	[241]
		Reduced graphene oxide (ERGO) modified GCE (ERGO–GCE)	SWASV	4.0	[242]
		Poly(diallyldimethylammonium chloride) functionalized graphene-modified GCE (PDDA– GR–Pd/GCE)	SWV	1	[243]
Isoflavone	Formononetin	Nitrogen-doped graphene (MIP/NG/ITO-PET)	CV	1140	[244]
	Puerarin	Reduced graphene oxide (rGO) and molecularly imprinted polymer (MIP) modified GCE (rGO–MIP–GCE)	CV	6	[245]
		Graphene/quantum dot-GCE (GR, QD-GCE)	DPV	6	[246]
		Polyarginine and reduced graphene oxide modified GCE	CV, DPV	40	[247]
		Poly(alizarin red *S*)-graphene-modified GCE	CV	34	[248]
		Poly(sodium 4-styrenesulfonate) functionalized graphene decorated with WO_3 nanorods (PSS–GN/WO_3) modified GCE (PSS–GN/WO_3—GCE)	CV, LCV	40	[249]
	Daidzein	Poly(sodium 4-styrenesulfonate)–reduced graphene oxide (PSS–rGO) functionalized with *o*-phenylenediamine (o-PD) on the surface modified GCE, PSS–rGO/o-PD–GCE	CV, DPV	0.5	[250]
		SnO_2-poly(diallyldimethylammonium chloride), graphene oxide (GO) (SnO_2–PDDA–GR/GCE)	CV	6.7	[251]

electrochemical biosensors for their detection. One approach involves incorporating CBNs with graphene, which greatly improves the application area of such sensors for flavonoid detection. For example, Lite and co-authors developed a highly sensitive electrode for quercetin detection by incorporating CNTs, molecularly imprinted polymer, Pd nanoparticles, and porous graphene [205]. The linear range of concentration was found to be between 0.01 and 0.50 μM, and the limit of detection determined using DPV analysis was 5 nM. Similarly, in another study, Rutin detection was achieved using a combination of Pt nanoparticles, GO, and multi-walled CNTs [206]. This approach provided highly electroanalytic activity for rutin with a detection limit of 5 nM. Others flavonoids were also studied using graphene with carbon-based materials such as luteolin [207], myricetin [206] and hesperetin [76]. Metal oxides have also been used to modify graphene-based nanomaterials for flavonoid detection. For instance, the flavonoid baicalein was studied through the functionalization of rGO with molybdenum trioxide (MoO_3) [208], and Peng *et al* used tin oxide (SnO_2) incorporated with nitrogen-doped graphene for rutin electroanalytical analysis [209]. GQDs are another promising approach for the development of graphene-based materials as sensors. GQDs exhibit excellent optical, electrical, and biocompatibility properties, as well as an enhanced surface-grafting due to the p–p conjugated network. However, due to their limited application as a single-electrode material, they are often combined with other conductive materials to expand their application. Li *et al* presented a sensitive electrode based on a composite of gold nanoparticles and GQDs for quercetin detection [210]. The modified electrode showed a linear concentration range of 0.01–6.0 μM for quercetin, with a detection limit of 2.0 nM. In a separate study, GQDs were incorporated with poly(3,4-ethylenedioxythiophene) to form bilayer films for the detection of rutin [211].

11.5 Conclusion

In this chapter, the well-known phenolic compounds that have been obtained from natural sources have been analyzed with electrochemical methods. Graphene is one of the most used materials in electrochemistry due to its excellent chemical, electronic and physical properties. Graphene-based sensors have been presented in this chapter for investigation of sensitive and selective electrochemical analyses of phenolic compounds. Quinones, glycosides and some flavonoids, which have been shown to have anti-inflammatory, antioxidant, antimicrobial, antibacterial and anti-carcinogenic effects, have been preferred as compounds by researchers in recent years. Hence, due to increasing interest, their sensitive analysis have been given in this chapter. The oxidation mechanism of these drugs can be investigated easily, and they also can be used as modifying agents to increase the sensitivity and selectivity.

References

[1] Homann K H and Wagner H G G 1967 Some new aspects of the mechanism of carbon formation in premixed flames *Symp. Combust.* **11** 371–9

[2] Bassham J A and Calvin M 1960 The path of carbon in photosynthesis *Die CO_2-Assimilation/The Assimilation of Carbon Dioxide* (Springer) pp 884–922

[3] Allen M J, Tung V C and Kaner R B 2010 Honeycomb carbon: a review of graphene *Chem. Rev.* **110** 132–45
[4] Adetayo A and Runsewe D 2019 Synthesis and fabrication of graphene and graphene oxide: a review *Open J. Compos. Mater* **09** 207–29
[5] Mao H Y, Laurent S, Chen W, Akhavan O, Imani M, Ashkarran A A and Mahmoudi M 2013 Graphene: promises, facts, opportunities, and challenges in nanomedicine *Chem. Rev.* **113** 3407–24
[6] Liu Y, Yu D, Zeng C, Miao Z and Dai L 2010 Biocompatible graphene oxide-based glucose biosensors *Langmuir* **26** 6158–60
[7] Zhen Z and Zhu H 2018 Structure and properties of graphene *Graphene Fabrication, Characterizations, Properties and Applications* (Elsevier) pp 1–12
[8] Avouris P and Dimitrakopoulos C 2012 Graphene: synthesis and applications *Mater. Today* **15** 86–97
[9] Maheshwari U, Varshney D, Deshpande D and Ghodke S 2023 Development and utilizing graphene oxide for the wastewater treatment: an experimental evaluation *Mater. Today Proc.* **78** 23–7
[10] Ostojić J, Herenda S, Bešić Z, Miloš M and Galić B 2017 Advantages of an electrochemical method compared to the spectrophotometric kinetic study of peroxidase inhibition by boroxine derivative *Molecules* **22** 1120
[11] Chekin F, Vasilescu A, Jijie R, Singh S K, Kurungot S, Iancu M, Badea G, Boukherroub R and Szunerits S 2018 Sensitive electrochemical detection of cardiac troponin I in serum and saliva by nitrogen-doped porous reduced graphene oxide electrode *Sensors Actuators* B **262** 180–7
[12] Guin P S, Das S and Mandal P C 2011 Electrochemical reduction of quinones in different media: a review *Int. J. Electrochem.* **2011** 1–22
[13] Lazanas A C and Prodromidis M I 2022 Electrochemical impedance spectroscopy—a tutorial *ACS Meas. Sci. Au.* **3** 162–93
[14] Niaz K, Khan F and Ajmal Shah M 2020 Analysis of quinonoids *Recent Advances in Natural Products Analysis* (Elsevier) pp 749–66
[15] Rafat Husain S, Cillard J and Cillard P 1987 Hydroxyl radical scavenging activity of flavonoids *Phytochemistry* **26** 2489–91
[16] Khoo H E, Azlan A, Tang S T and Lim S M 2017 Anthocyanidins and anthocyanins: colored pigments as food, pharmaceutical ingredients, and the potential health benefits *Food Nutr. Res.* **21** 1361779
[17] Özenver N and Efferth T 2020 Therapeutic targeting of SNAIL, RKIP, and YY1 in tumor metastasis and drug resistance *Prognostic and Therapeutic Applications of RKIP in Cancer* (Elsevier) pp 357–87
[18] Kumar S and Pandey A K 2013 Chemistry and biological activities of flavonoids: an overview *Sci. World J.* **2013** 16
[19] Panche A N, Diwan A D and Chandra S R 2016 Flavonoids: an overview *J. Nutr. Sci.* **5** 1–15
[20] Zhang Y, Cai P, Cheng G and Zhang Y 2022 A brief review of phenolic compounds identified from plants: their extraction, analysis, and biological activity *Nat. Prod. Commun.* **17** 1934578X2110697
[21] Alu'datt M H, Rababah T, Alhamad M N, Al-Mahasneh M A, Almajwal A, Gammoh S, Ereifej K, Johargy A and Alli I 2017 A review of phenolic compounds in oil-bearing plants: distribution, identification and occurrence of phenolic compounds *Food Chem.* **218** 99–106

[22] Evans W C 2009 *Trease and Evans' Pharmacognosy* (Elsevier) 6th edn pp 1–603

[23] Thomson R H 1996 *Naturally Occurring Quinones IV* (Springer)

[24] Wang Y, Ameer G A, Sheppard B J and Langer R 2002 A tough biodegradable elastomer *Nat. Biotechnol.* **20** 602–6

[25] Verrax J, Delvaux M, Beghein N, Taper H, Gallez B and Calderon P B 2009 Enhancement of quinone redox cycling by ascorbate induces a caspase-3 independent cell death in human leukaemia cells. An *in vitro* comparative study **39** 649–57

[26] El-Najjar N, Gali-Muhtasib H, Ketola R A, Vuorela P, Urtti A and Vuorela H 2011 The chemical and biological activities of quinones: overview and implications in analytical detection *Phytochem. Rev.* **10** 353–70

[27] Dandawate P R, Vyas A C, Padhye S B, Singh M W and Baruah J B 2010 Perspectives on medicinal properties of benzoquinone compounds *Mini Rev. Med. Chem.* **10** 436–54

[28] Vaverkova V, Vrana O, Adam V, Pekarek T, Jampilek J and Babula P 2014 The study of naphthoquinones and their complexes with DNA by using Raman spectroscopy and surface enhanced Raman spectroscopy: new insight into interactions of DNA with plant secondary metabolites *BioMed Res. Int.* **2014** 461393

[29] Verma R P 2008 Anti-cancer activities of 1,4-naphthoquinones: a QSAR study *Anticancer Agents Med. Chem.* **6** 489–99

[30] Papageorgiou V, Assimopoulou A and Ballis A 2008 Alkannins and shikonins: a new class of wound healing agents *Curr. Med. Chem.* **15** 3248–67

[31] Huu Tung N, Du G J, Wang C Z, Yuan C S and Shoyama Y 2013 Naphthoquinone components from *Alkanna tinctoria* (L.) Tausch showed significant antiproliferative effects on human colorectal cancer cells *Phytother. Res.* **27** 66

[32] Kong Y H, Zhang L, Yang Z Y, Han C, Hu L H, Jiang H L and Shen X 2008 Natural product juglone targets three key enzymes from *Helicobacter pylori*: inhibition assay with crystal structure characterization *Acta Pharmacol. Sin.* **29** 870

[33] López L I L, Flores S D N, Belmares S Y S and Galindo A S 2014 Naphthoquinones: biological properties and synthesis of lawsone and derivates – a structured review *Vitae* **21** 248–58

[34] Salunke-Gawali S, Pereira E, Dar U A and Bhand S 2017 Metal complexes of hydroxynaphthoquinones: lawsone, bis-lawsone, lapachol, plumbagin and juglone *J. Mol. Struct.* **1148** 435–58

[35] Witte P D 1993 Metabolism and pharmacokinetics of anthranoids *Pharmacology* **47** 86–97

[36] Duval J, Pecher V, Poujol M and Lesellier E 2016 Research advances for the extraction, analysis and uses of anthraquinones: a review *Ind. Crops Prod.* **94** 812–33

[37] Seigler D S 1998 *Plant Secondary Metabolism* (Berlin: Springer)

[38] Jiang H Y, Hu X D, Zhu J J, Wan J and Yao J B 2021 Studies on the photofading of alizarin, the main component of madder *Dye. Pigment.* **185** 108940

[39] Khan A, Ezati P and Rhim J W 2023 Alizarin: prospects and sustainability for food safety and quality monitoring applications *Colloids Surf.* B **223** 113169

[40] Su S, Wu J, Gao Y, Luo Y, Yang D and Wang P 2020 The pharmacological properties of chrysophanol, the recent advances *Biomed. Pharmacother.* **125** 110002

[41] Sanders B *et al* 2018 Anti-cancer effects of aloe-emodin: a systematic review *J. Clin. Transl. Res.* **3** 283

[42] Li X, Niu X, Chen W, Dong L, Niu Y, Shao T, Lai W and Sun W 2017 Electrochemical behavior and sensitive detection of luteolin by graphene-SnO_2 nanocomposite modified electrode *Int. J. Electrochem. Sci.* **12** 9774–83

[43] Guedes R C and Eriksson L A 2005 Theoretical study of hypericin *J. Photochem. Photobiol. A Chem.* **172** 293–9
[44] Kubin A, Wierrani F, Burner U, Alth G and Grunberger W 2005 Hypericin – the facts about a controversial agent *Curr. Pharm. Des.* **11** 233–53
[45] Liu X, Yang Y, Mao L, Li Z, Zhou C, Liu X, Zheng S and Hu Y 2015 SPR quantitative analysis of direct detection of atrazine traces on Au-nanoparticles: nanoparticles size effect *Sensors Actuators* B **218** 1–7
[46] Yu Y, Xu S, He R and Liang G 2023 Application of molecular simulation methods in food science: status and prospects *J. Agric. Food Chem.* **71** 2684–703
[47] Salehi B *et al* 2019 The therapeutic potential of apigenin *Int. J. Mol. Sci.* **20** 1305
[48] Fan Z, Zhou S, Garcia C, Fan L and Zhou J 2017 PH-responsive fluorescent graphene quantum dots for fluorescence-guided cancer surgery and diagnosis *Nanoscale* **9** 4928–33
[49] Mani R and Natesan V 2018 Chrysin: sources, beneficial pharmacological activities, and molecular mechanism of action *Phytochemistry* **145** 187–96
[50] Shaikh S, Nazam N, Rizvi S M D, Ahmad K, Baig M H, Lee E J and Choi I 2019 Mechanistic insights into the antimicrobial actions of metallic nanoparticles and their implications for multidrug resistance *Int. J. Mol. Sci.* **20** 1–15
[51] Zorzi G, Gambini S, Negri S, Guzzo F and Commisso M 2023 Untargeted metabolomics analysis of the orchid species oncidium sotoanum reveals the presence of rare bioactive C-diglycosylated chrysin derivatives *Plants* **12** 655
[52] Imran M *et al* 2019 Luteolin, a flavonoid, as an anticancer agent: a review *Biomed. Pharmacother.* **112** 108612
[53] Gendrisch F, Esser P R, Schempp C M and Wölfle U 2021 Luteolin as a modulator of skin aging and inflammation *BioFactors* **47** 170–80
[54] Seelinger G, Merfort I and Schempp C M 2008 Anti-oxidant, anti-inflammatory and anti-allergic activities of luteolin *Planta Med.* **74** 1667–77
[55] Nabavi S F, Braidy N, Gortzi O, Sobarzo-Sanchez E, Daglia M, Skalicka-Woźniak K and Nabavi S M 2015 Luteolin as an anti-inflammatory and neuroprotective agent: a brief review *Brain Res. Bull.* **119** 1–11
[56] Antika L D and Dewi R M 2023 Pharmacological aspects of fisetin *Asian Pac. J. Trop. Biomed.* **11** 1
[57] Kumar R M, Kumar H, Bhatt T, Jain R, Panchal K, Chaurasiya A and Jain V 2023 Fisetin in cancer: attributes, developmental aspects, and nanotherapeutics *Pharmaceuticals* **16** 196
[58] Pal H C, Pearlman R L and Afaq F 2016 Fisetin and its role in chronic diseases *Adv. Exp. Med. Biol.* **928** 213–44
[59] Li Q, Song F, Zhu M, Wang Q, Han Y, Ling Y, Qiao L, Zhong N and Zhang L 2022 Hyperoside: a review of pharmacological effects *F1000Research* **11** 635
[60] Zeng Z, Wu D, Tang L, Hu X, Zhang J and Geng F 2023 Exploring the binding effects and inhibiting mechanism of hyperoside to lipase using multi-spectroscopic approaches, isothermal titration calorimetry, inhibition kinetics and molecular dynamics *RSC Adv.* **13** 6507–17
[61] Kim S H *et al* 2023 Effect of isoquercitrin on free fatty acid-induced lipid accumulation in HepG2 cells *Molecules* **28** 1476
[62] Carlos Rebuglio Vellosa J, Regasini L O, Maissar Khalil N, da Silva Bolzani V, Khalil O A K, Alessandra Manente F, Pasquini Netto H and de Faria Oliveira O M M 2011 Antioxidant and cytotoxic studies for kaempferol, quercetin and isoquercitrin *Ecletica Quim.* **36** 07

[63] Gong G, Guan Y Y, Zhang Z L, Rahman K, Wang S J, Zhou S, Luan X and Zhang H 2020 Isorhamnetin: a review of pharmacological effects *Biomed. Pharmacother.* **128** 110301

[64] Akhter H, Murshed J, Rashed M A, Oshima Y, Nagao Y, Rahman M M, Asiri A M, Hasnat M A, Uddin M N and Siddiquey I A 2017 Fabrication of hydrazine sensor based on silica-coated Fe2O3magnetic nanoparticles prepared by a rapid microwave irradiation method *J. Alloys Compd.* **698** 921–9

[65] Mustafa S, Ijaz M U, Ul Ain Q, Afsar T, Almajwal A, Shafique H and Razak S 2022 Isorhamnetin: a flavonoid, attenuated doxorubicin-induced testicular injury via regulation of steroidogenic enzymes and apoptotic signaling gene expression in male rats *Toxicol. Res. (Camb).* **11** 475–85

[66] Sun J *et al* 2013 Isorhamnetin protects against doxorubicin-induced cardiotoxicity *in vivo* and *in vitro PLoS One* **8** e64526

[67] Devi K P, Malar D S, Nabavi S F, Sureda A, Xiao J, Nabavi S M and Daglia M 2015 Kaempferol and inflammation: from chemistry to medicine *Pharmacol. Res.* **99** 1–10

[68] Imran M *et al* 2019 Kaempferol: a key emphasis to its anticancer potential *Molecules* **24** 2277

[69] Imran M *et al* 2021 Myricetin: a comprehensive review on its biological potentials *Food Sci. Nutr.* **9** 5854

[70] Song X, Tan L, Wang M, Ren C, Guo C, Yang B, Ren Y, Cao Z, Li Y and Pei J 2021 Myricetin: a review of the most recent research *Biomed. Pharmacother.* **134** 111017

[71] Zhang Y, Sun J, Lin C C, Abemayor E, Wang M B and Wong D T W 2016 The emerging landscape of salivary diagnostics *Periodontol. 2000* **70** 38–52

[72] Al-Dhabi N A, Arasu M V, Park C H and Park S U 2015 An up-to-date review of rutin and its biological and pharmacological activities *EXCLI J.* **14** 59–63

[73] Agrawal P K, Agrawal C and Blunden G 2021 Pharmacological significance of hesperidin and hesperetin, two citrus flavonoids, as promising antiviral compounds for prophylaxis against and combating COVID-19 *Nat. Prod. Commun.* **16** 1934578X2110425

[74] Li L, Yan X, Chen F, Zheng L, Hu Y, He F, Ni H, Chen F and Li Q 2023 A comprehensive review of the metabolism of citrus flavonoids and their binding to bitter taste receptors *Compr. Rev. Food Sci. Food Saf.* **22** 1763–93

[75] Khezri M R, Ghasemnejad-Berenji M and Moloodsouri D 2022 Hesperetin and the PI3K/AKT pathway: could their interaction play a role in the entry and replication of the SARS-CoV-2? *J. Food Biochem.* **46** e14212

[76] Wu J, Wang L, Wang Q, Zou L and Ye B 2016 The novel voltammetric method for determination of hesperetin based on a sensitive electrochemical sensor *Talanta* **150** 61–70

[77] Zalpoor H, Bakhtiyari M, Shapourian H, Rostampour P, Tavakol C and Nabi-Afjadi M 2022 Hesperetin as an anti-SARS-CoV-2 agent can inhibit COVID-19-associated cancer progression by suppressing intracellular signaling pathways *Inflammopharmacology* **30** 1533–9

[78] Salehi B, Fokou P V T, Sharifi-Rad M, Zucca P, Pezzani R, Martins N and Sharifi-Rad J 2019 The therapeutic potential of naringenin: a review of clinical trials *Pharmaceuticals (Basel)* **12** 11

[79] Zhao Y and Liu S 2021 Bioactivity of naringin and related mechanisms *Pharmazie* **76** 359–63

[80] Jong B K, Stein R and O'Hare M J 2004 Three-dimensional *in vitro* tissue culture models of breast cancer – a review *Breast Cancer Res. Treat.* **85** 281–91

[81] Gaur R, Kumar S, Trivedi P, Bhakuni R S, Bawankule D U, Pal A and Shanker K 2010 Liquiritigenin derivatives and their hepatotoprotective activity *Nat. Prod. Commun.* **5** 1243–6

[82] Park B W, Oh J W, Kim J H, Park S H, Kim K S, Kim J H and Lee K S 2008 Preoperative CA 15-3 and CEA serum levels as predictor for breast cancer outcomes *Ann. Oncol.* **19** 675–81

[83] Tian F, Woo S Y, Lee S Y, Park S B, Im J H and Chun H S 2023 Plant-based natural flavonoids show strong inhibition of aflatoxin production and related gene expressions correlated with chemical structure *Food Microbiol.* **109** 104141

[84] Yugender Goud K, Sunil Kumar V, Hayat A, Vengatajalabathy Gobi K, Song H, Kim K H and Marty J L 2019 A highly sensitive electrochemical immunosensor for zearalenone using screen-printed disposable electrodes *J. Electroanal. Chem.* **832** 336–42

[85] Kurbanoglu S, Dogan-Topal B, Rodriguez E P, Bozal-Palabiyik B, Ozkan S A and Uslu B 2016 Advances in electrochemical DNA biosensors and their interaction mechanism with pharmaceuticals *J. Electroanal. Chem.* **775** 8–26

[86] Lu L, Wang H, Zhou Y, Xi S, Zhang H, Hu J and Zhao B 2002 Seed-mediated growth of large, monodisperse core–shell gold-silver nanoparticles with Ag-like optical properties *Chem. Commun.* **2** 144–5

[87] Maaliki D, Shaito A A, Pintus G, El-Yazbi A and Eid A H 2019 Flavonoids in hypertension: a brief review of the underlying mechanisms *Curr. Opin. Pharmacol.* **45** 57–65

[88] Pan H *et al* 2017 20-year risks of breast-cancer recurrence after stopping endocrine therapy at 5 years *N. Engl. J. Med.* **377** 1836–46

[89] Landete J M, Arqués J, Medina M, Gaya P, de Las Rivas B D and Muñoz R 2016 Bioactivation of phytoestrogens: intestinal bacteria and health *Crit. Rev. Food Sci. Nutr.* **56** 1826–43

[90] Ganai A A and Farooqi H 2015 Bioactivity of genistein: a review of *in vitro* and *in vivo* studies *Biomed. Pharmacother.* **76** 30–8

[91] Dwiecki K, Neunert G, Polewski P and Polewski K 2009 Antioxidant activity of daidzein, a natural antioxidant, and its spectroscopic properties in organic solvents and phosphatidylcholine liposomes *J. Photochem. Photobiol.* B **96** 242–8

[92] Alshehri M M *et al* 2021 Therapeutic potential of isoflavones with an emphasis on daidzein *Oxid Med Cell Longev* **2021** 6331630

[93] Masilamani M, Wei J and Sampson H A 2012 Regulation of the immune response by soybean isoflavones *Immunol. Res.* **54** 95–110

[94] Jiang D *et al* 2019 Potential anticancer properties and mechanisms of action of formononetin *BioMed Res. Int.* **2019** 5854315

[95] Kulczyński B, Gramza-Michałowska A, Suliburska J and Sidor A 2021 Puerarin-an isoflavone with beneficial effects on bone health *Front. Biosci. (Landmark Ed.)* **26** 1653–67

[96] Zhou Y X, Zhang H and Peng C 2021 Effects of puerarin on the prevention and treatment of cardiovascular diseases *Front. Pharmacol.* **12** 3616

[97] Baig N, Kammakakam I, Falath W and Kammakakam I 2021 Nanomaterials: a review of synthesis methods, properties, recent progress, and challenges *Mater. Adv.* **2** 1821–71

[98] Jeevanandam J, Barhoum A, Chan Y S, Dufresne A and Danquah M K 2018 Review on nanoparticles and nanostructured materials: history, sources, toxicity and regulations *Beilstein J. Nanotechnol.* **9** 1050–74

[99] Mourdikoudis S, Pallares R M and Thanh N T K 2018 Characterization techniques for nanoparticles: comparison and complementarity upon studying nanoparticle properties *Nanoscale* **10** 12871–934

[100] Yang C, Denno M E, Pyakurel P and Venton B J 2015 Recent trends in carbon nanomaterial-based electrochemical sensors for biomolecules: a review *Anal. Chim. Acta* **887** 17–37

[101] Hernandez-Vargas G, Sosa-Hernández J E, Saldarriaga-Hernandez S, Villalba-Rodríguez A M, Parra-Saldivar R and Iqbal H M N 2018 Electrochemical biosensors: a solution to pollution detection with reference to environmental contaminants *Biosensors* **8** 1–21

[102] Nemiwal M, Zhang T C and Kumar D 2022 Enzyme immobilized nanomaterials as electrochemical biosensors for detection of biomolecules *Enzyme Microb. Technol.* **156** 110006

[103] Castañeda M T, Alegret S and Merkoçi A 2007 Electrochemical sensing of DNA using gold nanoparticles *Electroanalysis* **19** 743–53

[104] Hasan M R *et al* 2022 Evaluating the electrochemical detection of nitrite using a platinum nanoparticle coated jute carbon modified glassy carbon electrode and voltametric analysis *J. Phys. Chem. Solids* **165** 110659

[105] Jahanbakhshi M 2019 *In situ* synthesis of rhodium nanoparticles – mesoporous carbon hybrid via a novel and facile nanocasting method for simultaneous determination of morphine and buprenorphine *Mater. Sci. Eng.* C **97** 479–85

[106] George J M, Antony A and Mathew B 2018 Metal oxide nanoparticles in electrochemical sensing and biosensing: a review *Microchim. Acta* **185** 358

[107] John A, Benny L, Cherian A R, Narahari S Y, Varghese A and Hegde G 2021 Electrochemical sensors using conducting polymer/noble metal nanoparticle nanocomposites for the detection of various analytes: a review *J Nanostruct Chem.* **11** 1–31

[108] Jiang C, Chen H, Yu C, Zhang S, Liu B and Kong J 2009 Preparation of the Pt nanoparticles decorated poly(N-acetylaniline)/MWNTs nanocomposite and its electrocatalytic oxidation toward formaldehyde *Electrochim. Acta* **54** 1134–40

[109] Yadav D K, Ganesan V, Sonkar P K, Gupta R and Rastogi P K 2016 Electrochemical investigation of gold nanoparticles incorporated zinc based metal-organic framework for selective recognition of nitrite and nitrobenzene *Electrochim. Acta* **200** 276–82

[110] Zhang S, Rong F, Guo C, Duan F, He L, Wang M, Zhang Z, Kang M and Du M 2021 Metal–organic frameworks (MOFs) based electrochemical biosensors for early cancer diagnosis *in vitro Coord. Chem. Rev.* **439** 213948

[111] Lim W Q and Gao Z 2015 Metal oxide nanoparticles in electroanalysis *Electroanalysis* **27** 2074–90

[112] Lee S, Oh J, Kim D and Piao Y 2016 A sensitive electrochemical sensor using an iron oxide/graphene composite for the simultaneous detection of heavy metal ions *Talanta* **160** 528–36

[113] Zhang Y, Cheng Y, Zhou Y, Li B, Gu W, Shi X and Xian Y 2013 Electrochemical sensor for bisphenol A based on magnetic nanoparticles decorated reduced graphene oxide *Talanta* **107** 211–8

[114] Ran G, Chen X and Xia Y 2017 Electrochemical detection of serotonin based on a poly (bromocresol green) film and Fe_3O_4 nanoparticles in a chitosan matrix *RSC Adv.* **7** 1847–51

[115] Madrakian T, Maleki S, Heidari M and Afkhami A 2016 An electrochemical sensor for rizatriptan benzoate determination using Fe_3O_4 nanoparticle/multiwall carbon nanotube-modified glassy carbon electrode in real samples *Mater. Sci. Eng.* C **63** 637–43

[116] Wang X, Yang T and Jiao K 2009 Electrochemical sensing the DNA damage *in situ* induced by a cathodic process based on Fe@Fe_2O_3 core–shell nanonecklace and Au nanoparticles mimicking metal toxicity pathways *in vivo Biosens. Bioelectron.* **25** 668–73

[117] Cui X, Sun Y Q, Ma R and Song X C 2014 A hydrogen peroxide electrochemical sensor based on co-doped Fe_3O_4 nanoparticles *Adv. Mater. Res.* **941–944** 377–80

[118] Luo S X, Wu Y H, Gou H and Liu Y 2014 A novel electrochemical sensor for the analysis of salbutamol in pork samples by using $NiFe_2O_4$ nanoparticles modified glassy carbon electrode *Adv. Mater. Res.* **850–851** 1279–82

[119] He X and Hu C 2011 Building three-dimensional Pt catalysts on TiO_2 nanorod arrays for effective ethanol electrooxidation *J. Power Sources* **196** 3119–23

[120] Chen Y, Liu B, Chen Z and Zuo X 2021 Innovative electrochemical sensor using TiO_2 nanomaterials to detect phosphopeptides *Anal. Chem.* **93** 10635–43

[121] Luo Z, Ma X, Yang D, Yuwen L, Zhu X, Weng L and Wang L 2013 Synthesis of highly dispersed titanium dioxide nanoclusters on reduced graphene oxide for increased glucose sensing *Carbon N. Y.* **57** 470–6

[122] Goda E S, Gab-Allah M A, Singu B S and Yoon K R 2019 Halloysite nanotubes based electrochemical sensors: a review *Microchem. J.* **147** 1083–96

[123] Yang Z, Zheng X and Zheng J 2016 Non-enzymatic sensor based on a glassy carbon electrode modified with Ag nanoparticles/polyaniline/halloysite nanotube nanocomposites for hydrogen peroxide sensing *RSC Adv.* **6** 58329–35

[124] Ghanei-Motlagh M and Taher M A 2018 A novel electrochemical sensor based on silver/halloysite nanotube/molybdenum disulfide nanocomposite for efficient nitrite sensing *Biosens. Bioelectron.* **109** 279–85

[125] Kour R, Arya S, Young S-J, Gupta V, Bandhoria P and Khosla A 2020 Review—Recent advances in carbon nanomaterials as electrochemical biosensors *J. Electrochem. Soc.* **167** 037555

[126] Power A C, Gorey B, Chandra S and Chapman J 2018 Carbon nanomaterials and their application to electrochemical sensors: a review *Nanotechnol. Rev.* **7** 19–41

[127] Shetti N P, Mishra A, Basu S and Aminabhavi T M 2021 Versatile fullerenes as sensor materials *Mater. Today Chem.* **20** 100454

[128] Madurani K A, Suprapto S, Machrita N I, Bahar S L, Illiya W and Kurniawan F 2020 Progress in graphene synthesis and its application: history, challenge and the future outlook for research and industry *ECS J. Solid State Sci. Technol.* **9** 093013

[129] Rahi A, Karimian K and Heli H 2016 Nanostructured materials in electroanalysis of pharmaceuticals *Anal. Biochem.* **497** 39–47

[130] Yang N, Swain G M and Jiang X 2016 Nanocarbon electrochemistry and electroanalysis: current status and future perspectives *Electroanalysis* **28** 27–34

[131] Ma P C, Siddiqui N A, Marom G and Kim J K 2010 Dispersion and functionalization of carbon nanotubes for polymer-based nanocomposites: a review *Composites* A **41** 1345–67

[132] Patolsky F, Weizmann Y and Willner I 2004 Long-range electrical contacting of redox enzymes by SWCNT connectors *Angew. Chem. – Int. Ed.* **43** 2113–7

[133] Georgakilas V, Gournis D, Tzitzios V, Pasquato L, Guldi D M and Prato M 2007 Decorating carbon nanotubes with metal or semiconductor nanoparticles *J. Mater. Chem.* **17** 2679–94

[134] Gao C, Guo Z, Liu J H and Huang X J 2012 The new age of carbon nanotubes: an updated review of functionalized carbon nanotubes in electrochemical sensors *Nanoscale* **4** 1948–63

[135] Ismail N H, Akindoyo J O and Mariatti M 2020 Solvent mediated dispersion of carbon nanotubes for glass fibre surface modification—suspensions stability and its effects on mechanical, interlaminar and dynamic mechanical properties of modified glass fibre reinforced epoxy laminates *Composites* A **139** 106091
[136] Lawal A T 2016 Synthesis and utilization of carbon nanotubes for fabrication of electrochemical biosensors *Mater. Res. Bull.* **73** 308–50
[137] Gergeroglu H, Yildirim S and Ebeoglugil M F 2020 Nano-carbons in biosensor applications: an overview of carbon nanotubes (CNTs) and fullerenes (C_{60}) *SN Appl. Sci.* **2** 603
[138] Tran T T and Mulchandani A 2016 Carbon nanotubes and graphene nano field-effect transistor-based biosensors *TrAC – Trends Anal. Chem.* **79** 222–32
[139] McKee G S B, Deck C P and Vecchio K S 2009 Dimensional control of multi-walled carbon nanotubes in floating-catalyst CVD synthesis *Carbon* **47** 2085–94
[140] Sehrawat M, Rani M and Singh B P 2022 One step fabrication of aligned carbon nanotube sheet via FC-CVD technique *J. Nanomater.* **2022** 8318217
[141] Kim S *et al* 2022 Reactor design for manufacturing carbon hybrid materials *Adv. Mater. Lett.* **13** 2201-1685
[142] Ribeiro H, Schnitzler M C, da Silva W M and Santos A P 2021 Purification of carbon nanotubes produced by the electric arc-discharge method *Surf. Interfaces* **26** 101389
[143] Chrzanowska J, Hoffman J, Małolepszy A, Mazurkiewicz M, Kowalewski T A, Szymanski Z and Stobinski L 2015 Synthesis of carbon nanotubes by the laser ablation method: effect of laser wavelength *Phys. Status Solidi Basic Res.* **252** 1860–7
[144] Pilehvar S and Wael K D 2015 Recent advances in electrochemical biosensors based on fullerene-C60 nano-structured platforms *Biosensors* **5** 712–35
[145] Yin R, Wang M, Huang Y Y, Landi G, Vecchio D, Chiang L Y and Hamblin M R 2015 Antimicrobial photodynamic inactivation with decacationic functionalized fullerenes: oxygen-independent photokilling in presence of azide and new mechanistic insights *Free Radic. Biol. Med.* **79** 14–27
[146] Peyghan A A and Yourdkhani S 2014 Exohedral functionalization of C60 by [4 + 2] cycloaddition of multiple anthracenes *Struct. Chem.* **25** 785–91
[147] Eizenberg M and Blakely J M 1979 Carbon monolayer phase condensation on Ni(111) *Surf. Sci.* **82** 228–36
[148] Lang B 1975 A LEED study of the deposition of carbon on platinum crystal surfaces *Surf. Sci.* **53** 317–29
[149] Bhuyan M S A, Uddin M N, Islam M M, Bipasha F A and Hossain S S 2016 Synthesis of graphene *Int. Nano Lett.* **6** 65–83
[150] Zhang Y, Small J P, Pontius W V and Kim P 2005 Fabrication and electric-field-dependent transport measurements of mesoscopic graphite devices *Appl. Phys. Lett.* **86** 1–3
[151] Geim A K and Novoselov K S 2007 The rise of graphene *Nat. Mater.* **6** 183–91
[152] Mubarak N M, Abdullah E C, Khalid M and Khalid M 2018 Recent trends in the synthesis of graphene and graphene oxide based nanomaterials for removal of heavy metals—a review *J. Ind. Eng. Chem.* **66** 29–44
[153] Novoselov K S *et al* 2016 Electric field effect in atomically thin carbon films *Science (80–)* **306** 666–9
[154] Prekodravac J R, Kepić D P, Colmenares J C, Giannakoudakis D A and Jovanović S P 2021 A comprehensive review on selected graphene synthesis methods: from electrochemical exfoliation through rapid thermal annealing towards biomass pyrolysis *J. Mater. Chem.* C **9** 6722–48

[155] Najafabadi A T and Gyenge E 2015 Synergistic production of graphene microsheets by simultaneous anodic and cathodic electro-exfoliation of graphitic electrodes in aprotic ionic liquids *Carbon* **84** 449–59
[156] Zhou Q, Lu Y and Xu H 2019 High-yield production of high-quality graphene by novel electrochemical exfoliation at air-electrolyte interface *Mater. Lett.* **235** 153–6
[157] Parvez K, Wu Z S, Li R, Liu X, Graf R, Feng X and Müllen K 2014 Exfoliation of graphite into graphene in aqueous solutions of inorganic salts *J. Am. Chem. Soc.* **136** 6083–91
[158] Bo Z, Yang Y, Chen J, Yu K, Yan J and Cen K 2013 Plasma-enhanced chemical vapor deposition synthesis of vertically oriented graphene nanosheets *Nanoscale* **5** 5180–204
[159] Mishra N, Boeckl J, Motta N and Iacopi F 2016 Graphene growth on silicon carbide: a review *Phys. Status Solidi Appl. Mater. Sci.* **213** 2277–89
[160] De Heer W A, Berger C, Ruan M, Sprinkle M, Li X, Hu Y, Zhang B, Hankinson J and Conrad E 2011 Large area and structured epitaxial graphene produced by confinement controlled sublimation of silicon carbide *Proc. Natl Acad. Sci. USA* **108** 16900–5
[161] Singh R, Ullah S, Rao N, Singh M, Patra I, Darko D A, Issac C P J, Esmaeilzadeh-Salestani K, Kanaoujiya R and Vijayan V 2022 Synthesis of three-dimensional reduced-graphene oxide from graphene oxide *J. Nanomater.* **2022** 8731429
[162] Smith A T, LaChance A M, Zeng S, Liu B and Sun L 2019 Synthesis, properties, and applications of graphene oxide/reduced graphene oxide and their nanocomposites *Nano Mater. Sci.* **1** 31–47
[163] Gao H and Duan H 2015 2D and 3D graphene materials: preparation and bioelectrochemical applications *Biosens. Bioelectron.* **65** 404–19
[164] Williams G, Seger B and Kamt P V 2008 TiO_2-graphene nanocomposites. UV-assisted photocatalytic reduction of graphene oxide *ACS Nano* **2** 1487–91
[165] Smirnov V A, Arbuzov A A, Shul'ga Y M, Baskakov S A, Martynenko V M, Muradyan V E and Kresova E I 2011 Photoreduction of graphite oxide *High Energy Chem.* **45** 57–61
[166] Stankovich S, Dikin D A, Piner R D, Kohlhaas K A, Kleinhammes A, Jia Y, Wu Y, Nguyen S B T and Ruoff R S 2007 Synthesis of graphene-based nanosheets via chemical reduction of exfoliated graphite oxide *Carbon* **45** 1558–65
[167] Chen W and Yan L 2011 *In situ* self-assembly of mild chemical reduction graphene for three-dimensional architectures *Nanoscale* **3** 3132–7
[168] Liu F, Wang C, Sui X, Riaz M A, Xu M, Wei L and Chen Y 2019 Synthesis of graphene materials by electrochemical exfoliation: recent progress and future potential *Carbon Energy* **1** 173–99
[169] Chen K, Chen L, Chen Y, Bai H and Li L 2012 Three-dimensional porous graphene-based composite materials: electrochemical synthesis and application *J. Mater. Chem.* **22** 20968–76
[170] Somani P R, Somani S P and Umeno M 2006 Planer nano-graphenes from camphor by CVD *Chem. Phys. Lett.* **430** 56–9
[171] Hiew B Y Z, Lee L Y, Lee X J, Thangalazhy-Gopakumar S, Gan S, Lim S S, Pan G T, Yang T C K, Chiu W S and Khiew P S 2018 Review on synthesis of 3D graphene-based configurations and their adsorption performance for hazardous water pollutants *Process Saf. Environ. Prot.* **116** 262–86
[172] Patil U M, Ghorpade R V, Nam M S, Nalawade A C, Lee S, Han H and Jun S C 2016 PolyHIPE derived freestanding 3D carbon foam for cobalt hydroxide nanorods based high performance supercapacitor *Sci Rep.* **6** 1–11

[173] Chen S, He G, Hu H, Jin S, Zhou Y, He Y, He S, Zhao F and Hou H 2013 Elastic carbon foam via direct carbonization of polymer foam for flexible electrodes and organic chemical absorption *Energy Environ. Sci.* **6** 2435–9

[174] Kishikawa N and Kuroda N 2014 Analytical techniques for the determination of biologically active quinones in biological and environmental samples *J. Pharm. Biomed. Anal.* **87** 261–70

[175] George P M, Mackay R J, Molyneux S L, Young J M, Lever M and Florkowski C M 2010 Coenzyme Q10 in health and disease *Clin. Chim. Acta* **411** 907–8

[176] Liebler D C, Burr J A, Philips L and Ham A J 1996 Gas chromatography-mass spectrometry analysis of vitamin E and its oxidation products *Anal. Biochem.* **236** 27–34

[177] Kasahara T and Kato T 2003 A new redox-cofactor vitamin for mammals *Nature* **422** 832–2

[178] Rejiniemon T S, Arasu M V, Duraipandiyan V, Ponmurugan K, Al-Dhabi N A, Arokiyaraj S, Agastian P and Choi K C 2014 In-vitro antimicrobial, antibiofilm, cytotoxic, antifeedant and larvicidal properties of novel quinone isolated from *Aegle marmelos* (Linn.) Correa *Ann. Clin. Microbiol. Antimicrob.* **13** 48

[179] Motoyama Y, Bekki K, Sang W C, Tang N, Kameda T, Toriba A, Taguchi K and Hayakawa K 2009 Oxidative stress more strongly induced by ortho- than para-quinoid polycyclic aromatic hydrocarbons in A549 cells *J. Heal. Sci.* **55** 845–50

[180] Rajendran M 2016 Quinones as photosensitizer for photodynamic therapy: ROS generation, mechanism and detection methods *Photodiagn. Photodyn. Ther.* **13** 175–87

[181] Li W-k, Ren P, Zhou Y-w, Feng J-t and Ma Z-q 2020 Europium(III) functionalized 3D covalent organic framework for quinones adsorption and sensing investigation *J. Hazard. Mater.* **388** 121740

[182] Kim R S and Chung T D 2014 The electrochemical reaction mechanism and applications of quinones *Bull. Korean Chem. Soc.* **35** 3143–55

[183] Chaisuksant R, Voulgaropoulos A, Mellidis A S and Papageorgiou V P 1993 Voltammetric determination of total alkannin using a glassy carbon electrode *Analyst* **118** 179–82

[184] Upan J, Reanpang P, Chailapakul O and Jakmunee J 2016 Flow injection amperometric sensor with a carbon nanotube modified screen printed electrode for determination of hydroquinone *Talanta* **146** 766–71

[185] Papageorgiou V P, Assimopoulou A N, Couladouros E A, Hepworth D and Nicolaou K C 1999 The chemistry and biology of alkannin, shikonin, and related naphthazarin natural products *Angew. Chem. – Int. Ed.* **38** 270–301

[186] An J, ping Li J, xia Chen W, xia Yang C, di Hu F and ming Wang C 2013 Electrochemical study and application on shikonin at poly(diallyldimethylammonium chloride) functionalized graphene sheets modified glass carbon electrode *Chem. Res. Chinese Univ.* **29** 798–805

[187] Kong F Y, Yao L, Li R F, Li H Y, Wang Z X, Lv W X and Wang W 2019 Synthesis of nitrogen-doped reduced graphene oxide loading with Au-Ag bimetallic nanoparticles for electrochemical detection of daunorubicin *J. Alloys Compd.* **797** 413–20

[188] Hashemi S A *et al* 2022 Bio-enhanced polyrhodanine/graphene oxide/Fe_3O_4 nanocomposite with kombucha solvent supernatant as ultra-sensitive biosensor for detection of doxorubicin hydrochloride in biological fluids *Mater. Chem. Phys.* **279** 125743

[189] Behravan M, Aghaie H, Giahi M and Maleknia L 2021 Determination of doxorubicin by reduced graphene oxide/gold/polypyrrole modified glassy carbon electrode: a new preparation strategy *Diamond Relat. Mater.* **117** 108478

[190] Zheng L, Xiong L, Li Y, Xu J, Kang X, Zou Z, Yang S and Xia J 2013 Facile preparation of polydopamine-reduced graphene oxide nanocomposite and its electrochemical application in simultaneous determination of hydroquinone and catechol *Sensors Actuators* B **177** 344–9
[191] Liu Z, Wang Z, Cao Y, Jing Y and Liu Y 2011 High sensitive simultaneous determination of hydroquinone and catechol based on graphene/BMIMPF6 nanocomposite modified electrode *Sensors Actuators* B **157** 540–6
[192] Wang J, Yin H, Meng X, Zhu J and Ai S 2011 Preparation of the mixture of graphene nanosheets and carbon nanospheres with high adsorptivity by electrolyzing graphite rod and its application in hydroquinone detection *J. Electroanal. Chem.* **662** 317–21
[193] Gil E S and Couto R O 2013 Flavonoid electrochemistry: a review on the electroanalytical applications *Rev. Bras. Farmacogn.* **23** 542–58
[194] Hidalgo M, Sánchez-Moreno C and de Pascual-Teresa S 2010 Flavonoid-flavonoid interaction and its effect on their antioxidant activity *Food Chem.* **121** 691–6
[195] Jucá M M *et al* 2020 Flavonoids: biological activities and therapeutic potential *Nat. Prod. Res.* **34** 692–705
[196] Jadreško D, Miličević A and Jovanović I N 2022 Reactivity of flavonoids toward superoxide radical: an electrochemical approach *Electrochim. Acta* **421** 140501
[197] Mohtar L G, Messina G A, Bertolino F A, Pereira S V, Raba J and Nazareno M A 2020 Comparative study of different methodologies for the determination the antioxidant activity of Venezuelan propolis *Microchem. J.* **158** 105244
[198] Hu J and Zhang Z 2020 Application of electrochemical sensors based on carbon nanomaterials for detection of flavonoids *Nanomaterials* **10** 1–14
[199] Yang L, Yang T, Li G, Ma J X and Yu Y 2022 Electrochemical studies of polyphenols, anthocyanins, and flavonoids extracted from blueberry fruit *Int. J. Electrochem. Sci.* **17** 22038
[200] Li Y, Tang J, Lin Y, Li J, Yang Y, Zhao P, Fei J and Xie Y 2022 Ultrasensitive determination of natural flavonoid rutin using an electrochemical sensor based on metal-organic framework CAU-1/acidified carbon nanotubes composites *Molecules* **27** 7761
[201] Ganesamurthi J, Shanmugam R, Chen T W, Chen S M, Balamurugan M, Gan Z W, Siddiqui M R, Wabaidur S M and Ali M A 2022 NiO/ZnO binary metal oxide based electrochemical sensor for the evaluation of hazardous flavonoid in biological and vegetable samples *Colloids Surf.* A **647** 129077
[202] Gao F, Tu X, Ma X, Xie Y, Zou J, Huang X, Qu F, Yu Y and Lu L 2020 NiO@Ni-MOF nanoarrays modified Ti mesh as ultrasensitive electrochemical sensing platform for luteolin detection *Talanta* **215** 120891
[203] Arvand M and Anvari M 2013 A graphene-based electrochemical sensor for sensitive detection of quercetin in foods *J. Iran. Chem. Soc.* **10** 841–9
[204] Arvand M, Shabani A and Ardaki M S 2017 A new electrochemical sensing platform based on binary composite of graphene oxide-chitosan for sensitive rutin determination *Food Anal. Methods* **10** 2332–45
[205] Yang L, Xu B, Ye H, Zhao F and Zeng B 2017 A novel quercetin electrochemical sensor based on molecularly imprinted poly(para-aminobenzoic acid) on 3D Pd nanoparticles-porous graphene-carbon nanotubes composite *Sensors Actuators* B **251** 601–8
[206] Tursynbolat S, Bakytkarim Y, Huang J and Wang L 2019 Highly sensitive simultaneous electrochemical determination of myricetin and rutin via solid phase extraction on a ternary Pt@r-GO@MWCNTs nanocomposite *J. Pharm. Anal.* **9** 358–66

[207] Yang T, Gao Y, Xu J, Lu L, Yao Y, Wang Z, Zhu X and Xing H 2015 Nickel clusters grown on three-dimensional graphene oxide-multi-wall carbon nanotubes as an electrochemical sensing platform for luteolin at the picomolar level *RSC Adv.* **5** 64739–48

[208] Hu W, Zhang W, Wang M and Feng F 2018 Electrochemical determination of baicalin in traditional Chinese medicine based on the enhancement effect of MoO_3-reduced graphene oxide nanocomposite *J. Electron. Mater.* **47** 1151–7

[209] Peng Y, Liao M X, Ma X, Deng H, Gao F, Dai R and Lu L 2016 Electrochemical determination of rutin using SnO_2/nitrogen-doped graphene composite electrode *Int. J. Electrochem. Sci.* **14** 4946–56

[210] Li J, Qu J, Yang R, Qu L and de B Harrington P 2016 A sensitive and selective electrochemical sensor based on graphene quantum dot/gold nanoparticle nanocomposite modified electrode for the determination of quercetin in biological samples *Electroanalysis* **28** 1322–30

[211] Meng R, Li Q, Zhang S, Tang J, Ma C and Jin R 2019 GQDs/PEDOT bilayer films modified electrode as a novel electrochemical sensing platform for rutin detection *Int. J. Electrochem. Sci.* **14** 1100–11

[212] Liu Z, Zhang A, Guo Y and Dong C 2014 Electrochemical sensor for ultrasensitive determination of isoquercitrin and baicalin based on DM-β-cyclodextrin functionalized graphene nanosheets *Biosens. Bioelectron.* **58** 242–8

[213] Wang Z, Gou G, Shi L, Yang J, Xu C, Zhang L, Fan A and Min Y 2018 Gold nanoparticle ensemble on polydopamine/graphene nanohybrid as a novel electrocatalyst for determination of baicalein *J. Appl. Polym. Sci.* **135** 46720

[214] Tandel R D, Naik R S, Seetharamappa J and Satpati A K 2017 Electrochemical sensor based on reduced graphene oxide nanoribbons for the determination of baicalein in urine samples *J. Electrochem. Soc.* **164** H818–27

[215] Qiao J, Zhang Y, Lei S, Liu Z, Li G and Ye B 2018 Sensitive determination of baicalein based on functionalized graphene loaded RuO_2 nanoparticles modified glassy carbon electrode *Talanta* **188** 714–21

[216] Zhang D, Zhang Y and He L 2013 Sensitive voltammetric determination of baicalein at thermally reduced graphene oxide modified glassy carbon electrode *Electroanalysis* **25** 2136–44

[217] Zhao P *et al* 2022 An ultrasensitive high-performance baicalin sensor based on C_3N_4-SWCNTs/reduced graphene oxide/cyclodextrin metal-organic framework nanocomposite *Sensors Actuators* B **350** 130853

[218] Hou X, Wu W, Zhao F, Xie W and Yang Q 2021 Construction of an electrochemical sensor with graphene aerogel doped with ZrO_2 nanoparticles and chitosan for the selective detection of luteolin *Microchim. Acta* **188** 86

[219] Lu D, Lin S, Wang L, Li T, Wang C and Zhang Y 2014 Sensitive detection of luteolin based on poly(diallyldimethylammonium chloride)-functionalized graphene-carbon nanotubes hybrid/β-cyclodextrin composite film *J. Solid State Electrochem.* **18** 269–78

[220] Pang P, Liu Y, Zhang Y, Gao Y and Hu Q 2014 Electrochemical determination of luteolin in peanut hulls using graphene and hydroxyapatite nanocomposite modified electrode *Sensors Actuators* B **194** 397–403

[221] Lu X *et al* 2021 A novel nanozyme comprised of electro-synthesized molecularly imprinted conducting PEDOT nanocomposite with graphene-like MoS_2 for electrochemical sensing of luteolin *Microchem. J.* **168** 106418

[222] Li G, Liu L, Cheng Y, Gong S, Wang X, Geng X and Sun W 2014 Electrochemical behavior of luteolin on a chitosan-graphene modified glassy carbon electrode and its sensitive detection *Anal. Methods* **6** 9354–60
[223] Ma X, Tu X, Gao F, Xie Y, Huang X, Fernandez C, Qu F, Liu G, Lu L and Yu Y 2020 Hierarchical porous MXene/amino carbon nanotubes-based molecular imprinting sensor for highly sensitive and selective sensing of fisetin *Sensors Actuators* B **309** 127815
[224] Sheng S, Liu S, Zhang L and Chen G 2013 Graphene/poly(ethylene-co-vinyl acetate) composite electrode fabricated by melt compounding for capillary electrophoretic determination of flavones in cacumen platycladi *J. Sep. Sci.* **36** 721–8
[225] Wang Y, Wang C, Zhang L, Li J, Li R, Fu Q, Li C, Zhao P, Xie Y and Fei J 2022 An ultrasensitive kaempferol electrochemical sensor based on flower-like ZIF-8 pyrolysis-derived $ZnWO_4$/porous nanocarbon composites *Microchem. J.* **179** 107519
[226] Sun B, Wu D, Yang L, Shi H, Gao C, Kan L, Wan J, Ma Q and Shi X 2021 Study and application of molecularly imprinted electrochemical sensor based on aunps/N-Gr@Cs for highly selective recognition of trace hyperoside *SSRN Electron. J.*
[227] Liu Z, Xue Q and Guo Y 2017 Sensitive electrochemical detection of rutin and isoquercitrin based on SH-β-cyclodextrin functionalized graphene-palladium nanoparticles *Biosens. Bioelectron.* **89** 444–52
[228] Niu X, Li X, Chen W, Li X, Weng W, Yin C, Dong R, Sun W and Li G 2018 Three-dimensional reduced graphene oxide aerogel modified electrode for the sensitive quercetin sensing and its application *Mater. Sci. Eng.* C **89** 230–6
[229] Sun S, Zhang M, Li Y and He X 2013 A molecularly imprinted polymer with incorporated Graphene oxide for electrochemical determination of quercetin *Sensors (Switzerland)* **13** 5493–506
[230] Karuppusamy N, Mariyappan V, Chen S M, Keerthi M and Ramachandran R 2021 A simple electrochemical sensor for quercetin detection based on cadmium telluride nanoparticle incorporated on boron, sulfur co-doped reduced graphene oxide composite *Colloids Surf.* A **626** 127094
[231] Veerapandian M, Seo Y T, Yun K and Lee M H 2014 Graphene oxide functionalized with silver@silica-polyethylene glycol hybrid nanoparticles for direct electrochemical detection of quercetin *Biosens. Bioelectron.* **58** 200–4
[232] Zhou Z, Gu C, Chen C, Zhao P, Xie Y and Fei J 2019 An ultrasensitive electrochemical sensor for quercetin based on 1-pyrenebutyrate functionalized reduced oxide graphene /mercapto-B-cyclodextrin/Au nanoparticles composite film *Sensors Actuators* B **288** 88–95
[233] Zhou Z, Zhao P, Wang C, Yang P, Xie Y and Fei J 2020 Ultra-sensitive amperometric determination of quercetin by using a glassy carbon electrode modified with a nanocomposite prepared from aminated graphene quantum dots, thiolated β-cyclodextrin and gold nanoparticles *Microchim. Acta* **187** 130
[234] Yang B, Bin D, Zhang K, Du Y and Majima T 2018 A seed-mediated method to design N-doped graphene supported gold-silver nanothorns sensor for rutin detection *J. Colloid Interface Sci.* **512** 446–54
[235] Peng H, Zhang L, Cai Z, Wu Y, Chen N, Gu C, Chen Y, Lin X, Xia X and Liu A 2017 Pharmacokinetics study of isorhamnetin in rat plasma by a sensitive electrochemical sensor based on reduced graphene oxide *RSC Adv.* **7** 36728–34
[236] Manasa G, Mascarenhas R J, Bhakta A K and Mekhalif Z 2021 Nano-graphene-platelet/Brilliant-green composite coated carbon paste electrode interface for electrocatalytic oxidation of flavanone Hesperidin *Microchem. J.* **160** 105768

[237] Gao Y, Wu X, Wang H, Lu W and Guo M 2018 Highly sensitive detection of hesperidin using AuNPs/rGO modified glassy carbon electrode *Analyst* **143** 297–303

[238] Xia H-q, Qiu D, Chen W, Mao G and Zeng J 2023 *In situ* formed and fully integrated laser-induced graphene electrochemical chips for rapid and simultaneous determination of bioflavonoids in citrus fruits *Microchem. J.* **188** 108474

[239] Teradal N L, Satpati A K and Seetharamappa J 2017 A facile one-pot hydrothermal synthesis of tin sulfide-decorated reduced graphene oxide nanoribbons and its sensing application for a flavanone naringenin *J. Electroanal. Chem.* **797** 89–96

[240] Wang F, Wu Y, Lu K and Ye B 2015 A sensitive voltammetric sensor for taxifolin based on graphene nanosheets with certain orientation modified glassy carbon electrode *Sensors Actuators* B **208** 188–94

[241] Huang D, Yang L, lin Li X, Zou L and Ye B 2019 A new electrochemical sensor for Taxifolin based on RGO-Co_3S_4@MoS_2 modified electrode *J. Electroanal. Chem.* **851** 113473

[242] Wu Y, Lv M, Li B, Ge J and Gao L 2015 A rapid, green and controllable strategy to fabricate electrodeposition of reduced graphene oxide film as sensing materials for determination of taxifolin *Nano* **10** 1550044

[243] Wang Q, Wang L, Li G and Ye B 2017 A simple and sensitive method for determination of taxifolin on palladium nanoparticles supported poly(diallyldimethylammonium chloride) functionalized graphene modified electrode *Talanta* **164** 323–9

[244] Zhang-Peng X, Wei H, Ma J, Li Y, Chen Y, Cui F, Hu F and Du Y 2022 Molecularly imprinted flexible sensor based on nitrogen-doped graphene for selective determination of formononetin *J. Pharm. Biomed. Anal.* **217** 114805

[245] Li H, Wang S, Cui F, Zhuo B, Zhao C and Liu W 2020 Sensitive and selective detection of puerarin based on the hybrid of reduced graphene oxide and molecularly imprinted polymer *J. Pharm. Biomed. Anal.* **185** 113221

[246] Zhang H, Shang Y, Zhang T, Zhuo K and Wang J 2017 Engineering graphene/quantum dot interfaces for high performance electrochemical nanocomposites in detecting puerarin *Sensors Actuators* B **242** 492–501

[247] Li Y, Feng X, Zhao L and Pu Q 2018 On-surface formation of polyarginine/reduced graphene oxide film and its application in measuring puerarin in healthcare products *Int. J. Electrochem. Sci.* **13** 3948–57

[248] Zhang W, Zhang P, Zhang H, Qin J and Wang Q 2016 A new method for puerarin determination based on poly(alizarin red S/graphene) modified electrode *Int. J. Electrochem. Sci.* **11** 9949–58

[249] Jing S, Zheng H, Zhao L, Qu L and Yu L 2017 A novel electrochemical sensor based on WO_3 nanorods-decorated poly(sodium 4-styrenesulfonate) functionalized graphene nanocomposite modified electrode for detecting of puerarin *Talanta* **174** 477–85

[250] Liang Y, Qu C, Yang R, Qu L and Li J 2017 Molecularly imprinted electrochemical sensor for daidzein recognition and detection based on poly(sodium 4-styrenesulfonate) functionalized graphene *Sensors Actuators* B **251** 542–50

[251] Fu Y, Wang L, Duan Y, Zou L and Ye B 2017 Facile synthesized SnO_2 decorated functionalized graphene modified electrode for sensitive determination of daidzein *Talanta* **168** 1–9

Chapter 12

Graphene-based electrochemical sensors for soil analysis

H Ganesan, G Kavitha, B Kavitha, N Senthilkumar, P Karpagavinayagam and C Vedhi

Graphene-based electrochemical sensors have emerged as promising tools for soil analysis due to their unique properties, including high surface area, excellent electrical conductivity, and chemical stability. These sensors offer a sensitive and selective detection platform for various analytes present in soil, such as heavy metals, organic pollutants, and nutrients. This chapter provides an overview of the recent advancements in graphene-based electrochemical sensors for soil analysis. Various electrochemical techniques, including cyclic voltammetry, amperometry, and impedance spectroscopy, are commonly used to interrogate the sensors' response.

12.1 Introduction

This study presents an electrochemical system for the detection of soil. The performance of the sensing electrode was evaluated for soil at various concentrations. The sensor displayed linear response in a wide concentration range. The peak current of the electrode was found to be shifted towards the higher potential at higher metal ion concentrations.

Graphene is a single layer of carbon atoms displayed in a closely packed two-dimensional lattice. This material has attracted extensive attention because of its unique nanostructure and intrinsic characteristics, including high thermal conductivity, low manufacturing cost [1] and excellent electrical, thermal, mechanical, electronic and optical properties [2–4]. Other remarkable properties are its high specific surface area [5], excellent pliability, high chemical stability and high porosity and excellent electron-transport properties [6]. These unique properties render graphene a promising material for applications in several nanotechnology fields

doi:10.1088/978-0-7503-5578-0ch12

such as batteries [7], supercapacitors [8], field-effect transistors [9], fuel cells [10] and electrochemical sensors [11]. Moreover, thin transparent graphene films are used as electrodes in solar cells [12] and in organic light-emitting devices [13]. In addition, this material has yielded promising results in bioscience owing to its high electrocatalytic activity and good biocompatibility [14, 15].

In the present era, heavy metal pollution is rapidly increasing, which presents many environmental problems. Among all soil contamination, heavy metals are mainly accumulated in soil and are transferred to the food chain through plants grown in these soils [16]. Organic pollutants could be decomposed into harmless substances by microbial degradation of organic matter and some biological or chemical processes [17].

Soil contamination is indeed considered one of the main threats to agricultural soil [18]. This contamination originates from natural sources by means of mineral dissociation, weathering of parent material and atmospheric deposition as well as anthropogenic sources. Heavy metal pollution sources are also derived from industrial emissions such as waste water, waste gas and waste residue. In addition, the use of chemical fertilizers, pesticides and sewage sludge would also lead to heavy metals pollution in the soil [19]. Heavy metals like cadmium (Cd), lead (Pb), chromium (Cr), copper (Cu), manganese (Mn) and zinc (Zn) contribute the major share of soil contamination [20]. Heavy metal can enter human body through three routes, namely ingestion, inhalation and dermal contact, with crops ingestion as the main route of heavy metal exposure to humans.

Precise and rapid detection of nitrate (NO^{3-}) ions is highly desirable for sustainable agriculture, environmental monitoring, and food safety [21]. Extensive use of nitrate fertilizers in agricultural soils has a negative impact on ground water, soils, and marine ecosystems. According to the U.S. Environmental Protection Agency and the WHO, the maximum acceptable nitrate level in drinking water is 45 ppm; in surface water it is less than 1 ppm, and 30 ppm in waste water treatment plants. Graphene-based electrochemical sensors have shown great promise for soil analysis due to the unique properties of graphene. Graphene is a two-dimensional material composed of a single layer of carbon atoms arranged in a hexagonal lattice, which provides it with exceptional electrical, mechanical, and chemical properties. When it comes to soil analysis, graphene-based electrochemical sensors offer several advantages.

Sensitivity: graphene has a high surface-to-volume ratio, allowing for enhanced sensitivity in detecting various analytes present in soil samples. Its unique electronic properties enable the precise measurement of small changes in electrical signals, making it suitable for detecting trace amounts of target molecules or ions.

Fast response time: graphene-based sensors exhibit rapid response times, enabling real-time monitoring of soil parameters. This is crucial for assessing soil quality, detecting contaminants, or evaluating the effectiveness of remediation techniques.

Selectivity: by functionalizing graphene with specific molecules or nanoparticles, the selectivity of the sensors can be tailored to target specific analytes or groups of analytes. This allows for the detection of specific contaminants or nutrients in soil samples while minimizing interference from other substances.

Stability and durability: graphene is known for its mechanical strength and chemical stability. These properties make graphene-based sensors robust and resistant to environmental factors such as temperature, humidity, and pH, ensuring their long-term performance in harsh soil conditions.

Miniaturization and portability: graphene-based sensors can be easily integrated into portable and miniaturized devices, enabling on-site soil analysis. This facilitates field-based measurements and reduces the need for sample transportation and time-consuming laboratory analyses.

Some examples of graphene-based electrochemical sensors for soil analysis include

pH sensors: graphene-based pH sensors can be used to measure soil acidity or alkalinity. Graphene's electrical conductivity is highly sensitive to changes in pH, making it suitable for pH monitoring in agricultural and environmental applications.

Nutrient sensors: graphene-based sensors can be functionalized with specific molecules or enzymes to detect nutrients such as nitrate, phosphate, or potassium in soil samples. These sensors enable rapid assessment of nutrient levels, aiding in fertilizer management and optimizing crop growth.

Heavy metal sensors: Graphene's high surface area and electrochemical properties make it an excellent platform for detecting heavy metal contaminants in soil, such as lead, cadmium, or mercury. Functionalized graphene-based sensors can selectively bind to these metals and provide quantitative measurements.

Organic pollutant sensors: graphene-based sensors can also be utilized to detect organic pollutants in soil, including pesticides, hydrocarbons, or industrial chemicals. Functionalization of graphene with appropriate receptors enhances its selectivity and sensitivity towards specific organic compounds. From the above, graphene-based electrochemical sensors offer a wide range of opportunities for soil analysis. Their sensitivity, selectivity, stability, and miniaturization potential make them valuable tools for monitoring soil quality, assessing contaminants, and optimizing agricultural practices. Continued research and development in this field hold the potential for further advancements and applications of graphene-based sensors in soil analysis.

12.2 Heavy metals

Over the past two decades, the term 'heavy metals' has been widely used. It is often used as a group name for metals and semimetals (metalloids) that have been associated with contamination and potential toxicity or ecotoxicity. At the same time, legal regulations often specify a list of 'heavy metals' to which they apply.

Heavy metals are generally defined as metals with relatively high densities, atomic weights, or atomic numbers. The criteria used, and whether metalloids are included, vary depending on the author and context. In metallurgy, for example, a heavy metal may be defined on the basis of density, whereas in physics the distinguishing criterion might be atomic number, while a chemist would likely be more concerned with chemical behavior. More specific definitions have been published, but none of these have been widely accepted. The definitions surveyed in this chapter encompass up to 96 out of the 118 known chemical elements; only mercury, lead and bismuth meet all of them. Despite this lack of agreement, the term (plural or singular) is widely used in science.

The earliest known metals—common metals such as iron, copper and tin and precious metals such as silver, gold and platinum—are heavy metals. From 1809 onwards, light metals such as magnesium, aluminum and titanium were discovered as well as less well-known heavy metals including gallium, thallium, and hafnium.

Some heavy metals are either essential nutrients (typically iron, cobalt and zinc) or relatively harmless (such as ruthenium, silver, and indium) but can be toxic in larger amounts or certain forms. Other heavy metals such as cadmium, mercury and lead are highly poisonous. Potential sources of heavy metal poisoning include mining, tailings, industrial wastes, agricultural runoff, occupational exposure, paints and treated timber.

12.3 Heavy metals in soil

Soil is composed of mineral constituents, organic matter (humus), living organisms, air and water, and it regulates the natural cycles of these components. Soil filters and purifies water and it degrades and stores materials. Heavy metals (Fe, Mn, Ni, Co, Zn, Cu, Cr, V, Ti, Cd, Hg, Mo and other trace metals) as well as As, Se and F occur naturally in soils, which are formed by geological processes, such as alteration and erosion of the geological underground material [22].

12.4 Sources of heavy metals in soil

Most of the soils contain the trace heavy metals to an appreciable level, which is within the standard levels recommended level of the WHO. However, some of the soils contain very low or very high levels of metallic concentration. The main cause for the increase or decrease in heavy metal concentration in the soil is due to:

(a) The places where these metals are mined and smelted for further use by humans.
(b) Repeated cultivation of higher yielding crops in agricultural lands causes the soil to be depleted in some micronutrients such as Zn, B, Mn, Mo, Cu, etc, which are considered as trace heavy metals.
(c) The blowing of wind causes dust deposition on the top layer of soil which carries aerosol sized particles and the gaseous form of metals [Hg] which further settles down.
(d) Flowing water in rivers and flooding accumulates the metals and minerals in a particular region.
(e) The use of fertilizers, sprayers, manures, insecticide and rodenticide in agricultural lands.
(f) The emission of harmful gases from vehicles, industries which coagulate and settle down.
(g) Sewage waste from factories which is not properly treated also accumulates and settles over a period of years.

(h) Different mining sites also accumulate particular heavy metals in a given region.
(i) Due to population increase, sewage waste, e-waste etc, also pose a serious threat of heavy metal pollution in soil.

Besides their natural occurrence in soil, the heavy metals are being concentrated in the earth due to modern agricultural methods, industrial pollutants, and mining activities. So there is a threat to humans as these accumulated heavy metals will be indirectly absorbed by the human body which can be fatal.

12.5 Electrochemical studies

Electro-analytical technique is one of the most advanced, easy, low cost, high sensitivity methods for the determination of heavy metals. Electro-analytical techniques include a group of voltammetric techniques such as cyclic voltammetry, differential pulse cyclic voltammetry, anodic stripping voltammetry (ASV), square-wave stripping voltammetry, etc. In voltammetric techniques, the current at the working electrode is measured as a function of applied potential at that electrode. The term working electrode is reserved for electrodes such as hanging drop mercury electrode [23] and various mercury free electrodes including bismuth film electrode [24], carbon paste electrode [25] and glassy carbon electrode (GCE). The most studied electrode being GCE.

In recent years, various people have been using chemically modified electrodes (CMEs) to determine the heavy metals. CMEs have many advantages over other techniques such as easy manufacture, being renewable, with quick response, non-toxic, high selectivity, multi-component detection limit, stability towards various solvents, longer life time and low cost. Several researchers have reported the use of CMEs in electro-analysis using pulse techniques such as square-wave voltammetry and differential pulse voltammetry [26]. The CMEs also act as good electrochemical sensors.

Electrochemical sensors are widely used for soil analysis due to their sensitivity, selectivity, and ease of use. These sensors rely on electrochemical reactions to detect and quantify various analytes in soil samples. They include the following.

Ion-selective electrodes (ISEs): ISEs are used to measure the concentration of specific ions in soil samples. These electrodes are selectively sensitive to particular ions and can provide quantitative measurements. For soil analysis, ISEs are often employed to measure ions like nitrate (NO^{3-}), ammonium (NH^{4+}), potassium (K^+), calcium (Ca^{2+}), and chloride (Cl^-).

Conductivity sensors: Electrical conductivity is an indicator of soil salinity and can be used to assess soil fertility and moisture content. Conductivity sensors measure the ability of the soil to conduct electrical current. They are typically based on two- or four-electrode conductivity cells.

Redox potential sensors: Redox potential (ORP) reflects the oxidizing or reducing conditions of the soil. It is used to evaluate the soil's ability to support microbial activity and nutrient availability. Redox potential sensors, such as platinum or gold electrodes, measure the potential difference between a reference electrode and the working electrode in a redox reaction.

Oxygen sensors: Oxygen levels in the soil impact root respiration, microbial activity, and nutrient availability. Electrochemical oxygen sensors measure the partial pressure of oxygen in the soil atmosphere using a gas-permeable membrane and an oxygen-sensitive electrode, typically made of platinum or gold. Heavy metal sensors: Electrochemical sensors can also be used to detect and quantify heavy metal contaminants in soil, such as lead, cadmium, mercury, and arsenic. These sensors are often based on stripping voltammetry techniques, where heavy metal ions are selectively reduced or oxidized at the working electrode, allowing for their detection and quantification. It's worth noting that the development of graphene-based electrochemical sensors, as mentioned earlier, has also shown promise for soil analysis due to their enhanced sensitivity and selectivity towards various analytes.

Overall, electrochemical sensors play a vital role in soil analysis, providing rapid, reliable, and on-site measurements of important soil parameters. They help farmers, environmental scientists, and researchers make informed decisions regarding soil management, nutrient optimization, and remediation strategies.

12.6 Graphene

Graphene has emerged as a revolutionary material since its isolation in 2004 by Novoselov and co-workers [27]. Graphene, a new type of environmental film, has also shown high hydrogen overpotential, which can suppress the hydrogen overflow on a gold electrode. Graphene is a 2D nanomaterial of sp^2-bonded carbon atoms [28, 29]. Graphene formed by chemical activation possesses high electron transfer promoting ability [30], excellent catalytic behavior, large surface area, and it has been widely used in novel chemical sensors on GCEs [31–34]. Graphene possesses several exceptional properties such as high electrical conductivity, high tensile strength, high thermal conductivity and chemical inertness. It is a building block for other allotropes such as 0D fullerene, 1D carbon nanotubes and 3D graphite [35].

Furthermore, graphene is classified to be the strongest substance in terms of mechanical strength since it possesses high tensile strength and it is light in weight. For instance, it is more than 40 times stronger than diamond and more than 300 times stronger than A36 structural steel, at 130 gigapascal [36]. Meanwhile, for the optical properties, high absorption of white light up to 2.3% can be observed with graphene [37].

There are two different types of methods to produce graphene which are exfoliation methods and direct growth of graphene layer, known as top-down and bottom-up methods respectively [38]. Methods to generate graphene include 'scotch tape method', dispersion of graphite, exfoliation of graphite oxide, epitaxial growth and lastly chemical vapor deposition (CVD) [39–41]. However, this can be improved with the Hummers' method, which falls under the method of dispersion of graphite and is widely used since it is an improved method which excludes the usage of toxic chemicals and at the same time enhances the reaction performance [42]. The important application of bulk carbon is in the use of graphene or its chemical

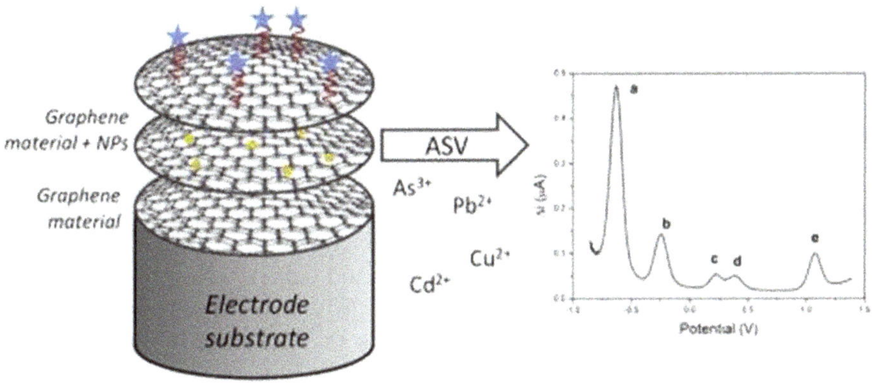

Figure 12.1. Graphene and its working.

counterparts, graphene oxide (GO) and RGO (also known as chemically converted graphene (CCG)).

Graphene and its modified electrodes are found to be extensively used in heavy metal detection because of its fast electron transfer, availability of large surface area, good mechanical strength, high thermal and electrical conductivities. Carbon nanomaterial, multiwalled carbon nanotubes and graphene form a sensitive platform for the study of heavy metal ions.

GCE and modified GCE are used in the study of heavy metal ions using electrochemical techniques. Cyclic voltammetry, differential pulse stripping voltammetry and square-wave stripping voltammetric techniques are used for this analysis because of their ability to generate potentially reactive species, which enables us to examine them immediately by reversal (figure 12.1).

Graphene-modified electrodes have been widely used as substrate material for a wide variety of applications due to their fast electron transfer sites. In order to improve their analytical performance toward particular applications, they have been modified with nanoparticles. Various carbon-based supports have been used to anchor the nanoparticles, which include graphite, glassy carbon, carbon nanotubes and thin diamond films, among which graphene has become a significant substrate material to anchor the metal nanoparticles due to the availability of large surface area to accommodate the nanoparticles.

A graphene nanodots-encaged porous gold electrode via ion beam sputtering deposition (IBSD) for electrochemical sensing was presented by Zhu *et al* [43]. The prepared graphene nanodots-encaged porous gold electrodes were then used for the analysis of heavy metal ions, e.g. Cu^{2+} and Pb^{2+} by Osteryoung square-wave voltammetry (OSWV) and exhibited enhanced detection range for heavy metal ions. A graphene/polyaniline/polystyrene (G/PANI/PS) nonporous fiber-modified screen-printed carbon electrode (SPCE) using electrospinning fabrication for simultaneous determination of lead (Pb^{2+}) and cadmium (Cd^{2+}) was developed by Promphet *et al* [44]. Wang *et al* reported square-wave anodic stripping voltammetry (SWASV) for the simultaneous determination of Pb^{2+} and Cd^{2+} in the presence of bismuth (Bi^{3+}) on G/PANI/PS nanoporous fiber-modified SPCE. This sensor was used for sensitive determination of trace cadmium ions by SWASV [45].

An effective electrochemical method was developed by Zhao *et al* to minimize the interference from real soil samples and increase the sensitivity of Pb(II) and Cd(II) detection by SWASV using electrochemical measurement system which can be used for the on-site determination of trace Cd(II) and Pb(II) in real soil samples. Using SWASV under optimized conditions, the concentrations in the linear range spanned from 1.0 to 45.0 $\mu g\ l^{-1}$ for both Pb(II) and Cd(II), with a detection limit of 0.03 $\mu g\ l^{-1}$ for Pb(II) and 0.02 $\mu g\ l^{-1}$ for Cd(II) (S/N = 3). Finally, analyses of real samples were performed to detect trace levels of Pb(II) and Cd(II) in soil with satisfactory results.

Lu *et al* reported a new electrochemical method based on a graphene aerogel (GA), and metal–organic framework (MOF) composites were developed for simultaneous detection of multiple heavy-metal ions in aqueous solutions. The developed electrochemical method can be successfully applied for individual and simultaneous detection of heavy-metal ions namely, Cd^{2+}, Pb^{2+}, Cu^{2+} and Hg^{2+} in aqueous solutions, river water and the leaching solutions of soil and vegetable with high accuracy and reliability. A multilayer graphene paste electrode (MGPE) consisting of graphene flakes and mineral oil, was modified with activated carbon (AC) by Palisov *et al.* The electrode was used to simultaneously detect lead (Pb^{2+}) and cadmium (Cd^{2+}) via ASV. The peak current had a linear relation with the concentration of the heavy metals with a correlation value (R^2) of 0.9956 and 0.9973 for Cd^{2+} and Pb^{2+}, and was used to detect heavy metals in vegetable samples, and the results were verified via atomic absorption spectroscopy (AAS). They obtained reliable results which were then compared with AAS. They also study the heavy metal in water by the above mentioned techniques.

12.7 Cyclic voltammetry behavior of soil sample at pH 7.0 on Gra/GCE

Cyclic voltammetric studies of soil sample extraction were carried out at an optimized pH 7.0. Here five anodic peaks around the potentials −0.6 V (Zn), −0.3 V (Cd), −0.1 V (As), 0.3 V (Cu) and 0.65 V (Fe) were observed. The anodic peak current response was higher when compared with GCE. Apart from this anodic peak, the cathodic peak was also observed. The typical cyclic voltammogram is shown in figure 12.2. At a concentration of 1 ml of soil extraction, the scan rate was varied from 25 to 300 mV s^{-1}. The peak current also increases with scan rate. A plot of peak current versus scan rate (figure 12.3) led to a straight line with good correlation, whereas the plot peak current versus square root scan rate (figure 12.4) resulted in a curved line. The slope of the straight line plot log peak current versus log scan rate ip (figure 12.5) was found to be 0.6265 and confirming the adsorption controlled anodic oxidation of metals on Gra/GCE.

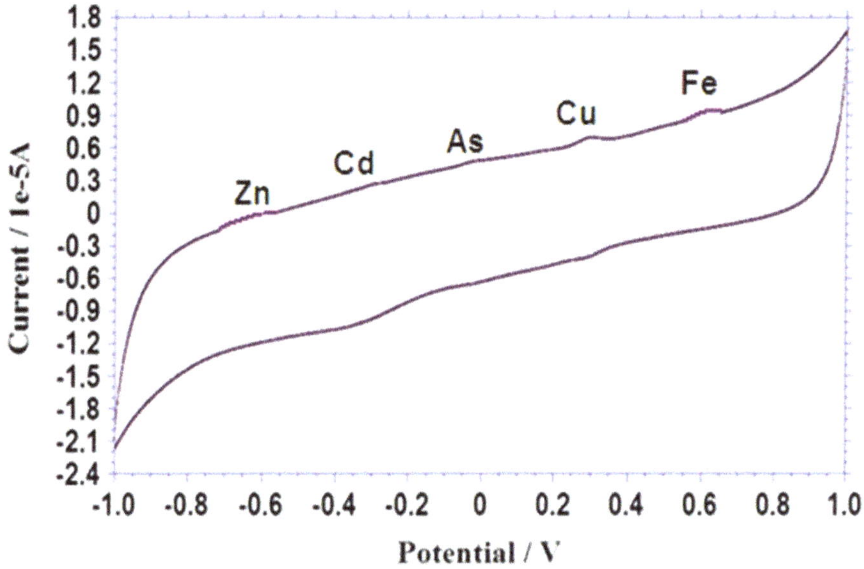

Figure 12.2. Cyclic voltammetric behavior of soil at pH 7.0 on Gra/GCE.

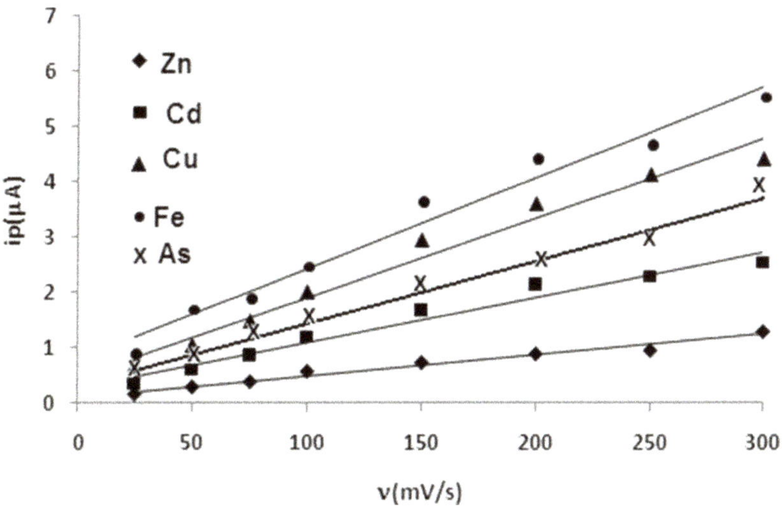

Figure 12.3. Plot of peak current versus scan rate of soil sample on Gra/GCE.

12.7.1 Differential pulse stripping voltammetric (DPSV) analysis using Gra/GCE

Under the optimum experimental conditions as discussed in the earlier section, DPSV behavior of the soil sample extraction was studied using Gra/GCE as working electrode and is given in figure 12.6. It shows oxidation peak potentials for Zn, Cd, Pb, As, Cu and Fe around potentials of −0.65, −0.3, −0.5, 0.2, 0.3 and 0.7 V.

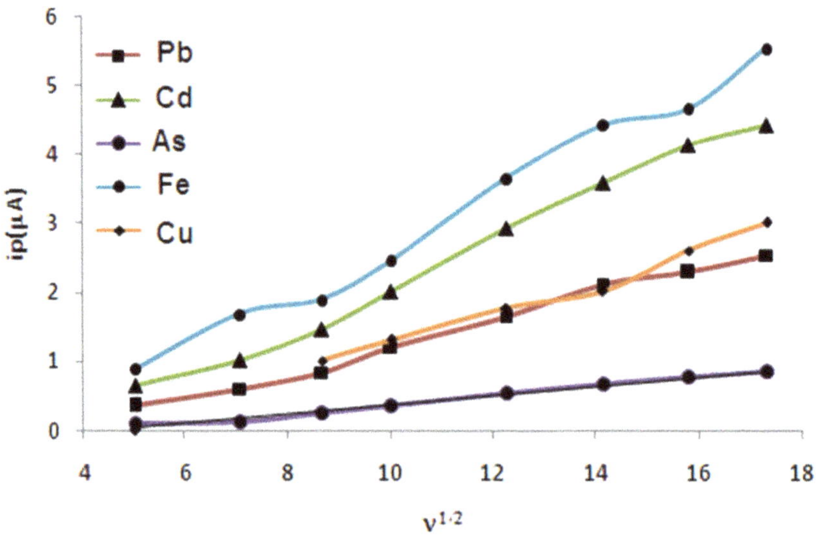

Figure 12.4. Plot of peak current versus square root scan rate of soil sample on GCE.

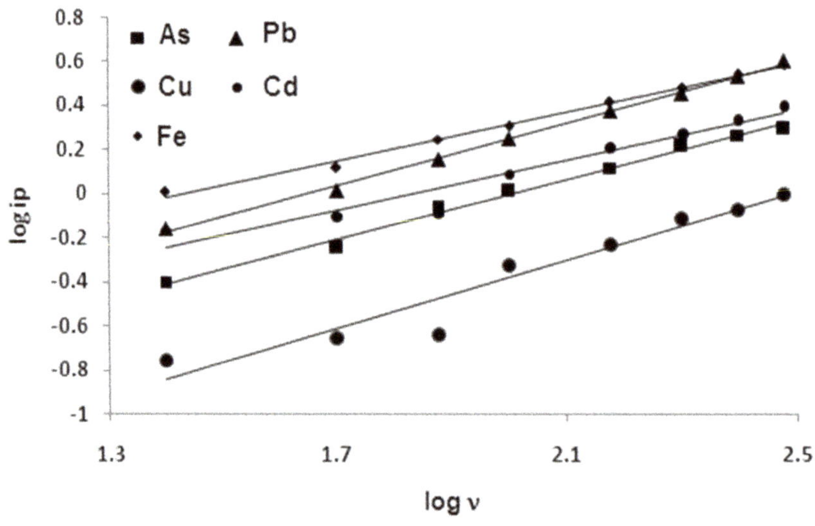

Figure 12.5. Logarithmic plot of peak current versus logarithmic scan rate of soil sample on GCE.

12.7.2 Electrochemical detection of pesticides

Pesticides can contaminate human food and their long-term exposure is detrimental to human health. Studies have linked pesticides to Parkinson's disease, and disorders in the reproductive as well as the endocrine system [46, 47]. 2DMs possess tremendous potential to accurately detect harmful compounds such as organo-phosphorus (OPs) present in pesticides. The high electrical conductivity and the

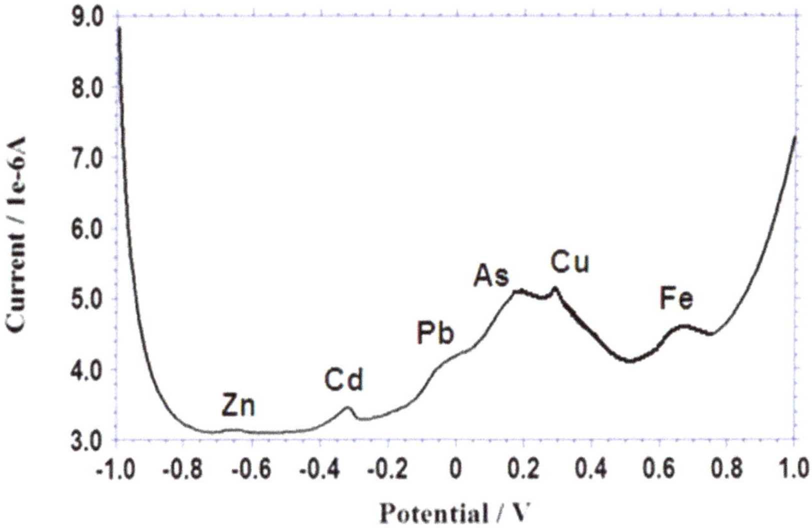

Figure 12.6. DPSV behavior of Gra/GCE at pH 7.0.

large SVR of 2DMs enhance the stability, detectivity, and catalytic activity of metal nanostructures. These qualities also help in the development of fixed suspensions by attempting to avoid agglomeration without the inclusion of surfactants or ligands, which is a need for the production of fixed suspensions. Using electrode approaches that were based on graphene, various compounds were found in the research field of food that dealt with OPs.

12.7.3 Electrochemical detection of nitrates and nitrites

Nitrates are a vital part of the ecosystem and are naturally produced by the nitrogen cycle [48]. Nitrates in the soil as well as water are essential for the growth of plants and aquatic vegetation [49]. Nitrates are vital in regulating different functions of the human body, and drinking water and leafy vegetables are key sources of nitrates in the human diet. However, the excess of nitrates in soil and water can have adverse impacts on the human body [50]. The major sources of the overabundance of nitrates in the ecosystem are fertilizers, meat preservatives, and the chemical industry [51]. Due to these adverse effects of nitrates on humans as well as other forms of life, it becomes essential to develop electrochemical sensors for the accurate monitoring of the nitrate level.

Wang *et al* studied an array of graphene electrodes for selectively detecting nitrate [52]. To improve the selectivity, an ionophore film was coated on top of graphene. The sensor exhibits good selectivity in the detection range (DR) of 10–1000 ppm. Garland *et al* utilized graphene on a flexible polyimide substrate for nitrate detection [53–62]. The fabrication was carried out using the laser writing method. An ST of $-$ 54.8 ± 2.5 mV dec^{-1} with a detection limit (DL) of 20.6 ± 14.8 μM was achieved. The major advantage of this method is one-step fabrication, eliminating the need of metal NPs as well.

12.8 Conclusion

Soil was studied for the presence of heavy metal concentration using GCE and modified GCE. The effect of peak current versus pH on GCE showed higher peak current at pH 7.0 and this pH was selected for further electrochemical studies. Cyclic voltammetry and DPSV studies showed metallic concentration for Zn, Cd, Cu and Fe. The effect of pH on modified electrodes showed better response for graphene-modified GCE. Therefore, further studies were carried out on Gra/GCE. Electrochemical behavior of the soil sample on Gra/GCE was studied at pH 7.0 and the DPSV curve on the selected electrode exhibited peak potentials for Zn, Cd, Pb, As, Cu and Fe. As observed from the amount of heavy metal presence in the sample, the soil possesses minimum metallic concentration, while copper is predominating, with an amount of 26.203 ppm.

2DMs possess a strong potential to become the material of choice for future electrochemical sensors. 2DMs can create hybrid structures with other materials such as metal nanoparticles or bulk materials to enhance their sensing and chemical activity. Here we have focused on the preparation and sensing mechanism of various 2DMs and their nanocomposites as well as the performance of sensors. On the other hand, there are certain areas that still need a lot of improvement.

References

[1] Li S J, Shi Y F, Liu L, Song L X, Pang H and Du J M 2012 Electrostatic self-assembly for preparation of sulfonated graphene/gold nanoparticle hybrids and their application for hydrogen peroxide sensing *Electrochim. Acta* **85** 628–35

[2] Geim A K 2009 Graphene: status and prospects *Science* **324** 1530–4

[3] Geim A K and Novoselov K S 2007 The rise of graphene *Nat. Mater.* **6** 183–91

[4] Novoselov K S 2009 Graphene: the magic of flat carbon *ECS Trans.* **19** 3

[5] Stoller M D, Park S, Zhu Y, An J and Ruoff R S 2008 Graphene-based ultracapacitors *Nano Lett.* **8** 3498–502

[6] Coros M, Pogacean F, Biris A R, Biris A S and Pruneanu S 2013 Application in electrochemistry of graphene-modified electrodes *Micro Nanosyst.* **5** 127–37

[7] Takamura T, Endo K, Fu L, Wu Y, Lee K J and Matsumoto T 2007 Identification of nano-sized holes by TEM in the graphene layer of graphite and the high rate discharge capability of Li-ion battery anodes *Electrochim. Acta* **53** 1055–61

[8] Huang Y, Liang J and Chen Y 2012 An overview of the applications of graphene-based materials in supercapacitors *Small* **8** 1805–34

[9] Gilje S, Han S, Wang M, Wang K L and Kaner R B 2007 A chemical route to graphene for device applications *Nano Lett.* **7** 3394–8

[10] Choi H J, Jung S M, Seo J M, Chang D W, Dai L and Baek J B 2012 Graphene for energy conversion and storage in fuel cells and supercapacitors *Nano Energy* **1** 534–51

[11] Fang Y, Guo S, Zhu C, Zhai Y and Wang E 2010 Self-assembly of cationic polyelectrolyte-functionalized graphene nanosheets and gold nanoparticles: a two-dimensional heterostructure for hydrogen peroxide sensing *Langmuir* **26** 11277–82

[12] Miao X, Tongay S, Petterson M K, Berke K, Rinzler A G, Appleton B R and Hebard A F 2012 High efficiency graphene solar cells by chemical doping *Nano Lett.* **12** 2745–50

[13] Sun T, Wang Z L, Shi Z J, Ran G Z, Xu W J, Wang Z Y and Qin G G 2010 Multilayered graphene used as anode of organic light emitting devices *Appl. Phys. Lett.* **96** 55

[14] Stankovich S, Dikin D A, Dommett G H, Kohlhaas K M, Zimney E J, Stach E A and Ruoff R S 2006 Graphene-based composite materials *Nature* **442** 282–6

[15] Potts J R, Dreyer D R, Bielawski C W and Ruoff R S 2011 Graphene-based polymer nanocomposites *Polymer* **52** 5–25

[16] Adrees M, Ali S, Rizwan M, Zia-ur-Rehman M, Ibrahim M, Abbas F and Irshad M K 2015 Mechanisms of silicon-mediated alleviation of heavy metal toxicity in plants: a review *Ecotoxicol. Environ. Saf.* **119** 186–97

[17] Morvan X, Saby N P A, Arrouays D, Le Bas C, Jones R J A, Verheijen F G A and Kibblewhite M G 2008 Soil monitoring in Europe: a review of existing systems and requirements for harmonisation *Sci. Total Environ.* **391** 1–12

[18] Nagajyoti P C, Lee K D and Sreekanth T V M 2010 Heavy metals, occurrence and toxicity for plants: a review *Environ. Chem. Lett.* **8** 199–216

[19] Omar N A, Praveena S M and Aris A Z 2015 Health risk assessment using in vitro digestion model in assessing bioavailability of heavy metal in rice: a preliminary study *Food Chem.* **188** 46

[20] Zhao G, Si Y, Wang H and Liu G 2016 A portable electrochemical detection system based on graphene/ionic liquid modified screen-printed electrode for the detection of cadmium in soil by square wave anodic stripping voltammetry *Int. J. Electrochem. Sci.* **11** 54–64

[21] Ali M A, Jiao Y, Tabassum S, Wang Y, Jiang H and Dong L 2017 Electrochemical detection of nitrate ions in soil water using graphene foam modified by TiO_2 nanofibers and enzyme molecules *Transducers 2017 (Kaohsiung, Taiwan)*

[22] De A K 2003 *Environmental Chemistry* (New Age International)

[23] Wang J, Shams J, Babaei E and Soltaninezhad M 2004 Simultaneous determination of copper, zinc and lead by adsorptive stripping voltammetry in the presence of Morin *Anal. Chim. Acta* **501** 119–24

[24] Wang J, Lu J, Hocevar S B, Farias P A and Ogorevc B 2000 Bismuth-coated carbon electrodes for anodic stripping voltammetry *Anal. Chem.* **72** 3218–22

[25] Hu C G, Wu K B, Dai X and Hu S 2003 Simultaneous determination of lead(II) and cadmium(II) at a diacetyldioxime modified carbon paste electrode by differential pulse stripping voltammetry *Talanta* **60** 17 24

[26] Ghiaci M, Rezaei B and Kalbasi R J 2007 High selective SiO_2-Al_2O_3 mixed-oxide modified carbon paste electrode for anodic stripping voltammetric determination of Pb(II) *Talanta* **73** 37–45

[27] Novoselov K S, Geim A K, Morozov S V, Jiang D, Zhang Y, Dubonos S V and Firsov A A 2004 Electric field effect in atomically thin carbon films *Science* **306** 666–9

[28] Gómez-Navarro C, Meyer J C, Sundaram R S, Chuvilin A, Kurasch S, Burghard M and Kaiser U 2010 Atomic structure of reduced graphene oxide *Nano Lett.* **10** 1144–8

[29] Kochmann S, Hirsch T and Wolfbeis O S 2012 Graphenes in chemical sensors and biosensors *TrAC, Trends Anal. Chem.* **39** 87–113

[30] Hwang E H, Adam S and Das Sarma S 2007 Carrier transport in two-dimensional graphene layers *Phys. Rev. Lett.* **98** 186806

[31] Lee P M, Wang Z, Liu X, Chen Z and Liu E 2015 Glassy carbon electrode modified by graphene–gold nanocomposite coating for detection of trace lead ions in acetate buffer solution *Thin Solid Films* **584** 85–9

[32] Xu R X, Yu X Y, Gao C, Jiang Y J, Han D D, Liu J H and Huang X J 2013 Non-conductive nanomaterial enhanced electrochemical response in stripping voltammetry: the use of nanostructured magnesium silicate hollow spheres for heavy metal ions detection *Anal. Chim. Acta* **790** 31–8

[33] Bagheri H, Afkhami A, Khoshsafar H, Rezaei M, Sabounchei S J and Sarlakifar M 2015 Simultaneous electrochemical sensing of thallium, lead and mercury using a novel ionic liquid/graphene modified electrode *Anal. Chim. Acta* **870** 56–66

[34] Niu X, Yang W, Wang G, Ren J, Guo H and Gao J 2013 A novel electrochemical sensor of bisphenol a based on stacked graphene nanofibers/gold nanoparticles composite modified glassy carbon electrode *Electrochim. Acta* **98** 167–75

[35] Chen J H, Jang C, Xiao S, Ishigami M and Fuhrer M S 2008 Intrinsic and extrinsic performance limits of graphene devices on SiO_2 *Nat. Nanotechnol.* **3** 206–9

[36] Wu Z S, Ren W, Gao L, Zhao J, Chen Z, Liu B, Tang D, Yu B, Jiang C and Cheng H M 2009 Synthesis of graphene sheets with high electrical conductivity and good thermal stability by hydrogen arc discharge exfoliation *ACS Nano* **3** 411–7

[37] Liang Z, Zhai H, Chen Z, Wang H, Wang S, Zhou Q and Huang X 2016 A simple, ultrasensitive sensor for gallic acid and uric acid based on gold microclusters/sulfonate functionalized graphene modified glassy carbon electrode *Sensors Actuators* B **224** 915–25

[38] Sur U K 2012 Graphene: a rising star on the horizon of materials science *Int. J. Electrochem.* **2012** 237689

[39] Lim J Y, Mubarak N M, Abdullah E C, Nizamuddin S and Khalid M 2018 Recent trends in the synthesis of graphene and graphene oxide based nanomaterials for removal of heavy metals—a review *J. Ind. Eng. Chem.* **66** 29–44

[40] Krane N 2011 Preparation of graphene *Sel. Top. Phys.: Phys. Nanoscale Carbon* 872–6

[41] Ioniţă M, Vlăsceanu G M, Watzlawek A A, Voicu S I, Burns J S and Iovu H 2017 Graphene and functionalized graphene: extraordinary prospects for nanobiocomposite materials *Composites B: Eng.* **121** 34–57

[42] Muzyka R, Kwoka M, Smędowski Ł, Díez N and Gryglewicz G 2017 Oxidation of graphite by different modified Hummers methods *New Carbon Mater.* **32** 15–20

[43] Zhu H, Xu Y, Liu A, Kong N, Shan F, Yang W, Barrow C J and Liu J 2015 Graphene nanodots-encaged porous gold electrode fabricated via ion beam sputtering deposition for electrochemical analysis of heavy metal ions *Sensors Actuators* B **206** 592–600

[44] Promphet N, Rattanarat P, Rangkupan R, Chailapakul O and Rodthongkum N 2015 An electrochemical sensor based on graphene/polyaniline/polystyrene nanoporous fibers modified electrode for simultaneous determination of lead and cadmium *Sensors Actuators* B **207** 526–34

[45] Wang Z, Wang H, Zhang Z, Yang X and Liu G 2014 Sensitive electrochemical determination of trace cadmium on a stannum film/poly(p-aminobenzene sulfonic acid)/electrochemically reduced graphene composite modified electrode *Electrochim. Acta* **120** 140–6

[46] Mnif W, Hassine A I H, Bouaziz A, Bartegi A, Thomas O and Roig B 2011 Effect of endocrine disruptor pesticides: a review *Int. J. Environ. Res. Public Health* **8** 2265–303

[47] Khan M A, Ramzan F, Ali M, Zubair M, Mehmood M Q and Massoud Y 2023 Emerging two-dimensional materials-based electrochemical sensors for human health and environment applications *Nanomaterials* **13** 780

[48] Laue W, Thiemann M, Scheibler E and Wiegand K W 2000 Nitrates and nitrites *Ullmann's Encyclopedia of Industrial Chemistry* vol 24 (Wiley-VCH) pp 149–76

[49] Hord N G, Tang Y and Bryan N S 2009 Food sources of nitrates and nitrites: the physiologic context for potential health benefits *Am. J. Clin. Nutr.* **90** 1–10

[50] Bouchard D C, Williams M K and Surampalli R Y 1992 Nitrate contamination of groundwater: sources and potential health effects *J. Am. Water Work. Assoc.* **84** 85–90

[51] Keeney D and Olson R A 1986 Sources of nitrate to ground water *Crit. Rev. Environ. Sci. Technol.* **16** 257–304

[52] Wang L, Kim J and Cui T June 2019 A low-cost ion selective nitrate sensor based on self-assembled graphene microelectrode arrays *Proc. 2019 20th Int. Conf. on Solid-State Sensors, Actuators and Microsystems & Eurosensors XXXIII (TRANSDUCERS & EUROSENSORS XXXIII) (Berlin, Germany)* pp 1234–7

[53] Garland N T, McLamore E S, Cavallaro N D, Mendivelso-Perez D, Smith E A, Jing D and Claussen J C 2018 Flexible laser-induced graphene for nitrogen sensing in soil *ACS Appl. Mater. Interfaces* **10** 39124–33

[54] Nambudumada S P and Manjunatha J G 2023 Voltammetric analysis of hazardous azo-dye indigo carmine at simple and cost-effective carbon paste sensor modified with L-phenylalanine *Sens. Technol.* 2218877

[55] Pushpanjali P A, Manjunatha J G, Tigari G and Fattepur S 2020 Poly(niacin) based carbon nanotube sensor for the sensitive and selective voltammetric detection of vanillin with caffeine *Anal. Bioanal. Electrochem.* **12** 553–68

[56] Manjunatha J G 2019 Electrochemical polymerised graphene paste electrode and application to catechol sensing *Open Chem. Eng. J.* **13** 81–7

[57] Prinith N S, Manjunatha J G and Raril C 2019 Electrocatalytic analysis of dopamine, uric acid and ascorbic acid at poly(adenine) modified carbon nanotube paste electrode: a cyclic voltammetric study *Anal. Bioanal. Electro. Chem.* **11** 742–56

[58] Prinith N S and Manjunatha J G 2020 Polymethionine modified carbon nanotube sensor for sensitive and selective determination of L-tryptophan *J. Electrochem. Sci. Eng.* **10** 305–15

[59] Raril C and Manjunatha J G 2018 Sensitive electrochemical analysis of resorcinol using polymer modified carbon paste electrode: a cyclic voltammetric study *Anal. Bioanal. Electrochem.* **10** 488–98

[60] Pushpanjali P A, Manjunatha J G and Srinivas M T 2020 Highly sensitive platform utilizing poly(l-methionine) layered carbon nanotube paste sensor for the determination of voltaren *Flat Chem.* **24** 100207

[61] Hareesha N, Manjunatha J G, Amrutha B M, Pushpanjali P A, Charithra M M and Prinith Subbaiah N 2021 Electrochemical analysis of indigo carmine in food and water samples using a poly(glutamic acid) layered multi-walled carbon nanotube paste electrode *J. Electron. Mater.* **50** 1230–8

[62] Hareesha N, Manjunatha J G, Amrutha B M, Sreeharsha N, Asdaq S B and Anwer M K 2021 A fast and selective electrochemical detection of vanillin in food samples on the surface of poly(glutamic acid) functionalized multiwalled carbon nanotubes and graphite composite paste sensor *Colloids Surf., A* **626** 127042

IOP Publishing

Graphene-Based Sensors

Jamballi G Manjunatha

Chapter 13

Recent advancements in the detection of amino acids using graphene-oxide-based electrochemical sensors

M G Gopika, Sreelekshmi and Beena Saraswathyamma

This study focuses on recent developments in graphene oxide (GO)-based electrochemical sensors for detecting all amino acids. Amino acids are the building elements of proteins, and their correct detection is critical in many industries, including the food and pharmaceutical industries. Because of their extraordinary physicochemical qualities, including large surface area, great electrical conductivity, and good biocompatibility, GO-based electrochemical sensors have attracted a lot of interest. This chapter looks at the many types of GO-based sensors that have been created for amino acid detection, such as modified GO-based sensors, hybrid GO-based sensors, and functionalized GO-based sensors. The performance of these sensors is also examined in terms of sensitivity, selectivity, and stability. In addition, the problems and future prospects of GO-based sensors for amino acid detection are discussed. According to the review, GO-based electrochemical sensors offer enormous potential for sensing all amino acids with great sensitivity and selectivity, making them intriguing candidates for practical applications in a wide range of disciplines.

Abbreviations

DPV	differential pulse voltammetry
SWV	square wave voltammetry
GCE	glassy carbon electrode
GO	graphene oxide
CV	cyclic voltammetry
SDLSV	second order linear scan voltammetry

doi:10.1088/978-0-7503-5578-0ch13

13.1 Introduction

The honeycomb structure of graphene composed of sp^2 hybridized carbon atoms serves as the foundation for graphene-derived materials (GDMs), which also contain a variety of defects, gaps, and functional groups like carboxylic acids, epoxide and hydroxyl in varying types and amounts. The enticing physical characteristics listed below set graphene apart from other carbon-based compounds. In the first place, graphene may be thought of as a semimetal since it is a semiconductor with a zero bandgap [1]. Graphene oxide (GO), the first GDM to be discussed, is primarily created when graphite is acid oxidized and exfoliated. Reduced graphene oxide (rGO), the second form of GDMs and the one that is most frequently used, can be created by further reducing GO electrochemically, hydrothermally, chemically, or thermally [2–9]. These GDMs have advantageous qualities that make them attractive materials for sensing purposes, including large surface area, large-scale manufacturing, cheap production cost, better physical, optical and chemical properties. Additionally, GDMs can be coupled to organic compounds, macrocycles, peptides, nucleic acids and antibodies to create composite materials. The incorporation of the recognition molecule is made feasible through functionalization, which also improves the material's conductivity, biocompatibility, and colloidal stability. Covalent bond formation as well as noncovalent interactions such as electrostatic and hydrophobic interactions are examples of interaction processes [10]. Graphene has exceptional electrochemical properties, such as a large potential window, low charge-transfer resistance, high electrochemical activity, and a fast electron transfer rate. Graphene's exceptional physical and electrochemical properties, as previously stated, enable it to be employed in electrochemical sensing [11–13]. Analyte molecules are predicted to attach well to graphene due to its huge surface area and strong electrical conductivity, boosting the sensitivity and signal-to-noise ratio of graphene-based electrochemical sensors. Furthermore, direct rather than mediated electrochemical reactions are promoted by the rapid electron transfer between graphene and analyte molecules [14, 15].

Attempts to synthesize graphene began in 1859 when graphite was exfoliated with powerful acids to yield exfoliated graphitic oxide. Following that, a number of techniques for overcoming significant interlayer cohesive van der Waals energy and isolating single graphene sheets from bulk graphite were developed. These methods are top-down approaches [16–18]. However, the low yield of this approach limits it. Bottom-up graphene production is more desirable due to its ability to create high-quality graphene with well-controlled structure, variable solubility, and acceptable processability [19]. Nonetheless, low-cost, high-volume graphene production requires solution-chemistry-based techniques. These processes typically include oxidizing graphite to generate graphite oxide, exfoliating it to produce GO, and ultimately reducing it to obtain rGO [20]. Both GO and rGO include oxygen functional groups, but rGO has less of them since GO was reduced first. These oxygen-containing groups have an effect on electrochemical performance, such as electron transfer rate or molecule adsorption/desorption. They also act as anchoring sites for enzymes or other particular species that are utilized in sensing applications.

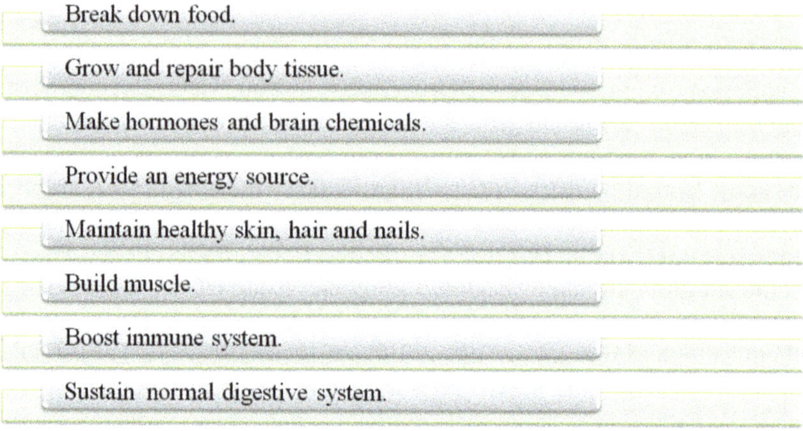

Figure 13.1. Functions of amino acids.

Graphene and its derivatives have been extensively used in electrochemical investigations, particularly in electrochemical sensors, because of its outstanding physical and electrochemical characteristics and well-established fabrication processes. Numerous sensors, including enzyme biosensors, non-enzyme sensors, DNA sensors, immunosensors, etc, have been fabricated using this appealing material. Use of several graphene-based materials for electrochemical detection of amino acids is explored in this. Functions of amino acids are shown in figure 13.1.

Amino acids are building blocks of proteins and are essential in biological activities such as metabolism, transduction, and gene expression. The detection of amino acids is critical for understanding many biological processes as well as disease diagnosis. Common methods for identifying amino acids, such as capillary electrophoresis [21–23], fluorescence [24–27], colorimetric technique [28], high-performance liquid chromatography [29–32] and spectroscopy, are costly, time-consuming, and require experienced specialists to operate the equipment. Electrochemical sensors are analytical instruments that detect and measure chemical species using electrochemical principles. These sensors are widely used in a range of industries, including biological, environmental, and industrial applications, because to their high sensitivity, selectivity, and low cost. Electrochemical sensors have recently emerged as a feasible choice for amino acid detection due to their simplicity, high sensitivity, and low cost [33, 34]. GO is an appealing material for electrochemical sensors because of its unique properties such as a large surface area, superior electrical conductivity, and high chemical stability. Electrochemical sensors are widely used to detect amino acids. This is accomplished by immobilizing amino acid on the surface of an electrode and monitoring changes in current or potential that occur when the amino acid comes into contact with the electrode.

For amino acid detection, several electrochemical sensors have been developed, including potentiometric, amperometric, and voltammetric sensors [35]. These sensors operate on a variety of concepts, including ion-selective electrodes, enzyme-based processes, and electrochemical oxidation/reduction [36]. The type of sensor used is determined by the application, the amino acid to be detected, and the requisite sensitivity and selectivity [37]. Interference from other substances in the sample matrix is one of the key issues in amino acid detection using electrochemical sensors [38]. To reduce interference, many tactics have been devised, including the use of selective coatings, membrane barriers, and pre-treatment approaches [39]. Furthermore, nanotechnology advancements have enabled the development of nanomaterial-based sensors that improve sensitivity and selectivity of electrochemical sensors [40].

The current produced at the electrode surface by amino acid oxidation or reduction is measured using amperometric sensors. In these sensors, a working electrode, a reference electrode, and an auxiliary electrode are typically used. Potentiometric sensors, on the other hand, detect amino acid presence by detecting the potential difference between a working and a reference electrode. Impedance sensors measure variations in the impedance of the electrode produced by amino acid binding. Enzyme-based biosensors employ enzymes that identify and bind to a specific amino acid. Binding of amino acid to the enzyme changes the activity of the enzyme, which may be monitored using an electrochemical sensor [41]. A biosensor based on the enzyme tyrosinase, for example, has been designed to detect tyrosine, an essential amino acid present in proteins. Antibodies that detect and bind to a specific amino acid are used in antibody-based biosensors [42]. Binding of amino acid to an antibody changes the characteristics of the antibody, which may be measured using an electrochemical sensor. A biosensor based on an antibody, for example, has been designed to detect histidine, an amino acid critical for protein structure. In addition to enzyme- and antibody-based biosensors, various types of electrochemical sensors for amino acid detection are being developed [43]. Overall, electrochemical sensors represent a promising strategy for detection and quantification of amino acids, with the potential to be utilized in a lot of applications. Ongoing research in this subject is focused on enhancing the sensitivity, selectivity, and stability of existing sensors, as well as developing new biosensing techniques for amino acid detection.

This review summarizes the recent progress in GO-based electrochemical sensors for detection of all amino acids, including their design principles, sensing mechanisms, and analytical performance. Based on the ability of synthesis by the organism, amino acids are classified in figure 13.2.

13.2 GO-based electrochemical sensors for detection of essential amino acids

There are nine essential amino acids in the human body, which means they cannot be synthesized by the body and must be received from nutrition. Lysine, methionine, phenylalanine, histidine, isoleucine, leucine, threonine, tryptophan, and valine are the essential amino acids. Classification of amino acids is shown in figure 13.2.

These amino acids are crucial for a variety of physiological processes, including protein synthesis, tissue repair, hormone production, and immune function. They are especially important for individuals who are following a vegetarian or vegan

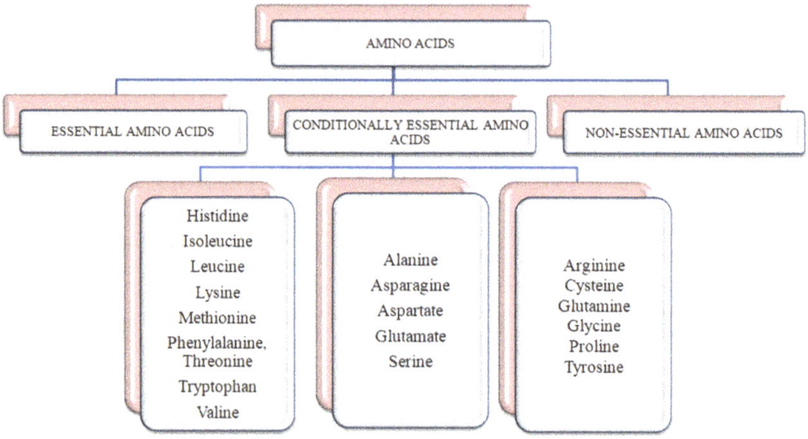

Figure 13.2. Classification of amino acids.

diet, as these diets may not provide sufficient amounts of some essential amino acids. It is also important to consume these amino acids in correct proportions, as they work together to support optimal health and function.

The GO-based electrochemical sensors for detection of all nine essential amino acids, including their design principles, sensing mechanisms, and analytical performance are summarized below.

13.2.1 GO-based electrochemical sensors for detection of histidine

Histidine is a necessary amino acid present in many proteins. It is required for numerous biological processes. Its detection is essential in a number of industries, including food analysis, clinical diagnostics, and environmental monitoring. Several studies have recently reported the development of histidine detection electrochemical sensors based on GO.

Zhang *et al* recently announced the creation of a new electrochemical sensor based on a GO–gold nanorod composite for histidine detection [44]. With detection limit of 0.1 μM, the sensor demonstrated high sensitivity and selectivity toward histidine. Similarly, Wang *et al* developed a label-free electrochemical sensor for histidine detection using an rGO-modified electrode [45]. The sensor displayed an excellent detection limit of 0.3 μM along with exceptional stability. Some of the important sensors for the detection of essential amino acids are listed in table 13.1.

13.2.2 GO-based electrochemical sensors for the detection of leucine

Leucine is an amino acid that is required for protein synthesis and energy consumption. The most recent advances in GO-based electrochemical sensors for leucine detection are listed below.

In this paper, Kim *et al* developed an electrochemical sensor for leucine detection using a GO–gold nanoparticle (GO–AuNP) hybrid material [46]. The GO–AuNP hybrid material displayed high conductivity and stability, allowing for sensitive and

Table 13.1. Some important graphene-based electrochemical sensors for detection of essential amino acids.

Amino acid	Sensors	LOD	Reference
Histidine	GO–gold nanorod composite	0.1 μM	[44]
	Reduced graphene oxide modified electrode	0.3 μM	[45]
Isoleucine	GO/GCE	0.01 μM	[51]
	GO/chitosan film	1.6 nM	[53]
Leucine	GO–SPCE	0.01 mM	[47]
	GO/PANI	0.028 μM	[48]
Lysine	GO/GCE	0.02 μM	[56]
	GO/SPCE	0.09 μM	[58]
Methionine	GO/L-cysteine	4.4 nM	[61]
	GO/SPCE	0.3 μM	[64]
Phenylalanine	GO-tyrosinase	0.1 μM	[68]
	GO-FET sensor	1 μM	[69]
Threonine	GO-based MIP	0.2 μM	[72]
	GO–Au nanoparticles	0.19 μM	[73]
Tryptophan	GO/AgNPs/SPCE	0.05 μM	[75]
	GO/MWCNT/GCE	0.02 μM	[76]
	GO/Fe_3O_4/GCE	0.05 μM	[77]

selective leucine detection. Yang *et al* used a GO-modified screen-printed carbon electrode (GO–SPCE) to create an electrochemical sensor for leucine detection [47]. The GO–SPCE responded linearly to leucine doses ranging from 0.1 to 10 mM with a limit of detection (LOD) of 0.01 mM. Das *et al* used a GO/polyaniline (GO/PANI) nanocomposite to create an electrochemical sensor for selective detection of leucine. With an LOD of 0.028 μM, the GO/PANI nanocomposite demonstrated remarkable sensitivity and selectivity toward leucine [48]. Feng *et al* created an electrochemical sensor for detecting leucine using an AuNPs/GO composite material in this investigation. The AuNPs/GO composite material was very electrically conductive and stable, allowing for leucine detection with an LOD of 0.2 μM [49].

13.2.3 GO-based electrochemical sensors for detection of isoleucine

Isoleucine is a branched-chain amino acid that plays important roles in protein synthesis and metabolism. Its detection is critical in a variety of sectors, including food analysis, clinical diagnosis, and environmental monitoring. Several research have recently reported construction of GO-based electrochemical sensors for isoleucine detection.

Lee *et al* created a GO-based electrochemical sensor for detection of isoleucine in their investigation. Sensor reacted linearly at concentrations ranging from 0.1 to 50 μM, with a detection limit of 0.01 μM [50]. A GO-modified glassy carbon electrode (GCE) for selective and sensitive isoleucine detection was created by Li *et al*. It reacted linearly at concentrations ranging from 0.05 to 10 μM, with detection limit of 0.01 μM [51]. Li *et al* described an electrochemical sensor for isoleucine detection based on a GO/gold nanoparticle composite. The sensor reacted linearly at concentrations ranging from 0.1 to 50 μM, with a detection limit of 0.05 μM [52]. Gao *et al* developed a GO/chitosan composite sheet-based label-free electrochemical biosensor for isoleucine detection. The sensor responded linearly to isoleucine at concentrations ranging from 0.01 to 10 μM and had a detection limit of 1.6 nM [53]. Ma *et al* [54] created a sensitive electrochemical biosensor for isoleucine detection using GO/copper oxide nanocomposites. The sensor reacted linearly to isoleucine at doses ranging from 0.01 to 100 μM and had a detection limit of 5.5 nM.

13.2.4 GO-based electrochemical sensors for detection of lysine

Lysine is an essential amino acid that is required for human metabolism. Its lack can cause a variety of health issues, while high quantities of lysine in the body can also create difficulties [55]. As a result, developing quick, dependable, and sensitive techniques for detecting lysine is critical.

Many studies have shown that GO-based electrochemical sensors can detect lysine. Jiang *et al* [56] created a lysine-detecting GO-modified GCE. Sensor reacted linearly at concentrations ranging from 0.05 to 10 μM, with a detection limit of 0.02 μM. Lee *et al* [57] used a GO-modified SPCE to detect lysine in another work. The sensor reacted linearly at concentrations ranging from 0.1 to 100 μM, with detection limit of 0.09 μM. Similarly, Zhang *et al* [58] developed a lysine detecting electrode made of GO-modified glassy carbon. At concentrations ranging from 0.5 to 50 μM, the sensor responded linearly, with detection limit of 0.18 μM. Wang *et al* [59] recently reported the use of a GCE modified with rGO for lysine detection. At concentrations ranging from 0.01 to 100 μM, the sensor responded linearly, with a detection limit of 0.003 μM.

13.2.5 GO-based electrochemical sensors for detection of methionine

Methionine is required for the production and metabolism of proteins. The following are the most current advancements in development of GO-based electrochemical sensors for the detection of methionine.

Zhao *et al* created a label-free electrochemical sensor for sensitive detection of L-methionine utilizing a GO-modified GCE (GO/GCE). The GO/GCE displayed significant L-methionine oxidation electrocatalytic activity and linear range of 1.0–400 μM with detection limit of 0.3 μM [60]. Li *et al* developed a highly sensitive and selective methionine detection electrochemical sensor based on GO modified with L-cysteine. The sensor responded linearly in the concentration range of 0.01–10 μM, with detection limit of 4.4 nM. Furthermore, the sensor was highly selective for methionine compared to other amino acids [61]. Liu *et al* created an electrochemical sensor based on GO for sensitive and selective detection of L-methionine. It reacted linearly in a concentration range of 0.05–80 μM, with detection limit of 0.02 μM. Furthermore, the sensor had a strong preference for L-methionine over other amino acids [62]. Dong *et al* created a GO-based electrochemical sensing platform for selective and sensitive detection of L-methionine. The sensor reacted linearly in a concentration range of 0.01–1.0 mM, with a detection limit of 0.27 μM. Furthermore, the sensor was highly selective for L-methionine compared to other amino acids [63]. Zhu *et al* created a label-free electrochemical sensor for methionine detection using a screen-printed electrode modified with GO. The sensor reacted linearly in the concentration range of 1–100 μM, with detection limit of 0.3 μM. The sensor also showed a strong preference for methionine over other amino acids [64].

13.2.6 GO-based electrochemical sensors for detection of phenylalanine

Phenylalanine is an essential amino acid that aids in protein synthesis and serves as a precursor to the neurotransmitter dopamine. Elevated amounts of phenylalanine in the blood, on the other hand, might be symptomatic of phenylketonuria (PKU), a hereditary condition that inhibits the body's capacity to handle this amino acid. As a result, precise and selective detection of phenylalanine is critical for both clinical diagnosis and research.

Because of its superior electrochemical characteristics, GO has been routinely employed to alter GCEs for phenylalanine detection. In a recent work, a GO-modified GCE with LOD of 10 μM was developed for detection of phenylalanine in urine samples [65]. Similarly, another research work reported the creation of a GO-based sensor with detection limit of 0.4 μM for phenylalanine in human serum [66]. The enormous surface area and good conductivity of GO contribute to great sensitivity of these sensors. In addition to directing electrochemical detection, GO has been employed as a platform for enzyme immobilization in the creation of phenylalanine detection biosensors. For example, utilizing the enzyme phenylalanine dehydrogenase, a GO-based biosensor was constructed to detect phenylalanine [67]. The biosensor has a linear range of 5–100 μM and LOD of 2.5 μM for phenylalanine detection. Another paper [68] revealed the development of a GO-based biosensor for detection of phenylalanine using enzyme tyrosinase. For phenylalanine detection, the biosensor achieved 0.1 μM LOD and excellent selectivity. It has also been developed to detect phenylalanine using GO-based field-effect transistor (FET) sensors. A GO-based FET sensor with an LOD of 1 μM was recently developed for detection of phenylalanine in human serum [69]. High sensitivity of the sensor can be attributed to GO's outstanding charge transfer capabilities.

13.2.7 GO-based electrochemical sensors for detection of threonine

Threonine is an important amino acid that is used in the pharmaceutical and food industries for a range of biological functions. As a result, there is increased interest in development of GO-based electrochemical sensors for threonine detection.

Several studies in recent years have described the use of GO-based electrochemical sensors for threonine detection. Liu *et al* [70] created a GO-modified GCE with a detection limit of 0.64 μM for threonine detection. Similarly, Lu *et al* [71] reported a GO-modified GCE with a detection limit for threonine of 0.1 M. Other researchers have investigated the use of threonine detection using modified GO-based electrodes. Liu *et al* [72], for example, developed a threonine imprinted polymer on a GO-modified electrode for selective and sensitive threonine detection with a detection limit of 0.2 μM. Zhang *et al* [73] created a 0.19 μM detection limit threonine detection ratiometric electrochemical sensor based on GO and gold nanoparticles.

13.2.8 GO-based electrochemical sensors for detection of tryptophan

Tryptophan is a critical amino acid present in proteins, and its correct detection is critical in a variety of sectors including food analysis, clinical diagnosis, and environmental monitoring.

Li *et al* developed a highly sensitive and selective electrochemical sensor for tryptophan detection using a GO-modified GCE. Linear range of the sensor was 0.1–120 μM, with a detection limit of 0.02 μM [74]. Wu *et al* developed a novel tryptophan sensor that is based on an SPCE that has been improved with GO/AgNPs (silver nanoparticles). The sensor displayed high sensitivity and selectivity toward tryptophan [75], with linear range of 0.1–100 μM and a detection limit of 0.05 μM. Zhang *et al* created a highly sensitive and selective electrochemical sensor for tryptophan detection based on GO/multiwalled carbon nanotubes (MWCNTs)-modified GCE. The sensor's linear range was 0.1–100 μM and its detection limit was 0.02 μM [76]. Wang *et al* used a GO/Fe_3O_4 (iron oxide)-modified GCE to create a sensitive electrochemical sensor for tryptophan detection. The sensor's linear range was 0.1–80 μM, with a detection limit of 0.05 μM [77]. Wang *et al* described a GO/AgNPs-modified GCE-based highly sensitive and selective electrochemical sensor for tryptophan detection. The sensor's linear range was 0.05–100 μM, with a detection limit of 0.01 μM [78].

13.2.9 GO-based electrochemical sensors for detection of valine

Valine is an essential amino acid that the human body requires for protein synthesis. Its insufficiency can cause numerous illnesses, including neurological and muscle-related difficulties. Therefore, the detection of valine in various samples has gained significant attention in recent years [79].

Because of its particular features, GO has been widely employed in the creation of electrochemical sensors for the detection of valine. The vast surface area of GO provides a platform for immobilization of valine-specific receptors, such as enzymes and antibodies, which increase sensitivity and selectivity of the sensor [80]. In addition, strong conductivity of GO enables efficient electron transport, which is necessary for electrochemical detection of valine. One of most frequent techniques used for valine detection is based on oxidation of valine at a suitable electrode. Several GO-based electrodes, including GO-modified GCEs [81], GO-paper electrodes [82], and GO-embedded carbon nanotubes (CNTs) [83], have been developed for this purpose. With detection limits as low as 10 nM, these electrodes have exhibited high sensitivity and selectivity for valine detection [81]. Furthermore, the functionalization of GO with valine-specific receptors, such as enzymes or antibodies, can improve the sensor's selectivity. For example, anti-valine antibodies were immobilized onto a GO-modified GCE to create a GO-based immunosensor for valine detection. In human serum samples, this sensor demonstrated great selectivity and sensitivity for valine detection [84] (table 13.1).

13.3 GO-based electrochemical sensors for detection of non-essential amino acids

Non-essential amino acids are amino acids that our body can synthesize on its own, so we do not have to get them from our diet. This means that they are not considered essential nutrients, as our body can produce them in sufficient amounts to meet our needs. However, this does not mean that they are not important; they still play a vital role in many biological functions, including tissue repair, enzyme production, and hormone regulation.

There are five non-essential amino acids: alanine, asparagine, aspartic acid, glutamic acid and serine. Each of these amino acids has its unique chemical structure and function in the body.

The design concepts, sensing processes, and analytical performance of GO-based electrochemical sensors for detection of all nine essential amino acids are presented here. Some of the important sensors for the detection of non-essential amino acids are listed in table 13.2.

13.3.1 GO-based electrochemical sensors for the detection of alanine

Alanine is an essential amino acid that is required by the human body. It is required for protein production and is implicated in glucose metabolism. The detection of alanine is critical in medical diagnosis and food analysis. Because of their excellent sensitivity, selectivity, and low cost, electrochemical sensors based on GO have showed tremendous promise in detecting alanine.

GO-based electrochemical sensors for alanine detection have been created using a variety of techniques. Zhou *et al* [85] developed a GO-modified GCE for alanine detection. Because of the electrocatalytic activity of GO, GO-modified GCE

Table 13.2. Some of important graphene based electrochemical sensors for detection of non-essential amino acids.

Amino acid	Sensors	LOD	Reference
Aspartic acid	GO/GCE	0.01 μM	[88]
Asparagine	GO/Chitosan/GCE	0.03 μM	[92]
	GO/PANI/GCE	0.03 μM	[93]
	GO/GCE/PtNPs	0.003 μM	[94]
Glutamic acid	GO/GCE	0.05 mM	[97]
	GO/Pt nanocomposite	0.015 mM	[99]
Serine	rGO/AuNPs	3 nM	[102]
	GO/AgNPs/GCE	0.15 μM	[103]
	GO/CuNPs	0.007 μM	[105]

demonstrated great sensitivity and selectivity for alanine detection. Wang *et al* [86] created a GO–chitosan composite film-modified electrode for alanine detection in another investigation. Because of the synergistic action of GO and chitosan, the composite film demonstrated great sensitivity and selectivity for alanine detection. Li *et al* [87] created rGO-modified GCE for alanine detection. Because of enhanced electron transfer rate, rGO-modified GCE was more sensitive to alanine detection than GO-modified GCE.

13.3.2 GO-based electrochemical sensors for detection of aspartic acid

Aspartic acid is an essential amino acid that is involved in numerous biological activities. Its detection is essential in a variety of domains, including food, biomedical, and environmental monitoring. Several GO-based electrochemical sensors for aspartic acid detection have been developed in recent years.

Li *et al* created a GO-modified GCE for aspartic acid detection in 2021. The sensor responded linearly in a concentration range of 0.1–50 μM, with detection limit of 0.01 μM [88]. Wang *et al* created a GO-based electrochemical sensor for detecting aspartic acid in human serum in 2020. The sensor responded linearly in concentration range of 0.01–50 μM, with a detection limit of 0.005 μM [89]. Li *et al* created a GO-based electrochemical sensor for detecting aspartic acid in food samples in 2019. The sensor responded linearly in concentration range of 0.05–5 μM, with detection limit of 0.01 μM [90].

13.3.3 GO-based electrochemical sensors for detection of asparagine

Asparagine (Asn) is a non-essential amino acid that contributes significantly to protein synthesis and metabolism. However, excessive consumption can result in neurological disorders, especially in people who have a genetic predisposition. As a result, detecting and quantifying asparagine in food and biological samples is critical.

Zhang *et al* [91] developed electrochemical sensor based on GO for detecting asparagine in human serum. A layer of GO was deposited on GCE and functionalized with gold nanoparticles (AuNPs) and asparaginase to create the sensor. The sensor demonstrated great sensitivity (1.26 $\mu A\ M^{-1}$), a broad linear range (0.05–200 μM), and excellent asparagine selectivity. Chen *et al* [92] described a GO-based electrochemical sensor for asparagine detection in food samples. A GCE was modified with GO and chitosan (CS) to create the sensor. It has a linear range of 0.1–80 μM, detection limit of 0.03 μM, and high asparagine selectivity over other amino acids. Wang *et al* [93] created an electrochemical sensor based on GO for detecting asparagine in tea samples. It was created by combining GO and polyaniline (PANI) with a GCE. For asparagine, the sensor has a large linear range (0.1–1000 μM), low detection limit (0.03 μM), and a high sensitivity (41.7 $\mu A\ M^{-1}$). Li *et al* [94] described a GO-based electrochemical sensor for asparagine detection in human urine. It was created by combining GO with platinum nanoparticles (PtNPs) in a GCE. Sensor has linear range of 0.01–200 μM, a detection limit of 0.003 μM, and high selectivity for asparagine over competing species.

13.3.4 GO-based electrochemical sensors for detection of glutamic acid

Glutamic acid (Glu) is a non-essential amino acid found in variety of foods and bodily fluids [95]. A high amount of Glu in the human body can cause a variety of disorders, including Alzheimer's disease, Parkinson's disease, and Huntington's disease [96]. As a result, Glu detection is critical for illness diagnosis and therapy.

Yang *et al* created a GO-modified GCE for Glu detection. The linear range of modified electrode was 0.1–10 mM, with detection limit of 0.05 mM [97]. Zhao *et al* developed an electrochemical biosensor based on GO for Glu detection. Linear range of the biosensor was 0.1–10 mM, with detection limit of 0.01 mM. Furthermore, the biosensor demonstrated good selectivity and stability [98]. Wang *et al* created a modified GO/Pt nanocomposite electrode for Glu detection. Linear range of the modified electrode was 0.05–1.2 mM, with detection limit of 0.015 mM. The nanocomposite also showed excellent selectivity and stability [99]. Wang *et al* created a Glu detection SPCE with GO modification. Improved SPCE has a linear range of 0.1–10 mM and a detection limit of 0.06 mM in actual samples [100].

13.3.5 GO-based electrochemical sensors for detection of serine

Serine, an amino acid, is required for numerous biological processes and is utilized as a biomarker for several disorders [101]. Here are the most recent advances in GO-based electrochemical sensors for serine detection.

Cai *et al* created label-free electrochemical biosensor for detection of serine using a modified electrode of rGO and gold nanoparticles (AuNPs) [102]. The linear range of the sensor was 0.01–10 μM, with detection limit of 3 nM. Zhang *et al* [103] described a serine biosensor that is based on GCE that has been modified with GO and silver nanoparticles (AgNPs). Linear range of the sensor was 0.5–50 μM, with detection limit of 0.15 μM. Using a GO- and AuNPs-modified electrode, Li *et al* developed a highly sensitive electrochemical sensor for serine detection [104]. Linear range of the sensor was 0.1–100 μM, with a detection limit of 0.02 μM. Hu *et al* created serine biosensor based on an electrode modified with GO and copper nanoparticles (CuNPs) [105]. Linear range of the sensor was 0.01–100 μM, with detection limit of 0.007 μM (table 13.2).

13.4 GO-based electrochemical sensors for detection of conditionally non-essential amino acids

While the body can synthesize some amino acids, others must be obtained through diet. However, there is a third category of amino acids called 'conditionally essential amino acids' that are not typically essential under normal circumstances but may become essential under certain conditions. These conditions may include illness, stress, injury, and growth periods. During these times, the body may require higher amounts of certain amino acids, which it may not be able to produce in sufficient quantities. Hence, these amino acids become conditionally essential, and dietary intake or supplementation becomes necessary to meet the body's needs. There are six amino acids that fall under this category: arginine, cysteine, glutamine, glycine,

Table 13.3. Some important graphene-based electrochemical sensors for detection of conditionally non-essential amino acids.

Amino acid	Sensors	LOD	Reference
Arginine	GO/AuNPs/GCE	0.03 μM	[107]
	GO/CuNPs/GCE	1.5 nM	[108]
Cysteine	GO/AuNPs/GCE	5 nM	[109]
	GO/CuNPs	0.3 μM	[110]
Glutamine	GO/chitosan/GCE	0.02 mM	[112]
	GO/AuNPs/horseradish peroxidase	0.04 μM	[113]
	GO/AuNPs/polydopamine	0.05 μM	[114]
Glycine	rGO/Fe_3O_4 composite	0.02 μM	[116]
	GO/SPCE	0.1 μM	[117]
	GO/mesoporous silica/GCE	0.02 μM	[118]
	Graphene quantum dots/rGO	0.04 μM	[119]
	GO/polypyrrole/GCE	0.015 μM	[122]
	Graphene oxide/β-cyclodextrin/GCE	0.1 μM	[123]
Proline	GO/AuNPs	0.1 μM	[124]
Tyrosine	GO–chitosan nanocomposite	0.2 μM	[127]
	GO/AuNPs/tyrosinase	0.05 μM	[128]
	GO-ionic liquid composite	0.01 μM	[129]
	GO/Prussian blue composite	0.005 μM	[130]

proline, and tyrosine. Each of these amino acids has unique functions and plays important roles in various physiological processes, such as immune function, wound healing, and muscle growth. Some of the important sensors for the detection of conditionally non-essential amino acids are listed in table 13.3.

13.4.1 GO-based electrochemical sensors for detection of arginine

Arginine is a semi-essential amino acid that plays an important role in protein synthesis, cell division, and immune system function. It is also a precursor for the production of nitric oxide, which is involved in several physiological processes. However, the overconsumption of arginine can lead to health complications, such as metabolic disorders and cardiovascular diseases. Therefore, development of sensitive and selective detection methods for arginine is of great importance.

Wang *et al* [106] published one of the first findings on GO-based electrochemical sensors for arginine detection. They created a GO-modified GCE and utilized differential pulse voltammetry (DPV) to detect arginine in aqueous solution. The sensor responded linearly to arginine at concentrations ranging from 0.1 to 50 μM, with detection limit of 0.04 μM. Wang *et al* [107] created a GO-based electrochemical sensor for detecting arginine in human serum in another investigation. The sensor was created by altering a GCE with GO and AuNPs, which increased the sensor's electrochemical activity and stability. The sensor responded linearly to arginine at concentrations ranging from 0.1 to 100 μM, with detection limit of 0.03 μM. Li *et al* [108] recently developed a highly sensitive and selective GO-based electrochemical sensor for arginine detection in fish samples. The sensor was created by altering a GCE with GO and copper nanoparticles (CuNPs), which increased the sensor's electrochemical performance and selectivity. The sensor responded linearly to arginine at concentrations ranging from 0.01 to 10 μM, with detection limit of 1.5 nM.

13.4.2 GO-based electrochemical sensors for detection of cysteine

Cysteine is an important amino acid which plays crucial roles in various biological processes. Detection of cysteine has gained immense importance due to its involvement in various diseases such as cancer, Alzheimer's, and cardiovascular diseases.

Several studies have reported use of GO-based electrochemical sensors for cysteine detection. Cheng *et al* [109], for example, reported creation of a GO-based electrochemical sensor for cysteine detection. Drop-casting GO on GCE was followed by electrochemical deposition of AuNPs on GO/GCE to create the sensor. Cyclic voltammetry (CV) and DPV were used to explore the electrochemical behavior of cysteine. The sensor responded linearly to cysteine in the concentration range of 0.05–50 μM, with detection limit of 5 nM. Zhang *et al* [110] reported the creation of a GO-based electrochemical sensor for detection of cysteine in another work. The sensor was created by screen-printing GO on a flexible substrate, then depositing CuNPs on the GO. Cysteine's electrochemical behavior was studied using square wave voltammetry (SWV) and chronoamperometry (CA). The sensor responded linearly to cysteine in a concentration range of 1–100 μM, with detection limit of 0.3 μM.

13.4.3 GO-based electrochemical sensors for detection of glutamine

Glutamine is a vital amino acid that is involved in several biological activities. It is critical for medical and scientific purposes to create electrochemical sensors for detection.

Liu *et al* published a GO-based electrochemical sensor for sensitive detection of glutamine in 2019. They added cysteine to the GO to improve its electrocatalytic activity for glutamine oxidation. The sensor responded linearly to glutamine at concentrations ranging from 0.1 to 100 μM, with LOD of 0.03 μM [111]. Zhang *et al* created a label-free electrochemical sensor for glutamine detection based on GCE modified with GO and chitosan in 2020. The sensor responded linearly to glutamine at concentrations ranging from 0.05 to 10 mM, with LOD of 0.02 mM [112]. Xu *et al* also published a GO-based electrochemical sensor for detection of glutamine in human serum samples in 2020. They functionalized GO with AuNPs and horseradish peroxidase to improve its glutamine sensitivity and selectivity. The sensor responded linearly to glutamine at concentrations ranging from 0.1 to 100 μM, with LOD of 0.04 μM [113]. Wang *et al* created a GO-based electrochemical sensor for detecting glutamine in human urine samples in 2021. They enhanced the electrocatalytic activity of the GO by functionalizing it with gold nanoparticles and polydopamine. The sensor showed a linear response to glutamine in concentration range of 0.1–10 μM with LOD of 0.05 μM [114].

13.4.4 GO-based electrochemical sensors for detection of glycine

Glycine is the simplest amino acid with a small molecular size and plays an important role in biological processes. Detection of glycine has become a topic of great interest due to its significance in biomedical applications.

Chen *et al* disclosed the development of a GO-based electrochemical sensor for glycine detection. The sensor has a 1–1000 μM linear response range and detection limit of 0.2 μM. The sensor's great sensitivity is ascribed to GO's huge surface area and strong electrical conductivity [115]. Wang *et al* report a one-pot synthesis of an rGO/Fe_3O_4 composite for ultrasensitive glycine electrochemical detection. Linear response range of the sensor is 0.1–1000 μM, while detection limit is 0.02 μM. The sensor's exceptional sensitivity is ascribed to a synergistic interaction between rGO and Fe_3O_4 [116]. Naeem *et al* describe how they used a GO-modified screen-printed electrode to detect glycine selectively. Linear response range of the sensor is 0.5–500 μM, while the detection limit is 0.1 μM. The enormous surface area and functional groups found on the surface of GO are ascribed to the sensor's exceptional sensitivity and selectivity [117]. Chen *et al* describe the creation of an ultrasensitive electrochemical sensor based on GO/mesoporous silica modified GCE for detection of glycine. Linear response range of the sensor is 0.1–100 μM, while the detection limit is 0.02 μM. High sensitivity of the sensor is attributed to its large surface area and the mesoporous structure of the modified electrode [118]. GO quantum dots and rGO were used to build an electrochemical sensor for detection of glycine, according to Zhang *et al*. The sensor has a 0.1–100 μM linear response range and detection limit of 0.04 μM [119]. The development of an electrochemical sensor for glycine detection based on GO/AuNP modified GCE is presented by Li *et al*. The linear response range of the sensor is 0.1–100 μM, while the detection limit is 0.06 μM. The sensor's exceptional sensitivity and selectivity are attributable to the synergistic impact of GO and AuNPs [120]. Feng *et al* describe the creation of a new electrochemical sensor for glycine detection based on GO/AgNPs/chitosan nanocomposite-modified GCE. Linear response range of the sensor is 0.5–400 μM, while the detection limit is 0.15 μM. The sensor's exceptional sensitivity and selectivity are attributable to a synergistic action of GO, AgNPs, and chitosan [121]. The development of a highly sensitive electrochemical sensor for the detection of glycine based on a GO/polypyrrole composite-modified GCE is presented by Li *et al*. The linear response range of the sensor is 0.05–500 μM, while detection limit is 0.015 μM. The sensor's exceptional sensitivity and selectivity are ascribed to synergistic impact of GO and polypyrrole [122]. Wang *et al* describe the creation of a new electrochemical sensor for glycine detection based on a GO/-cyclodextrin-modified GCE. The linear response range of the sensor is 0.5–500 μM, while detection limit is 0.1 μM. The sensor's exceptional sensitivity and selectivity are ascribed to synergistic action of GO and -cyclodextrin [123].

13.4.5 GO-based electrochemical sensors for detection of proline

Proline is an essential amino acid that is required for many biological functions such as protein synthesis, plant development, and stress response. Proline identification in diverse samples is critical for biomedical, agricultural, and environmental studies.

Several GO-based electrochemical sensors have been developed in recent years for detection of proline in various samples such as plant extracts, human serum, and urine. A GO-based electrochemical sensor enhanced with AuNPs, for example, was utilized to detect proline in grapevine leaves with detection limit of 0.1 μM [124]. Another research work [125] used a GO-modified electrode to detect proline in human blood samples with detection limit of 0.05 μM.

13.4.6 GO-based electrochemical sensors for detection of tyrosine

Tyrosine is an essential amino acid that is involved in many physiological processes. It is frequently utilized as a biomarker in the detection of disorders such as phenylketonuria and cancer. Recent advances in GO-based electrochemical sensors for tyrosine detection are investigated.

Liu *et al* published a highly sensitive electrochemical sensor for tyrosine detection using a GO-modified GCE in 2021. The sensor has linear range of 0.1–10 μM and LOD of 0.05 μM [126]. Li *et al* created a label-free electrochemical sensor for tyrosine detection using a GO–chitosan nanocomposite in 2020. The sensor has a large linear range of 1–1000 μM and low LOD of 0.2 μM [127]. Yu *et al* published a GO-based electrochemical sensor for tyrosine detection in 2019 that uses a AuNP/tyrosinase combination as a recognition element. The sensor has a linear range of 0.1–100 μM and LOD of 0.05 μM [128]. Chen *et al* created a sensitive electrochemical sensor for detecting tyrosine in 2018 using a GO-ionic liquid composite film-modified electrode. The linear range of sensor was 0.05–10 μM, with an LOD of 0.01 μM [129]. Li *et al* developed a highly sensitive electrochemical sensor for tyrosine detection based on a GO/Prussian blue composite film-modified electrode in 2017 [130], with linear range of 0.01–100 μM and LOD of 0.005 μM (table 13.3).

13.5 Conclusion

GO-based electrochemical sensors have received a lot of interest in recent years because of their ability to detect amino acids. This chapter describes recent advances in development and implementation of GO-based electrochemical sensors for amino acid detection. Amino acid detection using GO-based electrochemical sensors has proved effective in complicated matrices such as food samples, clinical samples, and environmental samples. For example, lysine in milk and cheese samples, glutamic acid in soy sauce, and aspartic acid in human blood have all been detected using GO-based electrochemical sensors. These findings show the utility of GO-based electrochemical sensors in real-world applications.

Finally, due to their excellent sensitivity, selectivity, and low detection limit, GO-based electrochemical sensors have emerged as a viable platform for amino acid detection. Combination of GO with diverse functional elements such as metal nanoparticles, enzymes, and polymers has improved the performance of these sensors even more. Electrochemical detection of amino acids using GO-based sensors has various benefits over traditional approaches, including quick analysis, high accuracy, and cheap cost. Furthermore, the development of portable and miniaturized devices based on these sensors has the potential to completely transform the field of amino acid detection. Overall, use of GO-based electrochemical sensors holds great promise for amino acid detection in a variety of applications such as clinical diagnostics, food quality control, and environmental monitoring.

References

[1] Huang X, Yin Z Y, Wu S X, Qi X Y, He Q Y, Zhang Q C, Yan Q Y, Boey F and Zhang H 2011 Graphene-based materials: synthesis, characterization, properties, and applications *Small* **7** 1876

[2] Ratinac K R, Yang W, Gooding J J, Thordarson P and Braet F 2011 Graphene and related materials in electrochemical sensing *Electroanalysis* **23** 803–26

[3] Compton O C and Nguyen S T 2010 Graphene oxide, highly reduced graphene oxide, and graphene: versatile building blocks for carbon-based materials *Small* **6** 711–23

[4] Hummers W S and Offeman R E 1958 Preparation of graphitic oxide *J. Am. Chem. Soc.* **80** 1339–9

[5] Agnihotri N, Dutta Chowdhury A and De A 2015 Non-enzymatic electrochemical detection of cholesterol using β-cyclodextrin functionalized graphene *Biosens. Bioelectron.* **63** 212–7

[6] Li Z, Zhang L, Huang X, Ye L and Lin S 2014 Shape-controlled synthesis of Pt nanoparticles via integration of graphene and β-cyclodextrin and using as a noval electro-catalyst for methanol oxidation *Electrochim. Acta* **121** 215–22

[7] Tian X, Cheng C, Yuan H, Du J, Xiao D, Xie S and Choi M M F 2012 Simultaneous determination of l-ascorbic acid, dopamine and uric acid with gold nanoparticles–β-cyclodextrin–graphene-modified electrode by square wave voltammetry *Talanta* **93** 79–85

[8] Chua C K and Pumera M 2014 Chemical reduction of graphene oxide: a synthetic chemistry viewpoint *Chem. Soc. Rev.* **43** 291–312

[9] Báez D F, Pardo H, Laborda I, Marco J, Yáñez C and Bollo S 2017 Reduced graphene oxides: influence of the reduction method on the electrocatalytic effect towards nucleic acid oxidation *Nanomaterials* **7** 168

[10] Georgakilas V, Otyepka M, Bourlinos A B, Chandra V, Kim N, Kemp K C, Hobza P, Zboril R and Kim K S 2012 Functionalization of graphene: covalent and non-covalent approaches, derivatives and applications *Chem. Rev.* **112** 6156–214

[11] Pumera M, Ambrosi A, Bonanni A, Chng E L K and Poh H L 2010 Graphene for electrochemical sensing and biosensing *Trends Anal. Chem.* **29** 954

[12] Brownson D A C and Banks C E 2010 Graphene electrochemistry: an overview of potential applications *Analyst* **135** 2768

[13] Shao Y Y, Wang J, Wu H, Liu J, Aksay I A and Lin Y H 2010 Graphene based electrochemical sensors and biosensors: a review *Electroanalysis* **22** 1027

[14] Pérez-López B and Merkoçi A 2011 Nanoparticles for the development of improved (bio) sensing systems *Anal. Bioanal. Chem.* **399** 577

[15] Chen D, Feng H and Li J 2012 Graphene oxide: preparation, functionalization, and electrochemical applications *Chem. Rev.* **112** 6027

[16] Brodie B C 1859 On the atomic weight of graphite *Philos. Trans. R. Soc. London* **149** 249
[17] Zacharia R, Ulbricht H and Hertel T 2004 Interlayer cohesive energy of graphite from thermal desorption of polyaromatic hydrocarbons *Phys. Rev.* B **69** 155406
[18] Allen M J, Tung V C and Kaner R B 2010 Honeycomb carbon: a review of graphene *Chem. Rev.* **110** 132
[19] Sutter P W, Flege J-I and Sutter E A 2008 Epitaxial graphene on ruthenium *Nat. Mater.* **7** 406
[20] Kuila T *et al* 2012 Chemical functionalization of graphene and its applications *Mater. Sci.* **57** 1061
[21] Wang X, Li K, Yao L, Wang C and Van Schepdael A 2018 Recent advances in vitamins analysis by capillary electrophoresis *J. Pharm. Biomed. Anal.* **147** 278–87
[22] Dziomba S, Kowalski P and Baczek T 2012 Field-amplified sample stacking-sweeping of vitamins B determination in capillary electrophoresis *J. Chromatogr.* A **1267** 224–30
[23] Shabangi M and Sutton J A 2005 Separation of thiamin and its phosphate esters by capillary zone electrophoresis and its application to the analysis of water-soluble vitamins *J. Pharm. Biomed. Anal.* **38** 66–71
[24] Li C, Yang Q, Wang X, Arabi M, Peng H, Li J, Xiong H and Chen L 2020 Facile approach to the synthesis of molecularly imprinted ratiometric fluorescence nanosensor for the visual detection of folic acid *Food Chem.* **319** 126575
[25] Wang M, Liu Y, Ren G, Wang W, Wu S and Shen J 2018 Bioinspired carbon quantum dots for sensitive fluorescent detection of vitamin B12 in cell system *Anal. Chim. Acta* **1032** 154–62
[26] Ding L, Yang H, Ge S and Yu J 2018 Fluorescent carbon dots nanosensor for label-free determination of vitamin B-12 based on inner filter effect *Spectrochim. Acta Mol. Biomol. Spectrosc.* **193** 305–9
[27] Yang K, Wang Y, Lu C and Yang X 2018 Ovalbumin-directed synthesis of fluorescent copper nanoclusters for sensing both vitamin B-1 and doxycycline *J. Lumin.* **196** 181–6
[28] Scudi J V J 1941 On the colorimetric determination of vitamin B6 *Biol. Chem.* **139** 707
[29] Jacobsen D W, Gatautis V J, Green R, Robinson K, Savon S R and Secic M 1994 Rapid HPLC determination of total homocysteine and other thiols in serum and plasma: sex differences and correlation with cobalamin and folate concentrations in healthy subjects *Clin. Chem.* **40** 873
[30] Yagi T, Iwamoto S, Mizuseki R, Furuya M and Nakayama K 2013 Contents of all forms of vitamin B6, pyridoxine-β-glucoside and 4-pyridoxic acid in mature milk of Japanese women according to 4-pyridoxolactone-conversion high performance liquid chromatography *J. Nutr. Sci. Vitaminol.* **59** 9–15
[31] Sampson D A, Eoff L A, Yan X L and Lorenz K 1995 Analysis of free and glycosylated vitamin B6 in wheat by high-performance liquid chromatography *Cereal Chem.* **72** 217–20
[32] Kall M A 2003 Determination of total vitamin B6 in foods by isocratic HPLC: a comparison with microbiological analysis *Food Chem.* **82** 315
[33] Saraswathyamma B, Grzybowska I, Orlewska C, Radecki J, Dehaen W, Kumar K G and Radecka H 2008 Electroactive dipyrromethene-Cu (II) monolayers deposited onto gold electrodes for voltammetric determination of paracetamol *Electroanalysis* **20** 2317–23
[34] Vadivaambigai A, Senthilvasan P A, Kothurkar N and Rangarajan M 2015 Graphene oxide-based electrochemical sensor for salicylic acid *Nanosci. Nanotechnol. Lett.* **7** 140–6

[35] Silva B V *et al* 2017 Electrochemical sensors and biosensors for determination of amino acids: a review *Microchem. J.* **130** 202–11
[36] Faridbod F *et al* 2019 Electrochemical sensors and biosensors for determination of amino acids: a review *TrAC, Trends Anal. Chem.* **112** 143–57
[37] Moulaee K and Neri G 2021 Electrochemical amino acid sensing: a review on challenges and achievements *Biosensors* **11** 502
[38] Govindasamy M *et al* 2018 Challenges and opportunities in the electrochemical detection of amino acids *Analyst* **143** 1576–90
[39] Karimi-Maleh H *et al* 2021 A review of recent developments in electrochemical sensors for amino acids: sensing mechanisms, materials, and applications *Anal. Chim. Acta* **1179** 338712
[40] Wang C *et al* 2020 Recent advances in electrochemical sensors for amino acid detection *TrAC, Trends Anal. Chem.* **124** 115783
[41] Nguyen H H, Lee S H, Lee U J, Fermin C D and Kim M 2019 Immobilized enzymes in biosensor applications *Materials* **12** 121
[42] Leca-Bouvier B and Blum L J 2005 Biosensors for protein detection: a review *Anal. Lett.* **38** 1491–517
[43] Conroy P J, Hearty S, Leonard P and O'Kennedy R J 2009 Antibody production, design and use for biosensor-based applications *Semin. Cell Dev. Biol.* **20** 10–26
[44] Zhang Y *et al* 2017 Graphene oxide-gold nanorod composite-based electrochemical sensor for ultrasensitive detection of histidine *Biosens. Bioelectron.* **87** 1123–8
[45] Wang L *et al* 2018 Reduced graphene oxide modified electrode for label-free electrochemical detection of histidine *Sensors Actuators* B **256** 780–6
[46] Kim J-H, Lee K B, Kang J-S, Lee W J and Shim J 2015 Highly sensitive and selective electrochemical detection of leucine using graphene oxide–gold nanoparticle hybrid materials *Anal. Bioanal. Chem.* **407** 8323–31
[47] Yang Z, Li H, Li C, Li Y and Li Z 2017 Electrochemical sensor for leucine detection based on graphene oxide modified screen-printed carbon electrode *Anal. Biochem.* **530** 7–12
[48] Das A K, Pal S and Maji S 2019 Graphene oxide/polyaniline based electrochemical sensor for selective detection of leucine *J. Electroanal. Chem.* **831** 101–8
[49] Feng X, Hu Q, Chen Z, Zhang Q and Chen X 2021 Fabrication of a highly sensitive and selective electrochemical sensor based on AuNPs/GO for detection of leucine *J. Electroanal. Chem.* **881** 114903
[50] Lee J H *et al* 2014 Graphene oxide-based electrochemical sensor for the detection of isoleucine *Biosens. Bioelectron.* **59** 222–6
[51] Li Y *et al* 2015 Highly selective and sensitive electrochemical detection of isoleucine using a graphene oxide-modified glassy carbon electrode *J. Mater. Chem.* B **3** 625–32
[52] Li Z *et al* 2016 A highly sensitive and selective electrochemical sensor based on graphene oxide/gold nanoparticle composites for the detection of isoleucine *Sensors Actuators* B **223** 178–85
[53] Gao F *et al* 2017 A label-free electrochemical biosensor based on graphene oxide/chitosan composite film for isoleucine detection *Anal. Methods* **9** 3975–82
[54] Ma Y *et al* 2018 A sensitive electrochemical biosensor based on graphene oxide/copper oxide nanocomposites for isoleucine detection *J. Electroanal. Chem.* **823** 14–21
[55] Wu G 2009 Amino acids: metabolism, functions, and nutrition *Amino Acids* **37** 1–17

[56] Jiang J *et al* 2015 Graphene oxide modified glassy carbon electrode for the sensitive detection of lysine *Sensors Actuators* B **208** 286–92

[57] Lee S H *et al* 2017 Graphene oxide modified screen-printed carbon electrode for the sensitive detection of lysine *Talanta* **164** 643–8

[58] Zhang X *et al* 2017 Graphene oxide modified glassy carbon electrode for the sensitive detection of lysine *J. Electroanal. Chem.* **799** 126–32

[59] Wang Y *et al* 2021 Reduced graphene oxide modified glassy carbon electrode for sensitive and selective detection of lysine *Anal. Chim. Acta* **1147** 63–70

[60] Zhao Y *et al* 2017 A label-free electrochemical sensor based on graphene oxide-modified glassy carbon electrode for sensitive detection of L-methionine *Talanta* **174** 458–63

[61] Li J *et al* 2017 A highly sensitive and selective electrochemical sensor for methionine detection based on graphene oxide modified with l-cysteine *Biosens. Bioelectron.* **95** 77–83

[62] Liu Y *et al* 2018 Graphene oxide-based electrochemical sensor for sensitive and selective detection of L-methionine *J. Electroanal. Chem.* **817** 17–23

[63] Dong Y *et al* 2018 Electrochemical sensing platform based on graphene oxide for selective and sensitive detection of L-methionine *Food Chem.* **256** 217–23

[64] Zhu Z *et al* 2021 Label-free electrochemical sensor for methionine detection based on graphene oxide-modified screen-printed electrode *Talanta* **221** 121684

[65] Kumar A *et al* 2019 Graphene oxide nanosheets modified glassy carbon electrode for detection of phenylalanine in urine samples *Sensors Actuators* B **282** 467–75

[66] Li M *et al* 2018 A graphene oxide-based electrochemical sensor for the sensitive detection of phenylalanine in human serum *RSC Adv.* **8** 20920–6

[67] Huang Y *et al* 2017 Graphene oxide-based biosensor for phenylalanine detection using phenylalanine dehydrogenase immobilized by glutaraldehyde *Sensors Actuators* B **253** 1240–6

[68] Zhang L *et al* 2015 A tyrosinase-graphene oxide-based biosensor for phenylalanine detection *Sensors Actuators* B **221** 891–6

[69] Wang X *et al* 2019 Graphene oxide field-effect transistor for sensitive and selective detection of phenylalanine in human serum *Biosens. Bioelectron.* **143** 111608

[70] Liu S *et al* 2012 Graphene oxide modified electrode for the electrochemical detection of threonine *Biosens. Bioelectron.* **36** 16–21

[71] Lu L *et al* 2014 Facile preparation of graphene oxide modified electrode and its application for electrochemical detection of threonine *Anal. Biochem.* **447** 1–6

[72] Liu L *et al* 2016 Threonine imprinted polymer on graphene oxide modified electrode for selective and sensitive detection of threonine *Electrochim. Acta* **212** 420–7

[73] Zhang J *et al* 2014 A ratiometric electrochemical sensor based on graphene oxide and gold nanoparticles for threonine detection *Biosens. Bioelectron.* **59** 207–13

[74] Li H, Li Y, Zhang S, Chen S and Chen X 2021 Highly sensitive and selective electrochemical sensor for tryptophan detection based on graphene oxide modified glassy carbon electrode *J. Electroanal. Chem.* **898** 115572

[75] Wu X, Zhang Y and Xie Q 2020 A novel tryptophan sensor based on graphene oxide/silver nanoparticles modified screen-printed carbon electrode *Food Chem.* **331** 127309

[76] Zhang Y, Gao F, Ma Y, Chen Y and Lin X 2019 Highly sensitive and selective electrochemical sensor for tryptophan detection based on graphene oxide/multi-walled carbon nanotubes modified glassy carbon electrode *Talanta* **196** 288–94

[77] Wang J, Xu Y, Wu Y and Zou H 2018 A sensitive electrochemical sensor for tryptophan detection based on graphene oxide/Fe_3O_4 modified glassy carbon electrode *Anal. Methods* **10** 2299–306
[78] Wang J, Lu C, Yang Y and Zou H 2017 A highly sensitive and selective electrochemical sensor for tryptophan detection based on graphene oxide/silver nanoparticles modified glassy carbon electrode *Electrochim. Acta* **248** 267–75
[79] Zhao Y *et al* 2018 Valine detection using a graphene oxide-based field-effect transistor *Sensors Actuators* B **259** 246–52
[80] Wang Y *et al* 2019 Graphene oxide-based electrochemical sensors for the detection of amino acids: a review *Microchim. Acta* **186** 320
[81] He Y *et al* 2016 Electrochemical sensor based on graphene oxide-modified glassy carbon electrode for sensitive detection of valine *Electroanalysis* **28** 1051–6
[82] Roushani M *et al* 2017 Graphene oxide/paper-based electrochemical sensor for determination of valine *Mater. Sci. Eng.* C **80** 50–5
[83] Shabani-Nooshabadi M *et al* 2018 Highly sensitive and selective detection of valine using a glassy carbon electrode modified with graphene oxide-embedded carbon nanotubes *Microchim. Acta* **185** 327
[84] Song Y *et al* 2016 Graphene oxide-based immunosensor for rapid and selective detection of valine in human serum *Anal. Bioanal. Chem.* **408** 2431–7
[85] Zhou X, Huang Y, Ma C, Luo X and Yang M 2017 A graphene oxide modified electrode for the electrochemical detection of L-alanine *Anal. Methods* **9** 1098–102
[86] Wang Y, Li X and Du D 2016 A graphene oxide-chitosan composite film-modified electrode for the detection of L-alanine *Electroanalysis* **28** 81–7
[87] Li M, Huang L and Wang Y 2016 A reduced graphene oxide modified glassy carbon electrode for the electrochemical detection of L-alanine *Mater. Sci. Eng.* C **62** 147–53
[88] Li H *et al* 2021 A graphene oxide modified glassy carbon electrode for electrochemical detection of aspartic acid *Microchem. J.* **166** 106177
[89] Wang Y *et al* 2020 Graphene oxide modified electrode for the electrochemical detection of aspartic acid in human serum *Anal. Chim. Acta* **1098** 56–63
[90] Li M *et al* 2019 Graphene oxide based electrochemical sensor for the detection of aspartic acid in food samples *Food Chem.* **276** 521–7
[91] Zhang X, Wang Y, Li J, Zhang Y, Chen H and Zhang Y 2020 Gold nanoparticle-decorated graphene oxide-based electrochemical biosensor for asparagine detection *Sensors Actuators* B **308** 127700
[92] Chen C, Yang S, Cai J, Zhang Y, Lu X and Liu Y 2019 Chitosan/graphene oxide modified electrode for selective detection of asparagine *Sensors Actuators* B **282** 890–6
[93] Wang Y, Yang L, Huang X, Luo L and Feng Y 2018 Graphene oxide-polyaniline composite modified electrode for the determination of asparagine in tea samples *Microchim. Acta* **185** 12
[94] Li H, He Y, Liu S, Huang X, Cao Y, Zhang X and Xu Z 2017 Platinum nanoparticle-decorated graphene oxide for ultrasensitive electrochemical detection of asparagine *Anal. Chem.* **89** 12006–12
[95] Lee M H *et al* 2014 Glutamate and glutamine in Alzheimer's disease *World J. Biol. Psychiatry* **15** 517–26
[96] Zhang Y *et al* 2016 Glutamate in Alzheimer's disease: a target for therapy *J. Alzheimers Dis.* **52** 367–78

[97] Yang Y *et al* 2017 A graphene oxide modified electrode for the detection of glutamic acid *J. Electroanal. Chem.* **791** 188–92
[98] Zhao Y *et al* 2017 Graphene oxide-based electrochemical biosensor for glutamic acid detection *Sensors Actuators* B **238** 411–6
[99] Wang J *et al* 2018 Fabrication of a graphene oxide/Pt nanocomposite modified electrode for glutamic acid detection *Anal. Lett.* **51** 1305–17
[100] Wang C *et al* 2017 Graphene oxide modified screen-printed carbon electrode for glutamic acid detection *Electrochim. Acta* **254** 91–7
[101] Zhang X and Lu Y 2017 Rapid and label-free electrochemical detection of serine in human serum using an iridium oxide nanoparticles/graphene oxide composite modified electrode *Biosens. Bioelectron.* **89** 559–64
[102] Cai J, Zhang G, Jiang X and Sun S 2015 Label-free electrochemical biosensor based on reduced graphene oxide and gold nanoparticles for ultrasensitive detection of serine *Biosens. Bioelectron.* **68** 175–81
[103] Zhang L, Zhou Y, Wu Y and Ma H 2016 A novel serine biosensor based on glassy carbon electrode modified with graphene oxide and silver nanoparticles *Sensors Actuators* B **223** 1–8
[104] Li Y, Zhang S, Zhang X and Du X 2018 A sensitive electrochemical sensor for serine detection based on graphene oxide and gold nanoparticles modified electrode *Anal. Biochem.* **558** 48–54
[105] Hu Y, Li X, Liu Y and Li J 2016 A label-free serine biosensor based on graphene oxide-copper nanoparticles modified glassy carbon electrode *Anal. Biochem.* **505** 1–6
[106] Wang L *et al* 2014 Electrochemical detection of arginine using a graphene oxide-modified glassy carbon electrode *Electrochim. Acta* **142** 230–6
[107] Wang L *et al* 2015 Gold nanoparticles-decorated graphene oxide-modified glassy carbon electrode for sensitive detection of arginine *Electrochim. Acta* **151** 465–71
[108] Li X *et al* 2021 Highly sensitive and selective graphene oxide-copper nanoparticles modified glassy carbon electrode for the detection of arginine in fish samples *Food Chem.* **343** 128529
[109] Cheng Y, Guo L, Huang Y, Yuan R and Chai Y 2016 A graphene oxide-based electrochemical sensor for sensitive determination of cysteine *Anal. Chim. Acta* **936** 240–6
[110] Zhang Y, Jiang D, Xie J, Liu M and Wei S 2019 Flexible screen-printed graphene oxide/copper nanoparticles modified electrode for cysteine detection *Anal. Chim. Acta* **1085** 102–9
[111] Liu H, Guo Y and Chen X 2019 Sensitive electrochemical sensor based on cysteine functionalized graphene oxide for the detection of glutamine *Anal. Chim. Acta* **1058** 81–7
[112] Zhang X, Wang Y, Liu Y and Qiu B 2020 Label-free electrochemical sensor based on chitosan modified graphene oxide for the detection of glutamine *Sensors Actuators* B **306** 127590
[113] Xu L, Wang Y, Huang X and Zhang Y 2020 Highly sensitive electrochemical sensor based on gold nanoparticles and horseradish peroxidase functionalized graphene oxide for detection of glutamine *Talanta* **214** 120871
[114] Jiang M, Tang J, Zhou N, Liu J, Tao F, Wang F and Li C 2022 Rapid electrochemical detection of domoic acid based on polydopamine/reduced graphene oxide coupled with in-situ imprinted polyacrylamide *Talanta* **236** 122885
[115] Chen J *et al* 2020 Graphene oxide-based electrochemical sensor for glycine detection *Sensors Actuators* B **309** 127763
[116] Wang L *et al* 2020 One-pot synthesis of reduced graphene oxide/Fe_3O_4 composite for ultrasensitive electrochemical detection of glycine *J. Electroanal. Chem.* **875** 114435

[117] Naeem M *et al* 2018 Graphene oxide modified screen-printed electrode for the selective detection of glycine *Mater. Sci. Eng.* C **91** 267–72
[118] Chen S *et al* 2018 Ultrasensitive electrochemical detection of glycine based on graphene oxide/mesoporous silica modified glassy carbon electrode *J. Electroanal. Chem.* **819** 58–64
[119] Zhang Y *et al* 2019 Graphene oxide quantum dots and reduced graphene oxide based electrochemical sensor for the detection of glycine *Anal. Methods* **11** 3695–700
[120] Li Y *et al* 2017 Enhanced electrochemical detection of glycine based on a graphene oxide/gold nanoparticles modified glassy carbon electrode *Anal. Chim. Acta* **961** 49–56
[121] Feng C *et al* 2020 A novel electrochemical sensor for glycine detection based on graphene oxide/silver nanoparticles/chitosan nanocomposite modified glassy carbon electrode *J. Electroanal. Chem.* **872** 114122
[122] Li Y *et al* 2017 A highly sensitive electrochemical sensor for glycine detection based on a graphene oxide/polypyrrole composite modified glassy carbon electrode *Sensors Actuators* B **242** 1302–9
[123] Wang H *et al* 2019 A novel electrochemical sensor for glycine detection based on graphene oxide/β-cyclodextrin modified glassy carbon electrode *Electroanalysis* **31** 1358–65
[124] Zarei M *et al* 2017 Detection of proline using a graphene oxide–gold nanoparticles modified glassy carbon electrode *Mater. Sci. Eng.* C **71** 198–204
[125] He Y *et al* 2017 A novel electrochemical sensor based on graphene oxide modified electrode for detection of proline *Talanta* **162** 277–82
[126] Liu Y *et al* 2021 Highly sensitive electrochemical sensor for tyrosine detection based on graphene oxide modified glassy carbon electrode *Mater. Sci. Eng.* C **118** 111502
[127] Li Y *et al* 2020 Label-free electrochemical sensor for tyrosine detection based on graphene oxide-chitosan nanocomposite *Mater. Sci. Eng.* C **112** 110891
[128] Yu J *et al* 2019 A graphene oxide-based electrochemical sensor for tyrosine detection using a gold nanoparticle/tyrosinase conjugate *Sensors Actuators* B **282** 723–9
[129] Chen Z *et al* 2018 Sensitive electrochemical sensor for tyrosine detection based on graphene oxide-ionic liquid composite film modified electrode *Biosens. Bioelectron.* **106** 58–64
[130] Li Y *et al* 2017 Highly sensitive electrochemical sensor for tyrosine detection based on graphene oxide/Prussian blue composite film modified electrode *Sensors Actuators* B **238** 1118

IOP Publishing

Graphene-Based Sensors

Jamballi G Manjunatha

Chapter 14

Application of graphene-based electrodes for water pollutants

M Stanleydhinakar, C Balaji, A Gowdhaman, J G Manjunatha and R Ramesh

14.1 Introduction

Water pollution is a pressing global issue that poses significant risks to ecosystems and human health. When all water-related illnesses, including cholera and arsenicosis, are taken into account, the number of fatalities attributable to diseases caused by the use of contaminated water is rapidly rising. To reduce them, a variety of actions must be performed, one of which is to conduct adequate monitoring of the drinking water quality throughout its life cycle. The detection and monitoring of water pollutants play a crucial role in safeguarding water resources and ensuring their sustainable use. However, it is not a simple process. Low-maintenance online sensors that can run constantly, independently, and reagent-free are the ideal kind to use for monitoring water quality. Electrochemical and electrical solid-state sensors have become a common class of sensors used for this purpose. Electrical sensors include chemoreceptive sensors and field-effect transistors (FETs), which track changes in a transducing material's conductance. Graphene-based materials are ideally suited to be the transducers in this class of sensors because of their outstanding electron transport characteristics and simplicity of functionalization. Graphene is a two-dimensional (2D) carbon material with extraordinary properties and has revolutionized many fields of science and technology [1]. Its distinctive electrical, mechanical, and chemical properties make it an ideal platform for the development of advanced sensing devices. Graphene electrodes, when used in water pollutant sensing, allow for the detection and analysis of a wide range of contaminants, including heavy metals, organic pollutants, microbial pathogens, and others. The integration of graphene-based electrodes in water pollutant sensing devices has several advantages. For starters, graphene's high electrical conductivity allows for efficient charge transfer, allowing for sensitive and accurate detection of target pollutants, and addressing the challenges associated with their complex

doi:10.1088/978-0-7503-5578-0ch14

chemical structures and low concentrations in environmental samples [2]. Second, the large surface area of graphene allows for ample analyte adsorption, increasing the detection limit and response of the sensor. Furthermore, graphene's exceptional chemical stability ensures the sensing device's long-term performance and durability.

In this comprehensive chapter, we explore the diverse applications of graphene electrodes in water pollutant sensing. We delve into the principles, mechanisms, and performance evaluation metrics of different sensing approaches, including graphene oxide (GO)-based sensors, graphene FETs (GFETs), and graphene-based optical sensors. Furthermore, we discuss specific examples and case studies showcasing the detection of heavy metals, organic pollutants, microbial pathogens, and other contaminants using graphene-based sensing platforms.

14.2 Electrode evolution: advancing the field of sensing electrode

The working/operational electrode holds paramount significance in electrochemical sensors, prompting researchers to concentrate on enhancing its performance due to its pivotal role in sensor operation. Traditionally, mercury electrodes were employed for electrochemical detection; however, the inherent volatility and toxicity of mercury have imposed significant limitations on its applications. Consequently, since the 1970s, researchers have shifted away from using mercury electrodes and instead focused on utilizing solid electrodes like gold and platinum. These solid electrodes possess desirable characteristics including high conductivity and stability, thereby circumventing toxicity concerns associated with heavy metals. Nevertheless, the use of heavy metal-based electrodes poses challenges due to their non-renewable nature and high costs. Consequently, a surge of interest has emerged in carbon-based solid electrodes, which offer promising alternatives. These carbon-based electrodes possess several advantages, including their renewable nature and cost-effectiveness, driving their increasing adoption in electrochemical sensing applications. Carbon electrodes exhibit notable attributes such as exceptional conductivity, affordability, minimal baseline current, expansive ranges of applied potentials and flexibility for modifications [3]. To enhance the performance of carbon electrodes, diverse techniques have been employed to modify their chemical composition and microstructure. These modifications can be achieved through two distinct approaches: the top-down and the bottom-up. Top-down modifications involve the utilization of techniques such as laser ablation, plasma treatment, ozone treatment, and other etching methods to selectively etch the surface of the electrode. This approach facilitates the attainment of desirable surface roughness and the introduction of specific surface functional groups, thereby enhancing the adsorption performance and electrochemical activity of the electrode [4–6]. Conversely, bottom-up modifications entail the incorporation of active materials deposition onto the electrode surface through diverse methodologies such as self-assembly, coatings, electrochemical polymerization, and electrochemical deposition [7]. These modifications enable the integration of versatile materials, including carbon nanomaterials, quantum dots, metal (oxide) nanoparticles, conductive organics, enzymes, and antibodies, onto the electrode surface [8–10].

14.3 Graphene-based sensing mechanism and performance metrics

14.3.1 Sensing mechanism in sensors based on graphene

Graphene-based sensors mainly depend on the process of adsorption and desorption of pollutants on the graphene surface, as well as charge transfer mechanisms. The interaction between the target contaminants and the functional groups on the graphene surface is at the heart of the adsorption and desorption processes. (Generally, GO or reduced graphene oxide (rGO) is employed). The high surface area of graphene allows for a large number of pollutant molecules to be adsorbed, increasing the sensitivity of the sensing device. Additionally, the conductivity of graphene enables the detection of adsorbed molecules through changes in electrical properties.

14.3.1.1 Adsorption and desorption processes

Adsorption and desorption are the most general process for the sensing of contaminants. Pollutants adsorb onto graphene via various mechanisms (figure 14.1) including electrostatic interactions, hydrogen bonding, van der Waals forces and π–π stacking interactions. For example, hydroxyl (–OH), carboxyl (–COOH), and epoxy (–O–) groups on GO provide an active site for the adsorption of many types of contaminants.

Various factors can be employed to enhance the adsorption capabilities of the electrode such as surface modification, doping and employing 2D heterostructures. For example, the unique properties of these heterostructures arise from the synergistic effects between the constituent materials. graphene combined with transition metal dichalcogenides (TMDs), such as molybdenum disulfide (MoS_2) or tungsten diselenide (WSe_2), can create heterojunctions with enhanced charge transfer, leading to improved adsorption capacities and selectivity.

Altering the ambient parameters such as pH, temperature, or the presence of competing ions can induce the desorption process. Adsorbed pollutants from GO

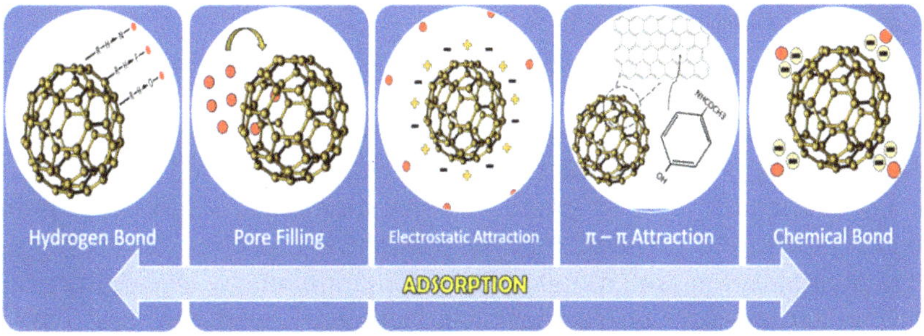

Figure 14.1. Types of adsorption.

can be tracked and quantified, providing important information regarding pollutant concentrations in water samples.

14.3.1.2 Charge transfer mechanisms

Graphene-based sensors also rely on charge transfer mechanisms for pollution detection. Pollutants can change the density and distribution of charge carriers in the graphene structure, affecting its electrical conductivity or electrochemical properties.

For Instance, adsorption of electron-donating contaminants onto GO, can result in an increase in electrical conductivity due to electron transfer from the adsorbed molecules to the GO surface. On the other hand, adsorption of electron-accepting contaminants can result in a decrease in electrical conductivity. By measuring the changes in electrical conductivity or current response, the presence and concentration of pollutants can be determined.

14.3.2 Sensing methods

In recent years, because of their exceptional properties and versatility, graphene electrodes have emerged as highly promising platforms for water pollutant sensing. Researchers have developed a variety of novel sensing methods for the detection and quantification of water pollutants by leveraging the unique properties of graphene. The most popular sensing method is the use of graphene electrodes based on the adsorption and desorption process. Another effective sensing method is the utilization of GFETs. In GFETs a graphene layer serves as the conducting channel, while a gate electrode regulates the charge carrier density.

Graphene-based optical sensors have also received a lot of attention for detecting water pollutants. Beyond these methods, there are several other graphene-based sensing approaches employed for water pollutant detection, including impedance spectroscopy, electrochemical impedance spectroscopy (EIS), capacitance sensing, and surface plasmon resonance (SPR) sensors. Each method has its own set of advantages and can be tailored to target specific pollutants or achieve desired levels of sensitivity.

In this chapter, we will explore in detail the various sensing methods utilized in the application of graphene electrodes for water pollutant detection. We will delve into each sensing method's underlying principles, performance evaluation metrics, and application examples.

14.3.2.1 Electrical sensing

The electrical sensing method based on graphene electrodes offers several advantages for detecting water pollutants. These methods take advantage of unique electrical responses exhibited by graphene when it interacts with various pollutants, allowing for precise and real-time analysis. Graphene-based sensors enable accurate detection, quantification, and monitoring of contaminants by leveraging changes in electrical properties such as electrical conductivity, impedance, and capacitance caused by the presence of water pollutants. One important approach in electrical sensing methods involves the use of GO, a derivative of graphene. Pollutants in

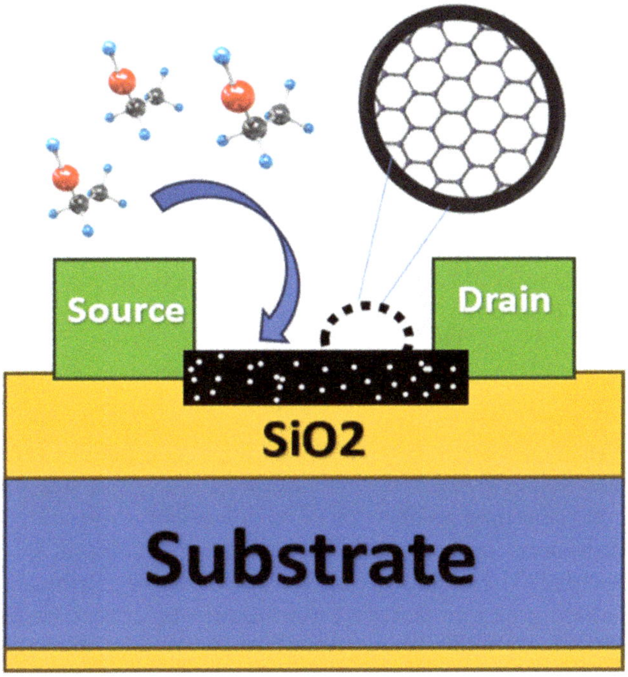

Figure 14.2. GFET.

water can interact with the GO surface, altering the electrical conductivity or impedance of the graphene-based sensor. This change in electrical response can be linked to pollutant concentrations, enabling their detection and quantification. Another prominent electrical sensing technique makes use of GFETs (figure 14.2), in which the graphene layer serves as the conducting channel. GFETs can selectively capture and bind target pollutants by functionalizing the graphene surface with specific receptors or sensing elements [11]. Biorecognition events occur when the specific biomolecular analyte of interest selectively binds to the active surface of the biosensor platform. This binding leads to a change in the carrier concentration within the gate channel. As a result, current signals are produced between the source and drain electrodes of an FET [12]. This binding process alters the GFET's conductance, resulting in measurable changes in electrical current that can be used to identify and quantify pollutants in water samples. The transfer characteristic equation can be used to describe the sensing mechanism:

$$I_{DS} = I_{DS0} \times (1 + \beta \quad \times (V_{GS} - V_T))$$

where I_{DS} is the drain current, I_{DS0} is the drain current in the absence of the analyte, β is the sensitivity factor, V_{GS} is the gate-to-source voltage, and V_T is the threshold voltage. Park *et al* developed a bio probe-immobilized GFET sensor for real-time monitoring of geosmin (an odorous substance) in water sources and its range of detection from 0.01 nM to 1 μM with a limit of detection (LOD) for 0.01 nM [13].

Other electrical sensing techniques, such as impedance spectroscopy and EIS [14], are used for water pollutant detection in addition to GO-based sensors and GFETs. When exposed to polluted water samples, these methods measure the impedance response of the graphene-based sensor. Impedance variations can provide useful information about the presence and concentration of specific contaminants.

14.3.2.2 Resistive and capacitive sensing

In resistive sensing, the change in the electrical resistance of graphene upon exposure to water pollutants is utilized for detection. The graphene resistance can be described using the following equation:

$$R = \rho \times \left(\frac{L}{W}\right)$$

where R is the resistance, ρ is the resistivity, L is the length, and W is the width of the graphene channel. When water pollutants are adsorbed onto the resistive material, they can alter its electrical properties. The adsorbed molecules may introduce barriers or perturbations to the electron flow within the material, leading to changes in electrical resistance. These changes in resistance can be attributed to a variety of mechanisms, including charge transfer, modulation of charge carrier mobility, or changes in the effective conducting paths within the material. It can operate through different sensing mechanisms, such as chemo-resistive sensing and FET sensing. Whereas capacitive sensing can operate based on the changes in the interfacial capacitance between the sensing material and the surrounding electrolyte. The sensing element can be a capacitive transducer, such as interdigitated electrodes or metal oxide films, which are designed to maximize the interfacial area with the adsorbate molecules. Adsorption of analyte molecules onto the surface of graphene can change the dielectric constant, resulting in capacitance changes that can be measured. Both resistive and capacitive sensing offer advantages and limitations depending on the specific application requirements. Resistive sensing is often preferred for its simplicity, cost-effectiveness, and compatibility with various materials. Capacitive sensing, on the other hand, provides high sensitivity, and low power consumption, and can be suitable for miniaturized sensor designs.

Yasin *et al* successfully synthesized a composite material consisting of graphene hydrogel and zinc oxide nanoparticles, incorporating activated carbon, known as AGHZ. The composite exhibited an impressive specific capacitance of 746.5 F g^{-1} at a scan rate of 10 mV s^{-1}, highlighting its exceptional energy storage capability. Furthermore, the AGHZ composite demonstrated a remarkable salt removal efficiency of 83.65% and an impressive electrosorption capacity of 9.95 mg g^{-1}, emphasizing its potential for efficient ion removal in various applications [15].

14.3.2.3 Optical sensing

Optical sensing techniques make use of changes in graphene's optical properties caused by interactions with water pollutants. The Kubo formula or the Drude model [16] can be used to describe the absorption spectrum of graphene, which exhibits

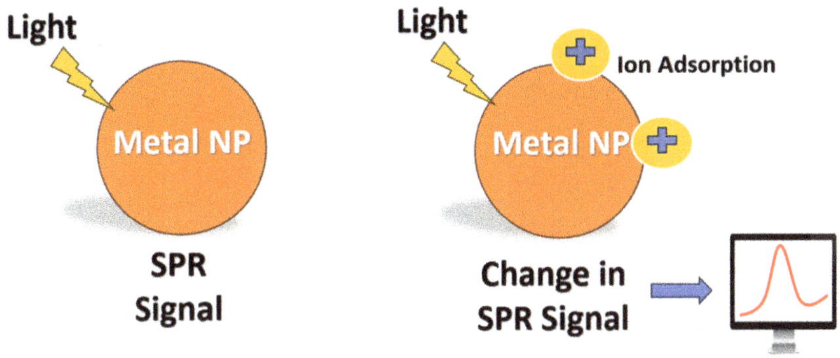

Figure 14.3. Surface plasmon resonance.

strong absorption in the visible to near-infrared spectral range. The adsorption of analyte molecules on the surface of graphene can change the optical absorption, which can be detected using techniques such as spectrophotometry (SPR) [17]. In general, methods like absorption spectroscopy, fluorescence spectroscopy, surface plasma resonance, and Raman spectroscopy are used for sensing. Advanced methods like surface-enhanced Raman spectroscopy (SERS), which involves the use of nanostructured surfaces such as metal nanoparticles or graphene, significantly increase the sensitivity and detection of trace amounts of pollutants. SPR (figure 14.3) is a phenomenon occurring when free electrons on the surface of a metal–dielectric interface collectively oscillate upon interaction with incident electromagnetic waves. This phenomenon is highly sensitive to the refractive index in the immediate vicinity of the thin film. Grating-based SPR, which allows for greater miniaturization and integration, offers several advantages. However, it is less commonly employed in sensing applications due to its lower sensitivity compared to prism-based SPR, which remains the most prevalent method in sensing applications.

In 2018, Fang *et al* conducted a study where they synthesized nitrogen-doped graphene quantum dots (N-GQDs) through the solvothermal process of GO using dimethylformamide. These N-GQDs were subsequently utilized as a fluorescent probe for the detection of Al^{3+} ions. The detection mechanism involved a photo-induced electron transfer (PET) process, which occurred due to the complexation between Al^{3+} ions and N-GQDs [18].

14.3.2.4 Electrochemical sensing

Electrochemical sensing is a widely used technique for water pollutant detection and monitoring due to its high sensitivity, selectivity, and compatibility with a wide range of target analytes [19]. The measurement of changes in electrochemical signals, such as current or potential, caused by the redox reactions of water pollutants at the graphene electrode surface is referred to as electrochemical sensing. The presence of analyte molecules can alter the electron transfer kinetics, resulting in quantifiable changes in the electrochemical response. There are three main electrochemical sensing modes: amperometry, potentiometry, and voltammetry. Amperometry involves the measurement of change in current, potentiometry measures the potential difference between the electrodes (indicator and reference), and voltammetry measures the current as a function of applied potential. Electrochemical sensors offer several advantages over other methods, including their compact size, ease of use, rapid electrical signal response, and elimination of the need for complex sample pretreatment. Additionally, they maintain high detection sensitivity. Due to these characteristics, electrochemical sensors hold significant potential for real-time and *in situ* analysis applications [20]. Banalata Maji *et al* developed a non-enzymatic glassy carbon electrode (GCE) sensor based on GO-g-C_3N_4-$MnCo_2O_4$ nanocomposite to detect chlorpyrifos (CPS), a well-known organophosphorus pesticide in water. This composite has been reported for the first time as a non-enzymatic sensor to detect CPS with superior sensing capability toward the detection of CPS [21].

14.3.2.5 Piezoelectric sensing

The piezoresistive sensing method with graphene electrodes is based on the idea of strain-induced changes in graphene's electrical conductivity. When a mechanical strain is applied to a graphene electrode as a result of interaction with water pollutants, the interatomic distances and overall lattice structure change, resulting in changes in electrical resistance. These changes in resistance can be measured and correlated to the concentration of pollutants in water samples, allowing for the detection and quantification of contaminants [22]. The piezoresistive sensing method has several advantages for detecting water pollutants. It has a high sensitivity and can detect trace amounts of pollutants in water samples. The method also has a broad dynamic range, allowing it to detect contaminants at both low and high concentrations. Additionally, the response time of piezoresistive sensing with graphene electrodes is typically fast, allowing for real-time monitoring of water quality [23].

Other graphene-based sensing approaches for water pollutant detection have been investigated in addition to the aforementioned techniques. These include impedance spectroscopy-based sensors, graphene-based SERS sensors, and graphene-based nanocomposite sensors. Ye *et al* presented a notable advancement in the realm of sensor fabrication by demonstrating the successful synthesis of flexible, robust, and porous sensors employing a combination of thermoplastic polyurethane (TPU) and rGO monolith. The resulting prototypes exhibited exceptional wettability characteristics, rendering them suitable for continuous oil/water separation applications in polluted environments. The hierarchical porous sensors developed during this study exhibited a remarkable combination of super-hydrophobic and super-oleophilic properties, selectively favoring interactions with specific liquids. Notably, the sensors exhibited an outstanding capability to absorb and remove oil pollution from aqueous environments, with an impressive efficiency of 15 000 times their weight. This feat was achieved through the implementation of an efficient pumping system, facilitating the seamless extraction of oil from water [24].

14.3.3 Performance evaluation metrics

Graphene-based sensor performance can be measured using a variety of metrics, including sensitivity, selectivity, detection limit, and response time. The sensor's sensitivity refers to its ability to detect and quantify small changes in pollutant concentrations. The ratio of the change in sensor response (e.g. electrical conductivity or current) to the change in pollutant concentration is commonly used to define it. The sensor's selectivity measures its ability to distinguish the target pollutant from other interfering species or background signals. High selectivity ensures that specific pollutants are detected accurately and reliably in complex water samples. The detection limit is the lowest concentration of a pollutant that the sensor can reliably detect. It is calculated by measuring sensor response at various concentrations of the target pollutant and determining the lowest concentration that can be distinguished from background noise. The time it takes the sensor to achieve a stable and measurable response after being exposed to the target pollutant is referred to as response time. For real-time monitoring applications, a quick response time is preferred.

14.3.4 Advantages of graphene-based sensors

Graphene-based sensors outperform traditional sensing platforms in several ways, making them particularly appealing for water pollutant detection.

High sensitivity: Because graphene has a high surface-to-volume ratio, it has a high density of active sensing sites, resulting in increased sensitivity and the ability to detect trace amounts of pollutants.

Rapid response time: Graphene's high carrier mobility and low mass enable rapid response times, allowing for real-time monitoring of water systems.

Selectivity: By functionalizing or modifying the graphene surface, the sensor's selectivity to specific analytes can be tailored, reducing interference from other compounds in the water sample.

Miniaturization and portability: Because graphene-based sensors can be fabricated on flexible substrates, portable and wearable sensing devices for on-site and in-field monitoring can be developed.

Low cost: Graphene production techniques like chemical vapor deposition (CVD) and mechanical exfoliation have become more scalable, resulting in lower manufacturing costs and greater accessibility to graphene-based sensors.

Graphene is compatible with existing microfabrication processes, allowing for the development of sensor arrays as well as the incorporation of on-chip signal processing and wireless communication capabilities.

Biocompatibility: Because graphene is biocompatible, it can be used in biological and environmental sensing applications such as water pollutant detection.

14.4 Case studies: graphene based water pollutant sensors

14.4.1 Detection of heavy metals

Heavy metals like mercury (Hg), lead (Pb), and arsenic (As) are major environmental pollutants that endanger water quality and human health. Graphene-based

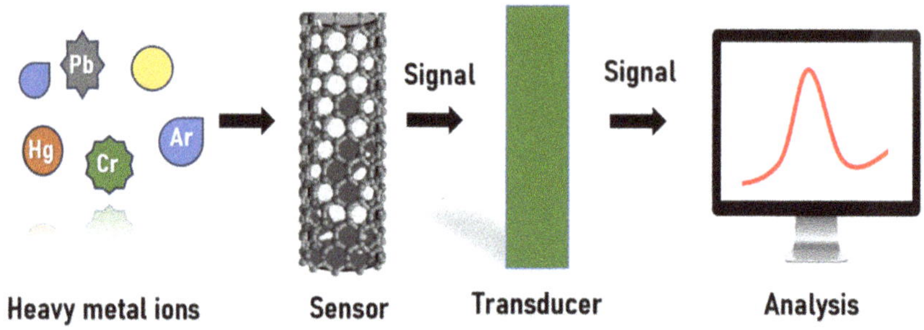

Figure 14.4. Heavy metal ion sensing.

sensors have proven effective in the sensitive and selective detection of these heavy metals [25] (figure 14.4). In this section, we will elaborate on the detection of mercury, lead, and arsenic using graphene-based sensors, providing the necessary equations and examples.

14.4.1.1 Mercury (Hg)

Mercury poisoning in water sources is a major problem due to its high toxicity and potential to accumulate in species across the food chain. Graphene-based sensors provide an effective platform for detecting mercury ions (Hg^{2+}) [26].

The redox reaction that occurs at the graphene electrode surface is one common approach for mercury detection.

$$Hg^{2+} + 2e^{-} \rightarrow Hg^{0}$$

The resulting change in current or potential can be measured and correlated to the concentration of mercury ions in the water sample. For example, consider a graphene-based electrochemical sensor used for mercury detection. When ions interact with the electrode, reduction of contaminant occurs, and the change in current is directly proportional to the concentration of Hg^{2+} ions. With respect to the change in current, the concentration of contaminant can be accurately quantified. The concentration of mercury in the sample can be accurately quantified by constructing a calibration curve.

Tang *et al* (2019) conducted a study aimed at achieving the lowest LOD for Hg^{2+} detection using an electrochemiluminescence (ECL) sensor. Their research demonstrated the successful implementation of a poly(5-formylindole)/rGO (P5FIn/erGO) nanocomposite that was employed as the ECL substrate, along with a gold nanoparticle-linked GQDs-DNA (GQDs-DNA-AuNP) as the signal probe. The synthesis of GQDs was carried out through the pyrolysis of citric acid. Remarkably, their experimental findings revealed a notably low limit of the detection value of 0.00248 nM for Hg^{2+} detection, showcasing the high sensitivity and accuracy achieved by their ECL sensor system. This advancement holds significant potential for enhancing the detection and monitoring of Hg^{2+} in various environmental and analytical applications [27].

14.4.1.2 Lead (Pb)

Lead contamination in water is a major public health concern, particularly because of its negative effects on neurodevelopment in children. For the sensitive detection of lead ions (Pb^{2+}), graphene-based sensors were used [28]. Similar to mercury detection, lead ion detection can be accomplished through electrochemical sensing. The following equation can be used to represent the oxidation of Pb^{2+} ions at the graphene electrode surface:

$$Pb^{2+} + 2e^{-} \rightarrow Pb^{0}$$

By measuring the current response at different concentrations of lead ions and constructing a calibration curve, the concentration of lead in water samples can be quantified using the graphene-based sensor. In the field of electrochemical sensing, anodic stripping voltammetry (ASV) has emerged as the most commonly employed technique over the past five years. By incorporating a pre-concentration step, during which Cd^{2+} ions are effectively accumulated on the electrode surface through adsorption or faradaic processes, the detection sensitivity can be significantly enhanced. Subsequently, the analytical signal is generated through the subsequent stripping voltammetry process. Akhar Shuai *et al* have successfully developed an electrochemical sensor for the real-time measurement of heavy metals including Pb^{2+}. Their innovative approach involved the utilization of rGO decorated with polyaniline (PANI) and alanine (Ala), creating a synergistic nanocomposite termed rGO/Ala/PANI. The proposed sensor exhibited exceptional sensitivity and robustness, enabling the detection of Pb^{2+} at a remarkably low detection limit of 0.045 nM. Moreover, the detection range spanned from 100 to 0.08 nM [29].

14.4.1.3 Arsenic (As)

Arsenic contamination in water sources is a worldwide concern because of its widespread occurrence and toxicity. One popular method for detecting arsenic is to functionalize graphene surfaces with specific receptors or chelating agents that selectively bind to arsenic ions (As^{3+} or As^{5+}). This enhancement improves the sensor's specificity for arsenic detection [30].

Arsenic ion binding to the functionalized graphene surface causes changes in electrical, optical, or electrochemical properties, which can be measured and correlated to arsenic concentrations in water samples. For example, a graphene-based optical sensor functionalized with an arsenic receptor can show changes in absorbance or fluorescence when exposed to arsenic-containing water samples. The concentration of arsenic in the water sample can be determined by measuring these changes and comparing them to a calibration curve obtained from known concentrations of arsenic. Li *et al* presented a GFET aptasensor array for the sensitive detection of arsenite (As^{3+}) ions. The sensing mechanism of the aptamer functionalized GFET relied primarily on the conformational changes occurring in the negatively charged aptamer upon interaction with As^{3+} ions. This approach exhibited an impressive linear detection range (LDR) spanning from 0.05 to 1000 ppb. Furthermore, the LOD achieved by the developed sensor was approximately 0.02 ppb [31].

14.4.1.4 Cadmium

Cadmium is a toxic heavy metal that is perilous to both human health and the environment. Graphene's unique properties, such as its large surface area and excellent electrical conductivity, allow for efficient cadmium ions sensing. Cadmium ions can be detected using graphene-based electrochemical sensors via a variety of methods, including stripping voltammetry and impedance spectroscopy. The application of a potential to the graphene electrode causes the reduction or oxidation of cadmium ions, resulting in detectable current peaks. The sensor had high sensitivity and a wide detection range, making it suitable for monitoring cadmium contamination in the environment [32]. Shan *et al* demonstrated an electrochemical metal-free sensor using GO for the detection of cadmium ions. The electrochemical Cd^{2+} sensor with GR/GO composite electrode exhibited superior analysis performances and exhibited a wide linear range (0–10 μM), low value of detection limit (0.087 μM), reliable stability and selectivity [33]. Additionally, impedance spectroscopy is another technique employed for cadmium ion detection using graphene-based electrodes. By monitoring the changes in impedance caused by the adsorption of cadmium ions onto the graphene surface, the concentration of cadmium can be quantified.

In summary, graphene-based sensors provide effective solutions for sensing heavy metals in water such as mercury, lead, and arsenic. The use of graphene electrodes in conjunction with appropriate functionalization or sensing mechanisms enables sensitive, selective, and rapid monitoring of these contaminants to ensure water safety and public health.

14.4.2 Organic pollutant detection

Organic contaminants in water, such as pesticides, medicinal compounds [34], and industrial chemicals, pose substantial threats to ecosystems and human health. These pollutants must be detected and quantified in order to maintain water quality and ensure the safety of drinking water sources. Because of their high surface area, excellent electrical conductivity, and molecular adsorption properties, graphene electrodes have shown great promise in the detection of various organic pollutants [35, 36].

14.4.2.1 Pesticides

Pesticides are widely used to protect crops, but their overuse and improper disposal can pollute water. Pesticides can be detected using graphene-based sensors via a variety of mechanisms, including electrochemical sensing and SERS. The graphene electrode is functionalized with specific receptors or enzymes that possess a strong affinity for target molecules in electrochemical sensing. Pesticide binding to the functionalized graphene surface causes changes in the sensor's electrical properties, such as impedance or current, allowing detection and quantification. In the case of SERS, graphene can serve as a substrate to amplify Raman signals emitted by pesticide molecules. Pesticide molecules adsorb on the graphene surface, amplifying their Raman scattering and allowing for highly sensitive detection. Pesticides' presence and concentration in water samples can be determined by analyzing their characteristic Raman spectra. Among organic pesticides, organophosphorus or organophosphates (OPs) are widely employed in commercial agriculture, household, human medications, and domestic animal medicines [37]. Malathion, parathion, dimethoate, ethion, chlorpyrifos, and diazinon are examples of OP pesticides. When OPs enter the human body, they cause pesticide poisoning by blocking active sites and inhibiting the mechanism of the acetylcholinesterase (AChE) enzyme, which regulates acetylcholine levels in the body [38] which still poses an active threat. Sundaresan *et al* developed an electrochemical sensor that detects the widely used organophosphorus pesticide 'fenitrothion' using one-dimensional Pr_2 $(WO_4)_3$ nanorods (PWO) integrated rGO. The sensor exhibited superior performance in the detection of fenitrothion [39]. Beduk *et al* (2020) reported a laser-scribed graphene-imprinted polymer for the detection of bisphenol A (BPA). It is sensitive with an LOD of 8 nm, which is accurate in the sensing of BPA in tap and mineral water [40].

14.4.2.2 Pharmaceutical compounds

Pharmaceutical compounds, such as antibiotics and hormone residues, can infiltrate water sources via wastewater disposal and pose ecological risks [41, 42]. In recent years, graphene-based sensors have been investigated for the sensing of pharmaceutical compounds in water [43]. The binding of pharmaceutical molecules to the functionalized graphene surface causes changes in the sensor's electrical properties or Raman signals, allowing them to be detected and quantified. The concentration of the pharmaceutical in the water sample can be correlated with the change in the electrical signal or Raman scattering intensity upon pharmaceutical binding.

Detecting industrial chemicals with graphene-based sensors is also possible with the integration of other sensing techniques. For example, a graphene-based sensor can be paired with optical techniques such as fluorescence spectroscopy or SERS to enable highly sensitive and specific detection of certain industrial chemicals [44]. Alagumalai *et al* evaluated a graphene composite Au@f-CNT/GO as a selector toward the electrochemical oxidation of naproxen. The fabricated GCE exhibited a low resistance and higher electrocatalytic activity toward NPZ sensing with lower limit detection of 14 nm [45].

14.4.2.3 Industrial chemicals

Due to their ubiquitous use in numerous manufacturing processes, industrial chemicals pose a substantial danger to water quality. Heavy metals, solvents, acids, alkalis, and other toxic compounds can contaminate water sources and have an adverse effect on the environment and human health [46]. Detecting and monitoring industrial chemicals in water is critical for ensuring water supply safety and avoiding pollution-related hazards [47]. To enhance the selectivity and sensitivity of graphene-based sensors for specific industrial chemicals, the graphene surface can be modified or functionalized in several ways [48]. Generally, chemical functionalization or nanoparticle decoration [49] are employed. To determine the concentration of industrial chemicals in water, calibration curves can be established. Topçu *et al* demonstrated the utilization of a paper electrode composed of CoMOF (cobalt-based metal–organic framework) and rGO for the electrochemical detection of resorcinol (RS). This flexible electrochemical sensor exhibited remarkable analytical performance, including an impressive wide LDR spanning from 0.1 to 800 μM and an exceptionally low LOD of 19 nM. Moreover, the sensor displayed superior sensitivity when applied to real water samples [50].

14.4.2.4 Endocrine disruptors

Endocrine disruptors are chemicals that disrupt the normal hormonal systems of living organisms, resulting in negative effects on development, reproduction, and overall health. BPA, phthalates, and polychlorinated biphenyls (PCBs) are examples of common endocrine disruptors [51]. The graphene electrode can be functionalized with receptors or antibodies specific to the target endocrine disruptors, allowing for their selective capture and detection. One example is the detection of BPA using a graphene-based sensor. These sensors can be modified with molecular receptors that bind to BPA molecules. When BPA binds to the receptor on the graphene surface, it causes changes in the sensor's electrical properties, which can then be measured and correlated to the BPA concentration in the water sample. In a study conducted by Alam and Deen, a remarkable advancement in sensor fabrication was achieved for the detection of BPA by employing a composite material comprising GO, β-cyclodextrin (β-CD)-functionalized multiwalled carbon nanotubes (MWCNTs), and a solid-phase electrode (SPE) [52]. Also, Tian *et al* (2020) developed a highly effective sensor for the determination of BPA through the integration of manganese dioxide nanowires (MnO_2 NWs) and rGO. The MnO_2 nanowires/GO suspension was drop-casted onto a GCE for testing. This sensor demonstrated a significantly lower detection limit (6 nM) compared to the widely used high-performance liquid chromatography (HPLC) technique, underscoring the remarkable sensitivity achieved by the electrode [53].

14.4.3 Microbial detection

Waterborne microbial contamination [54] is a serious threat to public health, emphasizing the importance of developing reliable and sensitive detection methods to ensure the safety of water resources. Traditional microbial detection methods frequently necessitate time-consuming laboratory procedures, making real-time monitoring difficult. Integration of graphene electrodes into sensing platforms has emerged as a promising approach for the rapid and sensitive detection of micro-organisms in water in recent years.

14.4.3.1 Bacteria

Bacterial contamination in water can result in waterborne illnesses. According to reports from the World Health Organization (WHO), approximately 1–2 million human deaths are attributed to diarrhea as an effect of bacterial encounters. Graphene-based sensors have been used to detect bacterial pathogens, providing both rapid and sensitive detection capabilities. One method for detecting bacteria is to functionalize the graphene surface with specific receptors or antibodies capable of

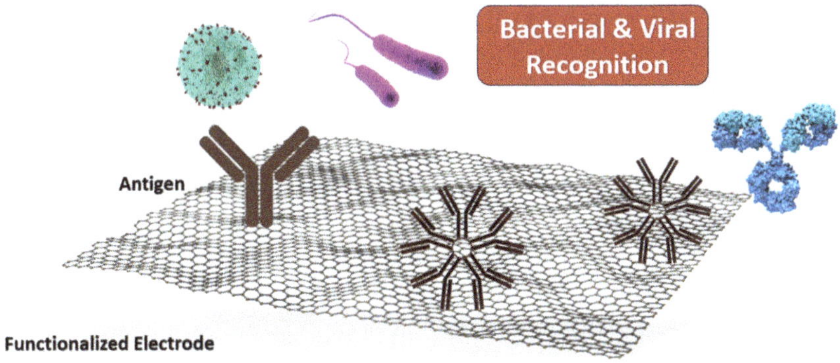

Figure 14.5. Bacterial and viral recognition using functionalized electrodes.

selectively capturing target bacteria (figure 14.5). Changes in the electrical conductivity or impedance of the graphene-based sensor occur when bacteria bind, allowing detection and quantification. Thakur *et al* published a study describing an rGO-FET biosensor that was modified with *Escherichia coli* antibodies (anti-*E. coli*) for the purpose of selectively detecting single *E. coli* bacteria. The sensor exhibited high sensitivity and selectivity, enabling the rapid detection of *E. coli* in water samples. When *E. coli* binds to the GFET, it induces a conformational change in the aptamer, causing the negatively charged *E. coli* to approach the surface of the graphene channel. This interaction leads to a noticeable right shift (p-doping effect) in the transfer characteristic curves [55]. Kim *et al* presented a novel approach where they developed a graphene micropattern FET (GMFET), a biosensor device combined with a microfluidic (MF) chip platform. This system aimed at enabling the early detection of both Gram-positive and Gram-negative bacteria. The integration of these technologies offered a portable platform with the potential for real-time on-site detection of pathogenic bacteria in environmental settings [56].

14.4.3.2 Viruses

Viruses, including waterborne viruses, have the potential to cause severe outbreaks and public health concerns. Graphene-based sensors have been investigated for virus detection, offering sensitive and specific detection capabilities [57]. One method for detecting viruses is to imbue the graphene electrode with aptamers or antibodies that specifically bind to the target virus. The virus's binding to the functionalized graphene surface causes changes in the electrical properties or optical signals, allowing them to be detected. Graphene-based sensors can also use techniques like FETs for label-free and real-time virus detection. Virus attachment to the graphene surface changes the conductance of the transistor, allowing for sensitive virus detection. In a pioneering study, Kim *et al* (2018) developed a graphene-based aptasensor specifically designed for the detection of norovirus GII. The guanine residues present in the single-stranded DNA aptamers underwent a reaction with 3,4,5-trimethoxylphenylglyoxal, resulting in the formation of a high-energy intermediate. This intermediate efficiently transferred energy to a fluorescent dye, such as fluorescein or 6-FAM, leading to the emission of detectable light. The aptasensor demonstrated an impressive detection limit of 80 ng ml^{-1} in tap water, showcasing its sensitivity for norovirus GII detection [58]. In addition to these, pathogen detection using graphene-based sensors can be accomplished through a variety of mechanisms, including impedance spectroscopy, electrochemical sensing, and optical sensing. These methods rely on contaminant interaction with graphene surfaces to produce measurable changes in electrical, electrochemical, or optical properties.

In impedance spectroscopy-based sensors, for example, the presence of a pathogen can change the electrical impedance of the graphene-based sensor. Furthermore, the combination of microfluidic devices and graphene-based sensors can provide a controlled and optimized environment for pathogen detection. Microfluidics enables the precise manipulation and analysis of small volumes of water samples, allowing for real-time monitoring and high-throughput protozoa screening. These sensors can provide sensitive, selective, and rapid detection of microbial organisms by leveraging the unique properties of graphene and various sensing mechanisms, ensuring the safety and quality of water resources. Continued research and development in this field have the potential to improve water quality monitoring techniques while also protecting public health.

14.4.4 Water quality monitoring

Ensuring the quality of water resources is critical for the wellbeing of both ecosystems and human populations. Water pollution caused by a variety of contaminants, including heavy metals, organic pollutants, and microbial pathogens, necessitates effective monitoring strategies to assess and maintain water quality. Graphene electrodes have emerged as a promising tool for water quality monitoring in recent years, offering numerous advantages for accurate and sensitive pollutant detection.

14.4.4.1 pH sensing with graphene electrodes

pH is a fundamental parameter that reflects a solution's acidity or alkalinity. pH-sensitive dyes or ion-selective electrodes are frequently used in traditional pH sensors. Because of their high sensitivity and stability, graphene-based electrodes, on the other hand, offer a promising alternative. The pH sensing mechanism based on graphene electrodes is based on changes in graphene electrical properties when exposed to different pH values. When graphene comes into contact with a solution, the adsorption of protons (H^+) or hydroxide ions (OH^-) from the solution changes the charge carrier density in graphene, resulting in changes in its electrical conductivity [59]. Changes in conductivity can be measured and correlated to the pH value of a solution using a graphene-based pH sensor. The sensor can be calibrated using standard pH solutions to create a calibration curve for precise pH measurement. The Nernst equation can be used to describe the relationship between the pH value and the electrical conductivity of the graphene-based pH sensor.

14.4.4.2 Conductivity and temperature sensing

Temperature and conductivity are critical parameters in water quality monitoring. The conductivity sensing mechanism is based on changes in graphene's electrical conductivity when exposed to various contaminants or conductive solutions. These sensors could be built as a microfluidic device, with graphene-based electrodes integrated into a microchannel. The solution's conductivity can be determined by measuring the electrical conductivity between the electrodes. The temperature-dependent electrical resistance of graphene is used in temperature sensing with graphene-based sensors. The temperature of the surrounding environment or the water sample can be determined by measuring the resistance of the graphene-based sensor. This is especially useful for monitoring water temperature in a variety of applications, such as aquaculture and environmental studies. Poma *et al* (2019) presented a graphene-based pH sensor as an integral component of an autonomous system designed for remote pH and temperature monitoring in marine environments. The pH measurement was achieved using a potentiometric sensor, equipped with a wireless, real-time acquisition system based on smartphone technology. The pH sensor underwent deployment at sea for eight days. Throughout this period, pH readings were recorded at hourly intervals. In both the laboratory and in-field experiments, a commercially available glass electrode pH meter served as the reference device. Remarkably, the versatile and cost-effective graphene-based system demonstrated comparable performance to the reference glass pH electrode, thus validating its efficacy. Notably, the graphene-based system exhibited superior attributes, including reduced energy consumption and enhanced calibration stability, further enhancing its appeal as a reliable alternative to conventional glass electrode-based systems [60].

14.5 Recent advancements in graphene-based sensors

In recent years, graphene-based sensors have seen significant advancements in improving their performance and expanding their applications. These advancements have concentrated on a variety of topics, including the creation of hybrid structures and composites, functionalization and surface modification techniques [61], integration with microfluidics and the Internet of Things (IoT), and the use of 3D printing technology to fabricate graphene electrodes.

14.5.1 Hybrid structures and composites

Researchers have investigated the integration of graphene with other materials to form hybrid structures or composites to improve the sensing capabilities of graphene-based sensors [62]. The performance and selectivity of the sensors can be improved by combining the unique properties of graphene with other materials such as metal nanoparticles, metal oxides, polymers [63] or carbon nanotubes [64]. In a study, Farooq *et al* successfully developed GO@Fe_3O_4 composites by encapsulating Fe_3O_4 nanoparticles with GO sheets, leading to outstanding adsorption capabilities for pollutants. The GO@ Fe_3O_4 composites offered two distinct advantages. First, the complete avoidance of restacking sheets of GO, enhanced the performance. Secondly, the combination of GO and Fe_3O_4 resulted in a highly stable structure due to increased connection areas, making them more extensive. This composite demonstrated exceptional adsorption capacity for pollutants, particularly heavy metals and organic contaminants [65].

14.5.2 Functionalization and surface modification

Surface modification and functionalization techniques are critical in tailoring the properties of graphene-based sensors for specific sensing applications. These processes involve introducing specific functional groups, molecules, or nanostructures onto the graphene surface [66] to tailor its properties and enable selective detection of target analytes. It is designed to increase its affinity for target analytes or to change its electrical and chemical properties [67]. Functionalization and surface modification techniques are commonly employed to enhance the sensitivity and selectivity of graphene electrodes. These techniques involve chemical functionalization, nanoparticle decoration, and the use of molecular probes or recognition elements. Chemical functionalization of graphene electrodes can be achieved using diazonium salts or organic reagents. Diazonium salts react readily with the graphene surface, forming covalent bonds and introducing specific functional groups. Organic reagents, such as organic acids or alkyl halides, can also be used to covalently modify the graphene surface, allowing for tailored functionalization. To improve sensitivity, molecular probes or recognition elements like antibodies and aptamers can be employed. Antibodies can be immobilized onto the graphene surface, facilitating interactions with target antigens through antibody–antigen recognition. This approach is commonly used in biosensing applications. On the other hand, aptamers, which are short single-stranded nucleic acids, can be attached to the graphene surface. Aptamers exhibit high affinity and selectivity for specific target molecules, enabling sensitive detection. Baghayeri *et al* demonstrated a GCE coated with poly (amidoamine) dendrimer functionalized magnetic GO (GO-Fe_3O_4-PAMAM) for the simultaneous detection of Pb(II) and Cd(II) in environmental waters. The functionalization improved the specificity of the sensor toward the target analytes, allowing for highly selective detection of water pollutants [68].

14.5.3 Integration with microfluidics and the Internet of Things (IoT)

The combination of graphene-based sensors with microfluidics and IoT technologies has enabled advanced functionality and real-time monitoring. Microfluidics enables precise fluid sample control and manipulation, whereas IoT enables remote data collection, analysis, and communication [69] (figure 14.6). Microfluidics enables the miniaturization and automatization of complex chemical processes by simplifying their use, which expands its application in the fabrication of more portable devices [70]. This process holds a promising future for the remote sensing and real-time monitoring of data [71]. In a study, Ma *et al* introduced a portable microfluidic electrochemical sensing platform for the quick detection of lead (Pb) utilizing thermocapillary convection. The platform was constructed using a three-dimensional (3D) architecture composed of silver (Ag)–rGO-functionalized nickel

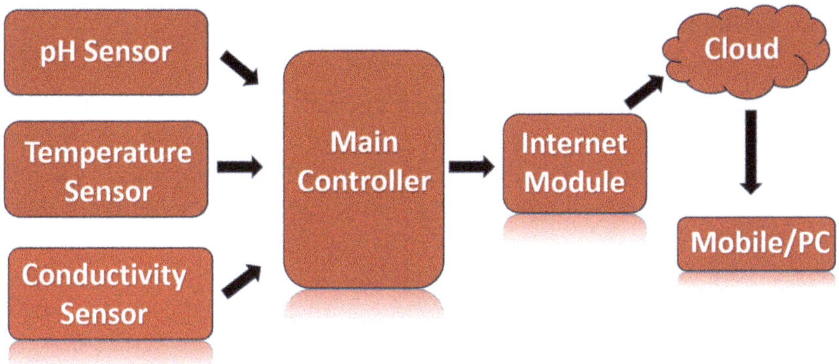

Figure 14.6. IoT enabled water sensing mechanism.

hydroxide (Ni $(OH)_2$) on a nanostructured flexible substrate. The developed sensor exhibited exceptional characteristics, including high selectivity, outstanding repeatability, good reproducibility, and remarkable stability, specifically tailored for the detection of Pb^{2+} [72]. In a pioneering study conducted by Alahi *et al* [73], a graphene-based IoT-enabled sensor was developed for the continuous and live monitoring of nitrate nitrogen in the water. The sensor system was equipped with Wi-Fi connectivity, allowing seamless data transmission to a cloud server for immediate and continuous monitoring. Through rigorous measurements, the sensor demonstrated excellent performance and enabled real-time data monitoring on a broader scale [74].

14.5.4 3D printing of graphene electrodes

With the advent of 3D printing technology, new possibilities for fabricating graphene electrodes with complex geometries and customized designs have emerged. The precise deposition of graphene ink or filaments enabled by 3D printing enables the fabrication of electrodes with controlled architectures and improved performance [75]. 3D printing techniques like direct ink writing (DIW), aerosol jet printing and inkjet printing are used to fabricate graphene electrodes. These advanced methods enable design flexibility, precise control and scalability over manufacturing complex electrodes in no time. Zhao *et al* integrated 3D-printed electrodes with graphene-based nanocomposite modified screen-printed electrodes for the detection of heavy metals. The accuracy and selectivity of the validated method were reported to be high and future promising [76].

14.6 Challenges and future perspectives

Despite significant progress in graphene-based sensors for water pollutant sensing, several challenges and future perspectives must be addressed before their full potential can be realized. Scalability and commercialization, environmental stability and durability, selectivity and sensitivity enhancement, multimodal sensing, and integration with water systems are among the challenges.

14.6.1 Scalability and commercialization

Scalability and cost-effectiveness for mass production are two major challenges in graphene-based sensor technology. While significant progress has been made in the synthesis and fabrication of graphene materials, large-scale production methods that preserve graphene's quality and properties are still being developed. Furthermore, further optimization and standardization are required for the incorporation of graphene-based sensors into practical devices and systems for commercial applications. Researchers are investigating scalable and cost-effective manufacturing processes such as roll-to-roll printing, CVD, and inkjet printing to address these challenges. These techniques enable the mass production of graphene-based sensors with consistent quality and performance on a large scale. Collaborations among academia, industry, and government agencies are also critical for the development of standardized protocols and regulations, as well as for fostering innovation, technology transfer and commercialization efforts.

14.6.2 Environmental stability and durability

Another challenge in the long-term environmental stability and durability of graphene-based sensors for water pollutant sensing is ensuring their long-term environmental stability and durability. When exposed to harsh environmental conditions such as high temperatures, humidity, and chemical interactions, graphene degrades. This can result in decreased performance and shorter sensor lifetimes. Researchers are investigating various protective coatings, encapsulation techniques, and functionalization approaches to improve the environmental stability and durability of graphene-based sensors. These strategies are intended to reduce the effects of environmental factors while also improving sensor resistance to fouling, corrosion, and mechanical stress.

14.6.3 Selectivity and sensitivity enhancement

Selectivity and sensitivity are important parameters in water pollutant sensing because they determine the sensors' ability to distinguish between target pollutants and interfering substances, as well as their detection limits. Improving the selectivity and sensitivity of graphene-based sensors is critical for detecting trace levels of pollutants in complex water samples. Researchers are investigating the functionalization of graphene surfaces with specific receptor molecules or functional groups that selectively interact with target pollutants to improve selectivity. This method improves the sensor's affinity for the desired analytes while reducing interference from other substances in the water matrix. Researchers are investigating various approaches to improve sensitivity, such as optimizing the electrode design, improving charge transfer kinetics at the electrode interface, and developing signal amplification strategies.

14.6.4 Multimodal sensing approaches

Multimodal sensing approaches that integrate graphene with other sensing modalities or combine multiple detection mechanisms have great potential for comprehensive and versatile water pollutant sensing. Multimodal sensors can provide complementary information about the target pollutants by utilizing different sensing principles such as electrochemical, optical, or mechanical sensing, resulting in improved detection accuracy and reliability. A graphene-based sensor, for example, can be used in conjunction with optical techniques such as SERS to enable label-free and highly sensitive detection of specific pollutants.

14.6.5 Integration with water treatment systems

Integrating graphene-based sensors with water treatment systems holds great promise for real-time pollutant monitoring and control during the treatment process. These sensors, which provide continuous and on-site sensing capabilities, can enable rapid feedback and adjustments to optimize treatment efficiency and ensure the safety of treated water. For example, graphene-based sensors can be included in filtration systems to monitor the removal effectiveness of certain contaminants or to identify breakthrough events, signaling the need for filter replacement or maintenance. They can also be integrated into disinfection systems to monitor disinfection byproduct formation and ensure regulatory compliance.

14.7 Conclusion

In summary, owing to their unique properties, graphene-based sensors have significant potential for water pollutant sensing. Researchers have made significant advances in graphene material synthesis and fabrication, and the integration of graphene-based sensors with other technologies. Promising strategies for improving the performance and functionality of graphene-based sensors, hybrid structures and composites, functionalization and surface modification, integration with microfluidics and IoT, and 3D printing of graphene electrodes have emerged. Advances in graphene-based water pollutant sensors have the potential to detect a wide range of pollutants quickly, sensitively, and selectively, enabling proactive pollution control strategies and efficient water treatment processes. The remaining challenges, such as scalability, environmental stability, selectivity and sensitivity enhancement, multimodal sensing approaches, and integration with water treatment systems, should be addressed in future research. By overcoming these obstacles, graphene-based sensors can be further optimized and tailored to specific pollutant detection needs, making them more practical for use in real-world scenarios.

References

[1] Novoselov K S and Geim A K 2007 The rise of graphene *Nat. Mater.* **6** 183–91

[2] Aboamera N M, Mohamed A, Salama A, Osman T A and Khattab A 2018 An effective removal of organic dyes using surface functionalized cellulose acetate/graphene oxide composite nanofibers *Cellulose* **25** 4155–66

[3] Niwa O, Ohta S, Takahashi S, Zhang Z, Kamata T, Kato D and Shiba S 2021 Hybrid carbon film electrodes for electroanalysis *Anal. Sci.* **37** 37–47

[4] Feng L, Xue Q, Liu F, Cao Q, Feng J, Yang L and Zhang F 2020 Voltammetric determination of ofloxacin by using a laser-modified carbon glassy electrode *Microchim. Acta* **187** 1–8

[5] Chen T, Liu Y, Lu J, Xing J, Li J, Liu T and Xue Q 2019 Highly efficient detection of ciprofloxacin in water using a nitrogen-doped carbon electrode fabricated through plasma modification *New J. Chem.* **43** 15169–76
[6] Ueda A, Kato D, Kurita R, Kamata T, Inokuchi H, Umemura S, Hironi S and Niwa O 2011 Efficient direct electron transfer with enzyme on a nanostructured carbon film fabricated with a maskless top-down UV/ozone process *J. Am. Chem. Soc.* **133** 4840–6
[7] Li J, Shao Y, Yin W and Zhang Y 2014 A strategy for improving the sensitivity of molecularly imprinted electrochemical sensors based on catalytic copper deposition *Anal. Chim. Acta* **817** 17–22
[8] Luo G, Zhu L, Li X, Zhou G, Sun J, Chen L, Chao Y, Jiang L and Zhu W 2022 Electrochemical lithium ions pump for lithium recovery from brine by using a surface stability Al_2O_3–ZrO_2 coated $LiMn_2O_4$ electrode *J. Energy Chem.* **69** 244–52
[9] Dang J, Cui H, Li X and Zhang J 2019 Determination of norfloxacin using a tetraoxocalix [2] arene [2]-triazine covalently functionalized multi-walled carbon nanotubes modified electrode *Anal. Sci.* **35** 979–85
[10] Annalakshmi M, Sumithra S, Chen S M, Chen T W and Zheng X H 2020 Facile synthesis of ultrathin $NiSnO_3$ nanoparticles for enhanced electrochemical detection of an antibiotic drug in water bodies and biological samples *New J. Chem.* **44** 10604–12
[11] Fu W, Jiang L, van Geest E P, Lima L M C and Schneider G F 2017 Sensing at the surface of graphene field-effect transistors *Adv. Mater.* **29** 1603610
[12] Kaisti M 2017 Detection principles of biological and chemical FET sensors *Biosens. Bioelectron.* **98** 437–48
[13] Park S J, Seo S E, Kim K H, Lee S H, Kim J, Ha S, Song H S, Lee S H and Kwon O S 2021 Real-time monitoring of geosmin based on an aptamer-conjugated graphene field-effect transistor *Biosens. Bioelectron.* **174** 112804
[14] Chang B Y and Park S M 2010 Electrochemical impedance spectroscopy *Annu. Rev. Anal. Chem.* **3** 207–29
[15] Yasin A S, Mohamed A Y, Kim D H, Doan T L L and Chougule S S 2021 Design of zinc oxide nanoparticles and graphene hydrogel co-incorporated activated carbon for efficient capacitive deionization *Sep. Purif. Technol.* **277** 119428
[16] Mak K F, Ju L, Wang F and Heinz T F 2012 Optical spectroscopy of graphene: from the far infrared to the ultraviolet *Solid State Commun.* **152** 1341–9
[17] Daniyal W, Saleviter S and Fen Y W 2018 Development of surface plasmon resonance spectroscopy for metal ion detection *Sens. Mater.* **30** 2023–38
[18] Fang B Y, Li C, Song Y Y, Tan F, Cao Y C and Zhao Y D 2018 Nitrogen-doped graphene quantum dot for direct fluorescence detection of Al3+ in aqueous media and living cells *Biosens. Bioelectron.* **100** 41–8
[19] Tigari G, Manjunatha J G and Raril C 2021 Electrochemical determination of indigotine based on poly(gibberellic acid)-modified carbon nanotube paste electrode *Environmental Applications of Carbon Nanomaterials-Based Devices* (New York: Wiley) pp 135–46
[20] Pienutsa N, Yannawibut K, Phattharaphongmanee J, Thonganantakul O and Srinives S 2022 Titanium dioxide-graphene composite electrochemical sensor for detection of hexavalent chromium *Int. J. Miner. Metall. Mater.* **29** 529–35
[21] Maji B, Achary L S K, Barik B, Sahoo S J, Mohanty A and Dash P 2022 $MnCo_2O_4$ decorated (2D/2D) rGO/g-C_3N_4-based non-enzymatic sensor for highly selective and

sensitive detection of chlorpyrifos in water and food samples *J. Electroanal. Chem.* **909** 116115
[22] Keser K, Mıhçıokur H and ÇağrıSoylu M 2020 Simple, rapid and sensitive detection of phenylarsine oxide in drinking water using quartz crystal microbalance: a novel surface functionalization technique *ChemistrySelect* **5** 2057–62
[23] Chen K Y, Xu Y T, Zhao Y, Li J K, Wang X P and Qu L T 2022 Recent progress in graphene-based wearable piezoresistive sensors: from 1D to 3D device geometries *Nano Mater. Sci.* **5** 247
[24] Ye S, Wang B, Pu Z, Liu T, Feng Y, Han W, Liu C and Shen C 2021 Flexible and robust porous thermoplastic polyurethane/reduced graphene oxide monolith with special wettability for continuous oil/water separation in harsh environment *Sep. Purif. Technol.* **266** 118553
[25] Raril C and Manjunatha J G 2020 Fabrication of novel polymer-modified graphene-based electrochemical sensor for the determination of mercury and lead ions in water and biological samples *J. Anal. Sci. Technol.* **11** 1–10
[26] Liu C, Chen X, Zong B and Mao S 2019 Recent advances in sensitive and rapid mercury determination with graphene-based sensors *J. Mater. Chem. A Mater.* **7** 6616–30
[27] Tang Y, Li J, Guo Q and Nie G 2019 An ultrasensitive electrochemiluminescence assay for Hg^{2+} through graphene quantum dots and poly (5-formylindole) nanocomposite *Sensors Actuators* B **282** 824–30
[28] Khawashi H P, Chavan V A, Bhagat D S and Deshmukh S U 2022 Recent advances in the detection of lead ions using nanoparticle-based sensors *Biointerface Res. Appl. Chem.* **13** 466
[29] Akhtar M, Tahir A, Zulfiqar S, Hanif F, Warsi M F, Agboola P O and Shakir I 2020 Ternary hybrid of polyaniline-alanine-reduced graphene oxide for electrochemical sensing of heavy metal ions *Synth. Met.* **265** 116410
[30] Xu X, Niu X, Li X, Li Z, Du D and Lin Y 2020 Nanomaterial-based sensors and biosensors for enhanced inorganic arsenic detection: a functional perspective *Sensors Actuators* B **315** 128100
[31] Li J *et al* 2022 Aptasensors based on graphene field-effect transistors for arsenite detection *ACS Appl. Nano Mater.* **5** 12848–54
[32] Yi Y, Zhao Y, Zhang Z, Wu Y and Zhu G 2022 Recent developments in electrochemical detection of cadmium *Trends Environ. Anal. Chem.* **33** e00152
[33] Shan Q, Tian J, Ding Q and Wu W 2022 Electrochemical sensor based on metal-free materials composed of graphene and graphene oxide for sensitive detection of cadmium ions in water *Mater. Chem. Phys.* **284** 126064
[34] Hareesha N and Manjunatha J G 2020 Surfactant and polymer layered carbon composite electrochemical sensor for the analysis of estriol with ciprofloxacin *Mater. Res. Innovations* **24** 349–62
[35] Fraga T J M, Carvalho M N, Ghislandi M G and da Motta Sobrinho M A 2019 Functionalized graphene-based materials as innovative adsorbents of organic pollutants: a concise overview *Braz. J. Chem. Eng.* **36** 1–31
[36] Karthik V, Selvakumar P, Senthil Kumar P, Vo D V N, Gokulakrishnan M and Keerthana P *et al* 2021 Graphene-based materials for environmental applications: a review *Environ. Chem. Lett.* **19** 3631–44
[37] Zhang L *et al* 2021 A highly sensitive and visible-light-driven photoelectrochemical sensor for chlorpyrifos detection using hollow Co9S8@ CdS heterostructures *Sensors Actuators* B **348** 130719

[38] World Health Organization 2015 *Human Biomonitoring: Facts and Figures* (World Health Organization. Regional Office for Europe)

[39] Sundaresan R *et al* 2023 One-dimensional rare-earth tungstate nanostructure encapsulated reduced graphene oxide electrocatalyst-based electrochemical sensor for the detection of organophosphorus pesticide *J. Nanostruct. Chem.* 1–14

[40] Beduk T, Lahcen A A, Tashkandi N and Salama K N 2020 One-step electrosynthesized molecularly imprinted polymer on laser scribed graphene bisphenol A sensor *Sensors Actuators* B **314** 128026

[41] Manjunatha J G 2020 A surfactant enhanced graphene paste electrode as an effective electrochemical sensor for the sensitive and simultaneous determination of catechol and resorcinol *Chem. Data Collect.* **25** 100331

[42] Hareesha N, Manjunatha J G, Amrutha B M, Pushpanjali P A, Charithra M M and Prinith Subbaiah N 2021 Electrochemical analysis of indigo carmine in food and water samples using a poly (glutamic acid) layered multi-walled carbon nanotube paste electrode *J. Electron. Mater.* **50** 1230–8

[43] Qian L, Durairaj S, Prins S and Chen A 2021 Nanomaterial-based electrochemical sensors and biosensors for the detection of pharmaceutical compounds *Biosens. Bioelectron.* **175** 112836

[44] Zhu C, Zhao Q, Wang X, Li Z and Hu X 2021 Ag-nanocubes/graphene-oxide/Au-nanoparticles composite film with highly dense plasmonic hotspots for surface-enhanced Raman scattering detection of pesticide *Microchem. J.* **165** 106090

[45] Alagumalai K, Babulal S M, Chen S M, Shanmugam R and Yesuraj J 2021 Electrochemical evaluation of naproxen through Au@ f-CNT/GO nanocomposite in environmental water and biological samples *J. Ind. Eng. Chem.* **104** 32–42

[46] Vallecillos L, Espallargas E, Allo R, Marcé R M and Borrull F 2019 Passive sampling of volatile organic compounds in industrial atmospheres: uptake rate determinations and application *Sci. Total Environ.* **666** 235–44

[47] Andre R S, Sanfelice R C, Pavinatto A, Mattoso L H C and Correa D S 2018 Hybrid nanomaterials designed for volatile organic compounds sensors: a review *Mater. Des.* **156** 154–66

[48] Manjunatha J G 2020 Poly(adenine) modified graphene-based voltammetric sensor for the electrochemical determination of catechol, hydroquinone and resorcinol *Open Chem. Eng. J.* **14** 52–62

[49] Hareesha N, Manjunatha J G, Alothman Z A and Sillanpää M 2022 Simple and affordable graphene nano-platelets and carbon nanocomposite surface decorated with cetrimonium bromide as a highly responsive electrochemical sensor for rutin detection *J. Electroanal. Chem.* **917** 116388

[50] Topçu E 2020 Three-dimensional, free-standing, and flexible cobalt-based metal-organic frameworks/graphene composite paper: a novel electrochemical sensor for determination of resorcinol *Mater. Res. Bull.* **121** 110629

[51] Yang F W, Li Y X, Ren F Z, Luo J and Pang G F 2019 Assessment of the endocrine-disrupting effects of organophosphorus pesticide triazophos and its metabolites on endocrine hormones biosynthesis, transport and receptor binding in silico *Food Chem. Toxicol.* **133** 110759

[52] Alam A U and Deen M J 2020 Bisphenol A electrochemical sensor using graphene oxide and β-cyclodextrin-functionalized multi-walled carbon nanotubes *Anal. Chem.* **92** 5532–9

[53] Tian Y, Deng P, Wu Y, Li J, Liu J, Li G and He Q 2020 MnO_2 nanowires-decorated reduced graphene oxide modified glassy carbon electrode for sensitive determination of bisphenol A *J. Electrochem. Soc.* **167** 046514

[54] Reta N, Saint C P, Michelmore A, Prieto-Simon B and Voelcker N H 2018 Nanostructured electrochemical biosensors for label-free detection of water-and food-borne pathogens *ACS Appl. Mater. Interfaces* **10** 6055–72

[55] Thakur B, Zhou G, Chang J, Pu H, Jin B, Sui X, Yuan X, Yang C-H, Magruder M and Chen J 2018 Rapid detection of single *E. coli* bacteria using a graphene-based field-effect transistor device *Biosens. Bioelectron.* **110** 16–22

[56] Kim K H *et al* 2020 High-performance portable graphene field-effect transistor device for detecting Gram-positive and-negative bacteria *Biosens. Bioelectron.* **167** 112514

[57] Gall A M, Mariñas B J, Lu Y and Shisler J L 2015 Waterborne viruses: a barrier to safe drinking water *PLoS Pathog.* **11** e1004867

[58] Kim B, Chung K W and Lee J H 2018 Non-stop aptasensor capable of rapidly monitoring norovirus in a sample *J. Pharm. Biomed. Anal.* **152** 315–21

[59] Avolio R *et al* 2022 Review of pH sensing materials from macro-to nano-scale: recent developments and examples of seawater applications *Crit. Rev. Environ. Sci. Technol.* **52** 979–1021

[60] Poma N *et al* 2019 Remote monitoring of seawater temperature and pH by low cost sensors *Microchem. J.* **148** 248–52

[61] Prinith N S and Manjunatha J G 2019 Surfactant modified electrochemical sensor for determination of anthrone – a cyclic voltammetry *Mater. Sci. Energy Technol.* **2** 408–16

[62] Ali Z, Ullah R, Tuzen M, Ullah S, Rahim A and Saleh T A 2022 Colorimetric sensing of heavy metals on metal doped metal oxide nanocomposites: a review *Trends Environ. Anal. Chem.* **37** e00187

[63] Charithra M M and Manjunatha J G 2021 Electroanalytical determination of acetaminophen using polymerized carbon nanocomposite based sensor *Chem. Data Collect.* **33** 100718

[64] Kanoun O *et al* 2021 A review of nanocomposite-modified electrochemical sensors for water quality monitoring *Sensors* **21** 4131

[65] Farooq N, Luque R, Hessien M M, Qureshi A M, Sahiba F, Nazir M A and ur Rehman A 2021 A comparative study of cerium-and ytterbium-based GO/g-C_3N_4/Fe_2O_3 composites for electrochemical and photocatalytic applications *Appl. Sci.* **11** 9000

[66] Charithra M M, Manjunatha J G, Prinith N S, Pushpanjali P A, Girish T and Hareesha N 2022 Electroanalytical determination of tinidazole by using surface modified carbon nano composite based sensor *Mater. Res. Innovations* **26** 285–94

[67] Mehta A, Mishra A, Basu S, Shetti N P, Reddy K R, Saleh T A and Aminabhavi T M 2019 Band gap tuning and surface modification of carbon dots for sustainable environmental remediation and photocatalytic hydrogen production – a review *J. Environ. Manage.* **250** 109486

[68] Baghayeri M, Alinezhad H, Fayazi M, Tarahomi M, Ghanei-Motlagh R and Maleki B 2019 A novel electrochemical sensor based on a glassy carbon electrode modified with dendrimer functionalized magnetic graphene oxide for simultaneous determination of trace Pb (II) and Cd (II) *Electrochim. Acta* **312** 80–8

[69] Jaywant S A and Arif K M 2019 A comprehensive review of microfluidic water quality monitoring sensors *Sensors* **19** 4781

[70] Adu-Manu K S, Tapparello C, Heinzelman W, Katsriku F A and Abdulai J D 2017 Water quality monitoring using wireless sensor networks: current trends and future research directions *ACM Trans. Sens. Networks* **13** 1–41

[71] Chen X, Zhang S, Han W, Wu Z, Chen Y and Wang S 2018 A review on application of graphene-based microfluidics *J. Chem. Technol. Biotechnol.* **93** 3353–63

[72] Ma S, Zhao W, Zhang Q, Zhang K, Liang C, Wang D, Liu X and Zhan X 2023 A portable microfluidic electrochemical sensing platform for rapid detection of hazardous metal Pb2+ based on thermocapillary convection using 3D Ag-rGO-f-Ni (OH) 2/NF as a signal amplifying element *J. Hazard. Mater.* **448** 130923

[73] Alahi M E E, Nag A, Mukhopadhyay S C and Burkitt L 2018 A temperature-compensated graphene sensor for nitrate monitoring in real-time application *Sensors Actuators* A **269** 79–90

[74] Li T, Xia M, Chen J, Zhao Y and De Silva C 2017 Automated water quality survey and evaluation using an IoT platform with mobile sensor nodes *Sensors* **17** 1735

[75] Ambaye A D, Kefeni K K, Mishra S B, Nxumalo E N and Ntsendwana B 2021 Recent developments in nanotechnology-based printing electrode systems for electrochemical sensors *Talanta* **225** 121951

[76] Zhao G, Tran T T, Modha S, Sedki M, Myung N V, Jassby D and Mulchandani A 2022 Multiplexed anodic stripping voltammetry detection of heavy metals in water using nano-composites modified screen-printed electrodes integrated with a 3D-printed flow cell *Front. Chem.* **10** 60

IOP Publishing

Graphene-Based Sensors

Jamballi G Manjunatha

Chapter 15

Biomedical applications of graphene-based electrochemical sensing devices

Maitri Shah, Pratik Kolhe, Drishya Prakashan and Sonu Gandhi

Graphene is a promising 2D nanomaterial that has attracted extensive interest because of its unique physiochemical features. Graphene-based nanomaterials are frequently employed in the manufacturing of biomedical devices and in a variety of clinical applications due to their optical, physical, and electrochemical capabilities. Because of its exceptional and superior physical and electric properties, graphene-based materials are commonly employed in the manufacture of high-sensitivity electrochemical sensors (ECSs). This chapter is intended to describe an overview on the biomedical uses, significance, and mechanisms of graphene-based ECSs. The electronic characteristics of graphene and graphene-derived materials that contribute to their exceptional sensing capacity are also explored in brief. Moreover, the chapter summarizes and critically evaluates recent breakthroughs in graphene-based electrochemical devices used in the biomedical area to detect a widespread range of analytes of interest, including small molecules such as nucleic acids, and extending to larger antibodies, proteins, microorganisms and drugs. Additionally, several prominent types of sensors and the combination with graphene are also listed in this chapter. Finally, a quick conclusion regarding in what way graphene and its variants may possibly impact the future of ECS applications is also discussed.

15.1 Introduction

In recent years, there has been a noteworthy rise in global importance since the invention of ECSs. ECSs have attracted attention from academics because of their great sensitivity and high selectivity (Kaushik *et al* 2021). The current rise in nanotechnology and nanomaterials has created new opportunities for developing electrodes modified with nanomaterials for use in ECSs (Yasun *et al* 2020, Byakodi *et al* 2022, Chouhan *et al* 2021). In this regard, there have been attempts to

doi:10.1088/978-0-7503-5578-0ch15

concentrate on improving catalytic performance by modifying electrodes using nanomaterials that are carefully assembled onto the surface of an electrode (Sumitha and Xavier 2023, Arduini *et al* 2017, Uçar *et al* 2023). Many different materials have been used as electrodes, including carbon-based nanomaterials (Shah *et al* 2023, Mhaisalkar *et al* 2010), conductive co-polymers, and metal/metal oxide nanoparticles (Shahdeo *et al* 2020, Chouhan *et al* 2023). Amongst which, carbon-based compounds are frequently utilized to create electrodes for electrochemical devices due to their generally wide potential range in water, cost effectiveness, and comparatively chemical dormancy in the majority of the electrolyte solutions (Novoselov *et al* 2012, Hareesha *et al* 2021a). Electrode modification has been done with a variety of carbon-based nanomaterials, including carbon nanotubes (CNTs), graphite, fullerenes, diamond, carbon dots, and graphene. A significant electrode material among these carbon nanostructures is graphene, the paramount nanostructure owing to its superior conductivity and large surface area (Eivazzadeh-Keihan *et al* 2022). In the following chapter, we briefly describe various graphene derivatives, their properties and fabrication processes. Later, the electronic, mechanical and optical properties are explained followed by different modification strategies for the electrodes. Lastly, the various biomedical applications are listed with respect to the detection molecule—antibodies, nucleic acid or enzyme.

Cutting-edge biosensors of various kinds have been created thanks to nanostructures made of graphene and its derivatives. In order to immobilize biomolecules like nucleic acids, antibodies, antigens, enzymes, etc for the creation of extraordinarily sensitive and selective biosensors, innovative sensing platforms have become extremely popular in recent years. The binding of analytes to the surfaces of graphene-based nanoparticles has been accomplished using a variety of techniques, including covalent coupling of biomolecules. Graphene as a two-dimensional (2D) material shows excellent physical, chemical, mechanical and electrochemical properties that has made it an emerging material for use in developing smart biosensors. The exceptional surface area, approximately 2630 m^2 g^{-1} for a single layer of graphene provides a milieu of applications in depositing various kinds of biological relevant materials such as antibodies, nucleic acids and enzymes. The fabricated biosensors can be further employed in early detection of diseases (Islam *et al* 2019b, Nandi *et al* 2020), or large biological microbes such as virus (Roberts *et al* 2022a), bacteria, proteins (Roberts *et al* 2019); or small molecules such as antibiotics, pesticides (Islam *et al* 2019a, Mahendra Wijaya *et al* 2010), nucleic acids, or biologically relevant materials such as dopamine (Pattar *et al* 2013, Manjunatha *et al* 2011, Manjunatha *et al* 2009, Charithra *et al* 2020, Madhavi and Reddy 2012), vanillin (Monnappa *et al* 2021, Hareesha *et al* 2021b) and curcumin (Pushpanjali *et al* 2020, Raril *et al* 2020) Furthermore, there are several wearable sensors developed in the recent times that can be used for real time detection of body fluids as well as physiochemical parameters like blood pressure.

In this chapter, we briefly discuss various graphene derivatives, their properties and fabrication processes. Later, the electronic, mechanical and optical properties are explained followed by different modification strategies for the electrodes. Lastly, the various biomedical applications are listed with respect to the detection molecule —antibodies, nucleic acid or enzyme.

15.2 Characteristics and types of graphene and its derivatives

The outstanding properties and stand-alone characteristics of graphene and its derivatives like graphene oxide (GO), graphene quantum dots (GQDs), reduced graphene oxide (rGO), and CNTs are discussed in the following section. Figure 15.1 represents the classification of graphene and its usage in various fields.

15.2.1 Graphene

Graphene is a one-atom thick atomic layer of 2D cyclohexane. There are a multitude of methods for production of graphene from graphite such as sonication, chemical, mechanical and photo exfoliation, liquid phase exfoliation, molecular beam epitaxy, etc. Graphene possesses some of the most promising features like high thermal conductivity, ultra-high strength, large surface area, optically transparent nature, light weight, as well as high electrical conductivity that make it a perfect candidate for a lot of biomedical applications (Yildiz *et al* 2021).

15.2.2 Graphene oxide

GO is the oxygen rich derivative of graphene. GO shows a substantially higher water solubility in comparison with graphene owing to the availability of oxygen atoms.

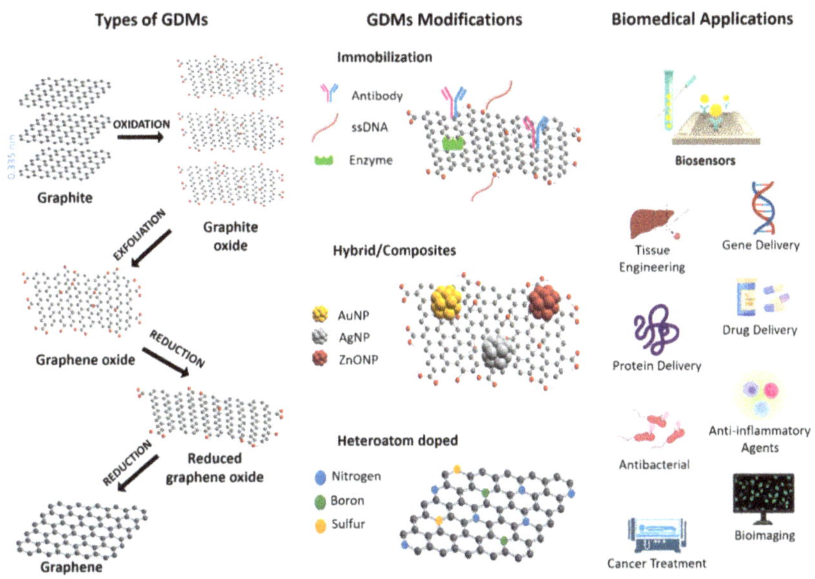

Figure 15.1. The varieties of graphene-derived materials, the route from graphite to graphene, its variations, and possible applications in many study disciplines are depicted schematically. Reproduced from (Badillo-Ramírez *et al* 2022), CC BY 4.0.

This makes it easier for GO to bind with positively or partially positively charged groups due to the presence of negatively charged oxygen. The impurities of oxygen in GO also result in formation of five ringed cyclic structures as compared to six ringed cyclic structures is graphene. This subsequently leads to an increase in resistance due to decrease in free ballistic electron transport (Chen *et al* 2010).

15.2.3 Reduced graphene oxide

rGO is made through electrochemical, thermal or chemical reduction of GO and has less oxygen as compared to GO. Since the overall oxygen group is reduced, rGO possesses a higher conductivity than GO and at the same time retains the water solubility that lacks in pure graphene. Due these characteristics, rGO has been widely used for fabricating ECSs (Pei and Cheng 2012).

15.2.4 Graphene quantum dots

GQDs are less than 100 nm sized photoluminescent particles having phenomenal biomimetic properties due to high permeability and low cytotoxicity. The high surface-to-volume ratio and effortlessly modifiable surface opens an extensive range of applications of GQDs. These advanced multifunctional materials can be used for applications in *in vitro* and *in vivo* research, enzymatic reactions, bioimaging, cancer therapeutics, drug delivery as well as photoluminescent studies. GQDs can further be doped with other transition metal and metal elements to increase the conductivity for further use in electrochemical detection (Li *et al* 2019).

15.2.5 Carbon nanotubes

CNTs are rolled-up allotropes of carbon with diameter in the range of nanometer. Even though not fabricated this way, CNTs can be envisioned as rolled-up cut-outs of 2D graphene nanosheet forming a hollow cylindrical structure that possesses exceptional surface-to-volume ratio. Multi-walled CNTs (MWCNTs) have more than one nanotube embedded one on top of the other either concentrically or eccentrically. CNTs have been employed in the biomedical field due to their high electrical, conductive and tensile properties as compared to other allotropes of carbon (Simon *et al* 2019).

15.2.6 Recent advancement in the catalytic activity of graphene

A recent study published in the *Proceedings of the National Academy of Science* (PNAS) suggested that the strong catalytic activity of graphene is due to the nano-rippled nature of graphene sheet as compared to the flat sheets of carbon in the precursor graphite. This wavy nature makes the splitting of hydrogen atoms from the surface of the graphene surface easily attainable. This recent breakthrough proves that 2D graphene demonstrates superior catalytic and electrochemical activity which is very different in comparison with their inert bulk material graphite (Sun *et al* 2023).

15.3 Properties of graphene

Graphene has numerous exceptional features, including optical transparency, mechanical strength, electric conductivity, and thermal conductivity. It is a 2D crystal having 0.77 mg m^{-2} planar density and is extremely light. Table 15.1 represents some of the properties of graphene and their values. Graphene's unit structure contains a hexagonal carbon ring of 0.052 nm^2 area. Because each atom at the vertices is shared by three-unit rings, such a ring only has two carbon atoms (Zhen and Zhu 2018, Kampouris and Banks 2010). Graphene is made up of only one atomic layer of carbon atoms, explaining its super thin and ultralight properties. If only one layer of graphite could be separated, it may be regarded as perfect graphene. Graphene with less than 10 layers typically retains its 2D crystalline qualities, however beyond that, it begins to exhibit 3D graphite-like electrical features. This metamorphosis may be seen in the alterations of its electrical structure (Abergel *et al* 2010). Graphene's exceptional electrical conductivity is because of its long-range conjugated network of sp^2-hybridized carbon atoms stacked in a honeycomb-like shape. Graphene sheets with resistances as small as 30/sq and conductivities as high as 1.6105 S cm^{-1} have been observed (Bae *et al* 2010).

15.3.1 Electronic properties

Graphene possesses two atoms per unit cell, resulting in two 'conical' sites, K and K′, in the Brillouin zone wherein band crossing happens. The electron energy is directly related on the wave vector that is close to the crossing places. Actually, this behavior arises from symmetry concerns and is hence resistant to long-range hopping processes. The following are the qualities of graphene which result from its electrical structure enabling them to be a great ECS construction material.

15.3.1.1 Ballistic electron transport

Ballistic electron transport attribute refers to graphene's outstanding electron mobility (EM). The typical EM of graphene is around 20 000 cm^2 $(V\ s)^{-1}$, while graphene in suspension has an EM of around 300 000 cm^2 $(V\ s)^{-1}$. Variations in EM

Table 15.1. Properties of graphene.

Sl no.	Properties	Values	Unit
1	Surface area	2360	$m^2\ g^{-1}$
2	Current density	$>10^8$	Acm^{-2}
3	Thermal conductivity	4000–5000	W mK^{-1}
4	Melting point	3800	K
5	Tensile strength	130.5	GPa
6	Fracture strength	125	GPa
7	Young's modulus	1	TPa
8	Light transmittance	97.7%	Percentage
9	Breaking strength	42	N m^{-1}

are caused by graphene inhomogeneity and homogeneity in the unsuspended and suspended states. These properties are mostly hampered by flaws, ripples, contaminants, and so on (Bolotin *et al* 2008).

15.3.1.2 Pseudospins and chirality

In the single unit cell state, graphene has two sublattices. Two wave functions are essential to explain the role of two graphene sublattices in the quasiparticle production method. This is analogous to quantum electrodynamics (QED), except that in this situation, the wave functions only correspond to the sublattices rather than the actual spin of electrons. As a result, these are known as pseudospins. Furthermore, the Dirac massless Fermion equation suggests that graphene electrons possess an electronic wavelength that is significantly greater than that for merely interatomic distances, implying that they possess a virtually infinite Bohr diameter. Nonetheless, at such scales, some unit cell features are maintained in all of these two sublattices, that may be interpreted by chirality in QED (Katsnelson and Novoselov 2007, Katsnelson 2007).

15.3.1.3 Ambipolarity

Ambipolarity describes the capacity of a conducting substance to tune the concentration of charge carriers among negative electrons and positive holes. In graphene, this value can reach 10^{13} (Allen *et al* 2010).

15.3.1.4 Klein paradox and zitterbewegung type behavior

Typically, a species' ability to communicate information via quantum tunneling is highly dependent on the potential height barrier. The outcome of applying the Dirac equation to scattering of electron by a potential barrier is the Klein paradox. According to investigational findings, the possibility of transmission for electrons in graphene approaches unity regardless of barrier potential. The German word for 'trembling motion' is zitterbewegung. The electron wave function exhibits frenetic movement in a confining potential because of the interference induced by the electron and positron energy levels. This is referred to as Zitterbewegung motion. These properties in graphene electrons are because of the defects (Castro Neto *et al* 2009).

15.3.1.5 The absence of weak localization

The weak localization theory describes how materials' conductivity decreases at low temperatures as a result of quantum interferences induced by electrons staying coherent throughout long distances. However, no substantial weak localization effect was detected in graphene, at liquid helium temperatures. This has been demonstrated using the chirality of graphene electrons (Bacon *et al* 2014, Morozov *et al* 2006).

15.3.2 Mechanical properties

The major reason that graphene stands out as a distinct component and as a strengthening element in composites is its outstanding mechanical capabilities.

The extraordinary mechanical capabilities of graphene are due to the stability of the sp^2 bonds that create the hexagonal lattice and resist a wide range of in-plane deformations (Papageorgiou *et al* 2017).

15.3.2.1 Elastic and fracture properties

Graphene possesses outstanding elastic properties, and its strength, elasticity, and intrinsic stress have been studied using both theoretical modeling and lab-based tests. Considering graphene is a nanoscale substance, the majority of its physical properties differ in comparison to bulk graphite (Liu *et al* 2012). Because fracture toughness is so crucial in engineering applications, it is one of graphene's most important mechanical qualities. Zhang *et al* used a scanning electron microscope to create an *in situ* micromechanical testing apparatus and a nanoindenter to determine the fracture toughness of chemical vapor deposition (CVD)-synthesized graphene (Zhang *et al* 2014a). Buehler and colleagues investigated the fracture mechanics of polycrystalline graphene, that is greatly influenced by the mean grain size, using full atomistic simulations. Because of the greater energy release rates, the fracture toughness of polycrystalline graphene was determined to be 20%–35% higher than that of pristine graphene (which reduce with growing grain size) (Jung *et al* 2015). Theoretical results agreed well with experimental data such as those obtained by Zhang *et al.*

15.3.2.2 Thermo-mechanical properties

The most significant issues for micro and nano technologies are thermal control. Because graphene has high thermal conductivity, it will effectively transfer heat from circuit hot spots to heat sinks. At room temperature, the thermal conductivity of defect-free monolayer suspended graphene ranges from 4840 to 5300 W m K^{-1}, according to Balandin *et al* (2008). Because Lee *et al* estimated the intrinsic strength to be 42 N m^{-1} for the monolayer membrane that amounts to an intrinsic strength of 130 GPa, monolayer graphene defect-free is regarded to be the strongest material yet verified (Lee *et al* 2008).

15.3.3 Optical properties

Graphene has excellent optical transitions and it absorbs light at a variety of frequencies. Graphene's capability to absorb radiation from many diverse regions of the electromagnetic spectrum is due to its band structure, the absence of a bandgap, and the interaction between the electromagnetic radiation and the Dirac fermions in the graphene sheet (Obraztsova *et al* 2021). The ability of graphene to absorb 2.3% of white light is yet another unique and intriguing attribute, especially given the fact that it is just one atom thick. This is because of the electrons' aforementioned electronic capabilities, which act as massless charge carriers through extreme high mobility. A few years ago, it was demonstrated that proportion of white light absorbed is determined by the fine structural constant rather than by material properties. Adding an additional layer of graphene raises the quantity of white light absorbed by the same amount (2.3%) (Falkovsky 2008). Graphene can also produce photoluminescence. While its zero bandgap prevents it from forming relaxed states, that is the usual mechanism of photoluminescence, in which an electron is excited to a higher energy band prior to actually liberating a photon as the electron returns to its electronic ground state, pristine graphene has been shown to emit light when excited by a near-infrared laser (Gangwar *et al* 2018, Kavitha and Jaiswal 2016). Because graphene has certain exclusive optical properties and can absorb an extensive range of electromagnetic radiation, it has several potential optical and photonics applications, ranging from saturable absorbers to transparent conductors in photonic devices and high-bandwidth photodetectors.

15.4 Preparation of graphene-based electrochemical electrodes

Graphene and its derivatives have been extensively used in electrochemical researches, particularly in the development of ECSs, owing to its high surface area, high electron transfer rate, electrical conductivity, and capability to immobilize diverse molecules. Graphene has been applied in the designing of numerous sensors for various transduction modes. Additionally, graphene's conjugated structure assists in the transfer of electrons amongst the biorecognition element and transducer, that will result in high signal sensitivity for these sensors (Chauhan *et al* 2017, Pumera 2011, Park *et al* 2016). The development of graphen-based electrochemical electrodes and the use of various graphene-based materials for the development of ECS for chemical and biomolecule detection will be presented in the content that follows.

15.4.1 Electrode fabrication using graphene oxide

GO is extensively used for the development of graphene-based electrodes. The well-known chemical modification method used for the conversion of graphene-to-GO is the Hummers method (Li *et al* 2008). Fabrication can be done by simple procedures, i.e. drop-casting on the surface of the electrode; for this purpose, Pt, Au (Shan *et al* 2010), and glassy carbon electrode (GCE) surfaces can be employed. The electrode serves as a modulating unit in electrochemical biosensors, but the main reaction occurs between the recognition of the probe and the biological molecule. For efficient electrode fabrication, the recognition unit should be attached to the electrode by chemical or physical processes. For chemical conjugation, various functional group conjugation techniques have been utilized by scientists, but in the case of GO or carboxylic acid-containing groups, the widely used technique is zero-length conjugation (Kolhe *et al* 2023), also known as carbodiimide chemistry (Shah *et al* 2022) (figure 15.2). In carbodiimide chemistry, 1-ethyl-3-(3-dimethylamino-propyl) carbodiimide hydrochloride (EDC) and *N*-hydroxysuccinimide (NHS) are used to convert the carboxylic group present in the GO to an amine-reactive *O*-acylisourea intermediate that instinctively reacts with the primary amine group present in the bioreceptor, which could be an enzyme (Liu *et al* 2010), antibody, aptamer, etc. Due to the abundance of oxygen functional groups, GO could be easily functionalized with a variety of substrates, including polymers, nanoparticles, polymers, organometallic compounds, to make electrodes. Qiu and co-workers

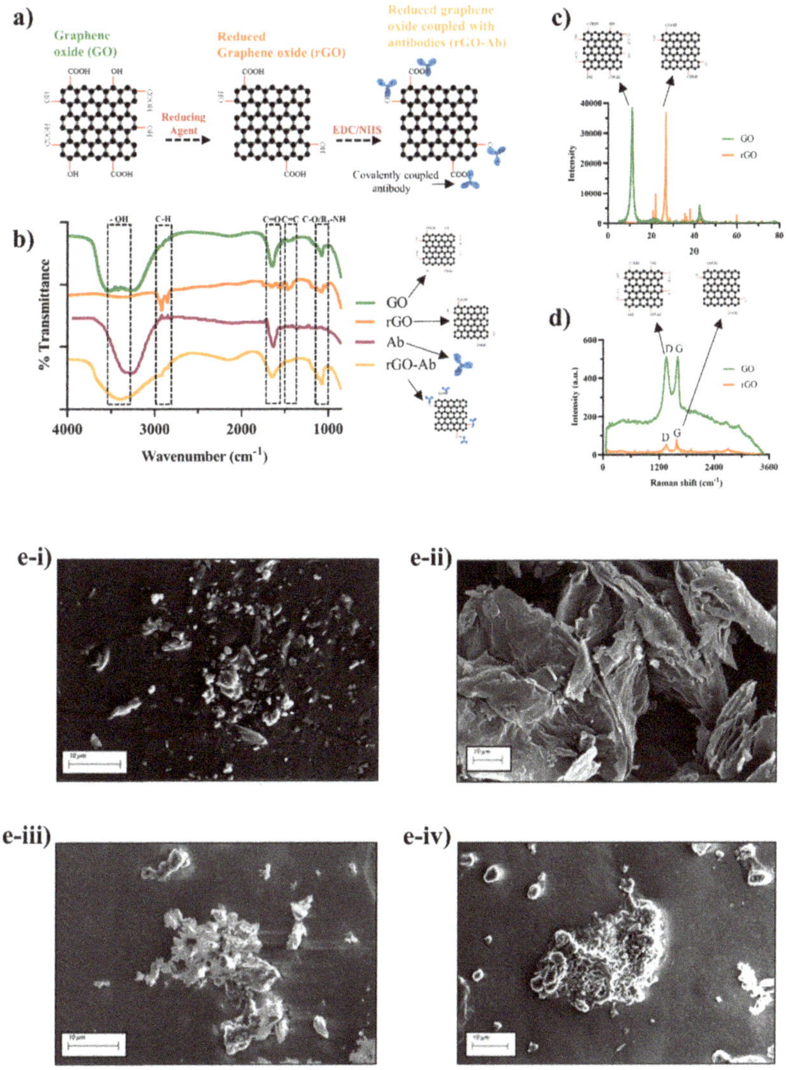

Figure 15.2. Various characterization of rGO. (a) Diagram showing the conversion of GO to rGO, followed by rGO carbodiimide activation and labeling with antibody; (b) FT-IR analysis; (c) XRD analysis; (d) Raman spectroscopy analysis; (e) SEM micrographs of: (i) GO, (ii) rGO, (iii) rGO–Ab and (iv) rGO–Ab–Ag. Reproduced from (Shah *et al* 2022) copyright (2022) with permission from Elsevier.

showed enhanced stability of immobilized enzyme with GO. For immobilization of the enzyme, they combined GO onto GCE with ferrocene branched chitosan (Fc-chitosan) (Qiu *et al* 2011). Other groups used direct adsorption of silver nanoparticles (AgNPs) or *in situ* chemical reduction of Ag to deposit AgNPs onto NH_2-functionalized silica (F-SiO_2)-covered GO nanosheets (Lu *et al* 2011).

15.4.2 Electrode fabrication using reduced graphene oxide

rGO has a heterogeneous structure made up of a graphene-like 2D planner, and also populated with regions containing oxidized chemical groups. rGO has properties that are comparable to those of pristine graphene in the mechanical, optoelectronic, or conductive. Indeed, GO is used as primary derivative of graphene but chemical conversion of graphene-to-GO needs a strong and hazardous oxidizing agent like potassium permanganate ($KMnO_4$), sulfuric acid (H_2SO_4) which creates a notable defect in the crystalline arrangement and ultimately results in a drop in the conductive properties of GO over graphene.

GO can again acquire graphene-like properties by chemical reduction (Iliut *et al* 2013), and hence rGO can be considered an evolving material between graphene and GO. The most commonly used chemical reagents for reduction are hydrazine (Ntsendwana *et al* 2012), borohydrate (Gao *et al* 2009), ascorbic acid (Fernández-Merino *et al* 2010, Kim *et al* 2010), etc. For enhanced covalent bonding or immobilization of bioreceptors, processes like encapsulation and polymerization were performed. rGO-encapsulated polymer matrix used for the enhanced attachment of the probe and conductivity of the working electrode (Bo *et al* 2011). Shah and group used a zero-length conjugation process for the fabrication of an electrochemical immunosensor from rGO (Shah *et al* 2022). rGO can be hybridized with 2D nanomaterials like MXenes and metallic nanostructures to improve the electrochemical performance and activity of bioreceptors like enzymes. Metallic nanoparticles are intertwined like confined nanostructures in the planar structure of rGO. Further, functionalizing rGO in three steps can increase its dissolvability: pre-reducing it with $NaBH_4$, reacting with salt of diazonium to attach sulfonic groups to the partially reduced GO, and then post-reducing it with hydrazine. There are some other modifications that were performed while fabricating the electrode over rGO on GCE, like treatment with ionic liquid (Shan *et al* 2009, Zhang *et al* 2011b), silanization (Hou *et al* 2010), and acid functionalization (Hu *et al* 2011).

As is well known, the regularly used drop-casting approach for fabricating electrodes typically results in an inconsistent layer of rGO thin film on the surface of electrodes and exhibits reduced control over the film thickness. As a result, Zeng and his group established an unusual layer-by-layer (LbL) assembly technique to create multilayer rGO-modified electrodes. Intermolecular and intramolecular forces such as hydrophobic, hydrophilic, and electrostatic forces act on layers in LbL assembly, holding layers together and assisting in self-assembly. These LbL-modified rGO electrodes can be prepared by simultaneously immersing the electrode in polyacid and polybase solutions (Zeng *et al* 2010).

15.4.3 Assembly of graphene-based electrode produced by non-conventional method

A graphene-based electrode can be made with the help of some non-conventional treatments on graphene. Synthesis and functionalization of graphene-based materials using methods such as solvothermal assisted synthesis (Cao *et al* 2010), electrochemical exfoliation (Cao *et al* 2017), microwave-assisted synthesis (Iskandar *et al* 2017), and photo-assisted synthesis (Zhang *et al* 2010), and thermal synthesis (Zhan *et al* 2011). In the current scenario, electrochemically reduced graphene has been utilized for the fabrication purpose. Unnikrishnan and co-workers made an electrochemically reduced electrode using a combination of GO and enzyme on GCE (Unnikrishnan *et al* 2013), while another group of scientists used a suspension of nanofiber (Zhang *et al* 2012) and polymer (Du *et al* 2012) for the preparation of a glucose sensing electrode using electrochemically reduced graphene.

In contrast to the methods discussed above, Shang and his group created multilayer graphene nanoflakes (MGNFs) using microwave-assisted plasma chemical vapor deposition (MPCVD) on a silicon material. These MGNFs are usually used as a working electrode for ECSs (Shang *et al* 2008). Analog to this, using epitaxial graphene as the working electrode, which was initially produced on silicon carbide materials was then given electrochemical treatment for anodization (Lim *et al* 2010).

In the most advanced developments, the electrochemical electrode was made from 3D graphene foam by CVD and electrostatic spray deposition. By changing the metallic substrate, like Ni foam (Dong *et al* 2012c), Co_3O_4 (Dong *et al* 2012b), carbon micropillars (Penmatsa *et al* 2012), and nanowires, scientists increased the active surface area of the electrode, created multiplexed charge transport channels within the electrode, and increased the sensitivity of the ECS.

Moreover, different morphologies of graphene have been used to create ECS, such as 0D GQDs and 2D graphene nanoribbons (GNRs). 0D GQD has advantageous properties over 2D and 3D pristine graphene. MWCNTs were first oxidized, and then they were reduced using a reducing agent like hydrazine (Sinitskii *et al* 2010) to produce GNR. The as-obtained GNRs featured a lot of groups containing oxygen, which helped in finding thiols using the 1,4-Michael addition as well as cycloaddition reaction amongst the thiols and carbonyl groups on the GNRs (Wu *et al* 2012).

15.5 Graphene and graphene-derived materials for the fabrication of electrochemical sensors

Because oxidation and reduction of various compounds occurs at particular potential range, the selection sensitivity is highly dependent on the electrochemically active species/molecule. Owing to the minimal residual current, rapidly renewed exterior, and broad potential range, carbon-based nanomaterials (graphene, graphite, fullerene, and CNTs) are frequently utilized in electrochemical analysis. The high potential for reducing O_2 and the high H_2 mass at the site of its deformed plane provides several active sites for transport of electron to biomolecules. Graphene is an efficient electrical charge conductor, and its heterogeneous electron transport occurs at the graphene's edges or at flaws in the basal plane (Ratinac *et al* 2011, Ananda Murthy *et al* 2021). Hence different graphene-based ECSs for various biomedical applications have been developed (figure 15.3).

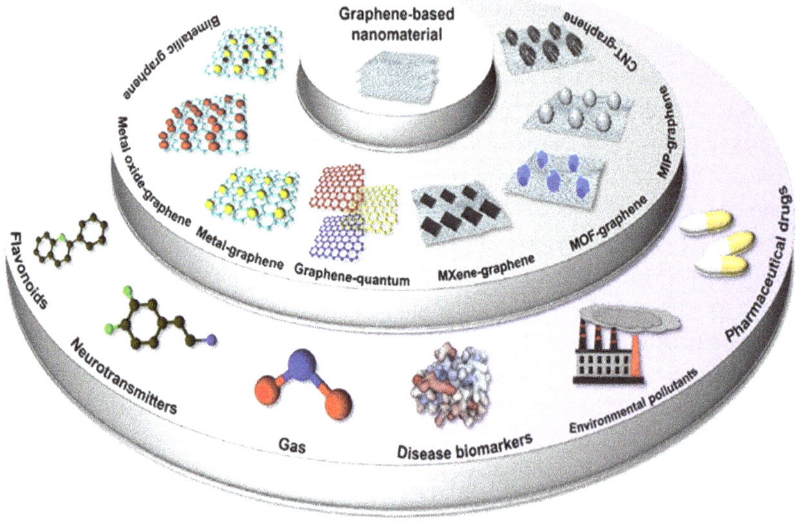

Figure 15.3. Applications of graphene-based ECSs in various fields. Reproduced from (Reddy *et al* 2022), copyright (2022) with permisssion from Elsevier.

15.5.1 Graphene-based DNA sensor

Aptamers are nucleotide sequences, including ribonucleic acid (RNA) and single-stranded deoxyribonucleic acid (ssDNA), or polypeptide molecules that have unique 3D topologies that enable them to attach to their targets with elevated selectivity and efficiency. Even though aptamers can attach to the same target, RNA and ssDNA aptamers might differ from one another in nucleotide sequence and folding arrangement. DNA is considered as a tool for biosensing because of its unique properties of flexibility, easy synthesis, simplistic chemical approach of immobilization on diverse platforms and finally high specificity due to distinctive nucleotide sequences (Pedersen *et al* 2014). Aptamers have been explored as a biocompatible material in several studies pertaining to their usage as a biosensing probe, a diagnostic and therapeutic tool, and in the creation of new pharmaceuticals, drug delivery systems, etc.

DNA shows π–π bonding interaction with graphene and derivatives, that can be affected by the polarizability of the nucleic acid bases and also physical adsorption (Gowtham *et al* 2007, Narlawar and Gandhi 2021). It was discovered that the physical adsorption of nucleic acid bases on the graphene surface can cause base-dependent modification in the (Min *et al* 2011, Bobadilla and Seminario 2013) generated transduced electric signals. Other than physisorption and π–π bonding, there are also covalent immobilization approaches for the fabrication of a DNA-based sensor. It is well known that the GO sheet usually has a variety of functional groups such as carboxylic acid (–COOH), alcohol (–OH), oxo (–C=O), and epoxide. With these groups, carbodiimide chemistry simplifies the immobilization of aptamers over graphene and derivatives.

Electrochemical sensing is dependent on factors like fluctuation in the applied voltage (v), current (I), or impedance (resistance in alternative current), which are based on resultant modifications in electrochemical aspects, like electron-loss, conductance, or change in ability to store current, due to the DNA hybridization or the oxidation of purines and pyrimidine bases of the DNA. Biorecognition event signals can be measured by differential pulse voltammetry (DPV), cyclic voltammetry (CV), and electrochemical impedance spectroscopy (EIS) (Song *et al* 2016). The continuous and ongoing research aims to increase the sensitivity and effectiveness of the transducer unit of the sensor.

The group of researchers proposed the decoration of graphene molecules as a DNA biosensor via through π–π bonding interactions, with positively charged nitrogen species (Zhang *et al* 2014b), quaternary nitrogen salts (Hu *et al* 2012), or nitrogen-containing aromatic molecules (Zheng *et al* 2015) owing elevated sensitivity and specificity. Modified properties of the rGO make the sensor more sensitive and selective. Using rGO, researchers developed the DNA sensor for methicillin-resistant *Staphylococcus aureus* DNA (Wang *et al* 2011) and the amelogenin gene (Benvidi *et al* 2014). The best material for dsDNA detection has been found to be 2D graphene nanosheets in previously reported studies, which is capable of sensing concentrations as low as 0.1 fM. rGO sensors for ssDNA can be modified with labeled ssDNA and gold nanoparticles (ssDNA–AuNPs–ERGO) to boost recognition capability to a lowered limit 0.005 fM (Dong *et al* 2012a). Graphene and its derivatives act as quencher species; this property is utilized in the fabrication of graphene/DNA fluorescence biosensors. Fluorescence quenching and emission depend on the DNA hybridization molecule with the analyte or target species. A gold nanoparticle GO composite electrode with a carbon dot as quencher is used for the detection of HIV-1. The detection is based on FRET (fluorescence resonance energy transfer), with a limit of detection 15 fM (Qaddare and Salimi 2017). The extensive outer layer and conductivity of the nanoparticles are taken into consideration by the electrochemical detection technique. The basis for the detection is deflection in electrical properties due to the presence of purines and pyrimidines in the DNA sequence. For precise measurements, it depends on the uniform deposition of the graphene substance. Moreover, the sensor's effectiveness may be impacted by DNA length and electrostatic potential. Nonetheless, regardless of the DNA's length, fluorescence detection can be carried out on ssDNA or dsDNA. Quenching and optical properties of nanomaterials based on graphene are the foundation of this technique. For the detection of glycated human serum albumin (GHSA), a label-free ECS combined with a GO-based aptamer-based biosensor was created by Waiwinya *et al* (figure 15.4).

15.5.2 Graphene-based immunosensor sensor

Analytical platform that is based on the specific reaction of antigen–antibody known as an immunosensor. The specific nature and biocompatibility of the antibody (Ab) towards the antigen (Ag) make immunosensors useful not only in the diagnostic but also in the therapeutic fields. However, this technique has some drawbacks,

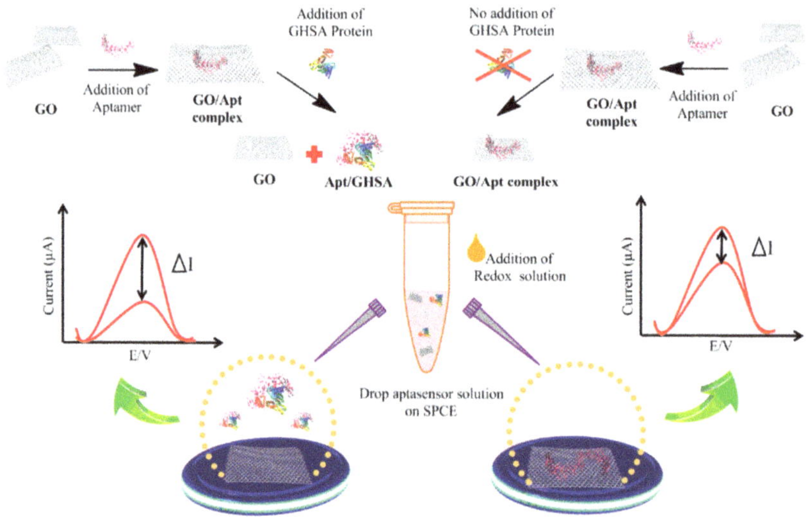

Figure 15.4. Label-free electrochemical aptasensor paired with GO based for detection of glycated human serum albumin (GHSA). When GHSA is present, the aptamer separates from the surface of GO and specifically binds to GHSA, resulting in free GO adsorption on the electrode surface. The aptamer remains attached on the surface of GO in the absence of GHSA, resulting in GO-aptamer complex adsorption on the electrode surface. Following the determination of square wave voltammetry (SWV), change in current (*I*) is calculated and displayed against the concentration of GHSA (Waiwinya *et al* 2021).

including binding specificity and irretrievable Ag–Ab interaction (Singh *et al* 2016). The antibody (Ab) is composed of an arm and a body; in the arm, polypeptide structure of the antibody variable region is responsible for the specificity for epitopes, while in the body, crystallizable fragment (Fc) and antigen-binding fragment (Fab) are present. In the body, carboxylic and amine functional groups are present, which help in the binding of Ab to the fabricated electrode surface (Wujcik *et al* 2014).

The versatile chemical nature of the graphene derivatives makes them easier for conjugation with Ab using variable chemical approaches. These approaches comprise micro-contact printing, biotin–streptavidin binding, direct spotting, electrostatic bonding, or via a 1-pyrenebutanoic acid succinimidyl ester (PASE) linker adsorption to a conductive polymer matrix like polypyrrole (Guo *et al* 2013) or polyaniline (Cosnier and Holzinger 2011), as well as covalent binding. In covalent bonding, carbodiimide bioconjugation is the most well-known and adapted technique.

In recent times, immunosensors have been created for a variety of bacterial and viral infections. Further, graphene and its derivative antibody biosensors have been now widely employed in clinical applications for early detection of disease caused by pathogens like *Escherichia coli* (Huang *et al* 2011b), *Salmonella* (Mahari and Gandhi 2022b, 2022a), *JEV* (Roberts *et al* 2020, 2022b). Graphene-anchored gold particles have been employed for cancer diagnosis, assessment, and forecasting of therapy effectiveness and recurrence (Elshafey *et al* 2016). Coupled graphene particles (Demeritte *et al* 2015) and graphene derivatives (Dey *et al* 2022) are also able to be used for the detection of Alzheimer's. In comparison to rGO, graphene and GO as sensors provide the minimal detection unit (ten times less) for bacterial pathogen sensing. While for viruses, when an antibody is covalently attached to graphene, it can be modified with gold and silver nanoparticles at concentrations of pictograms per milliliter (pg ml^{-1}).

15.5.3 ECSs of graphene based on enzyme and other biologically important molecules

Designing biosensors should pay special attention to enzymes because they are highly stable and easy to modify. Figure 15.5 represents the graphene-based paper device for electrochemical detection of various biomolecules. Moreover, these catalysts are reusable and extremely selective, able to distinguish between L and R enantiomers in various compounds, and they are engaged in the metabolism of all animals. Along with enzymes there is one application of graphene-based sensors

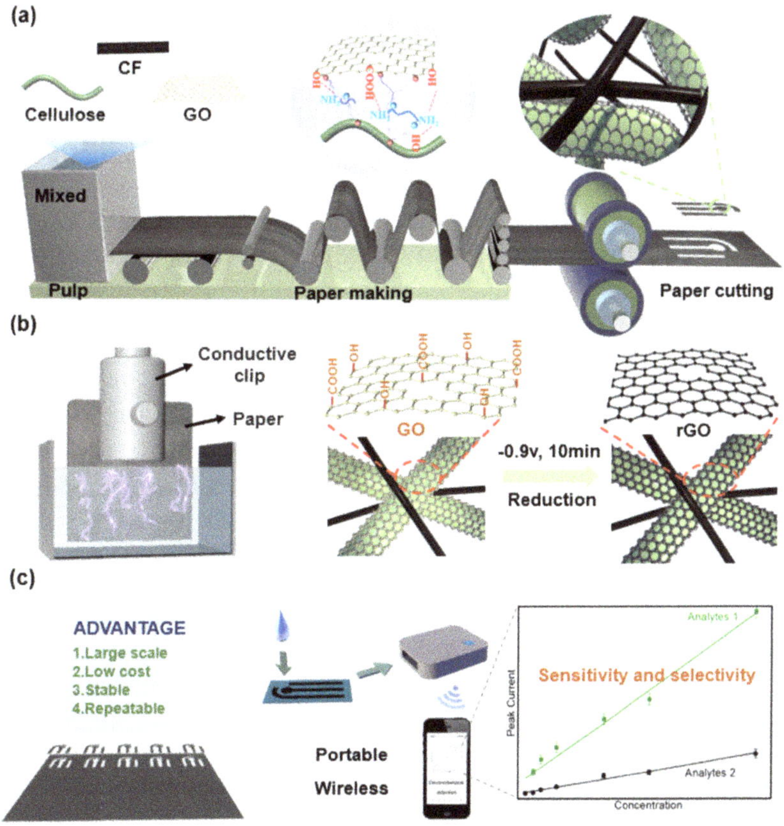

Figure 15.5. (a) Papermaking on a large scale. Composite paper comprising of cellulose, carbon fibers (CF, confers conductivity), and graphene oxidize was created by manipulating interfacial interactions; (b) GO reduction through electrochemistry. Electrochemistry is an easy and scalable method for reduction of GO; (c) illustration of an ECS made of paper. Reproduced from (Wang *et al* 2022), CC BY 4.0.

for recognition of small molecules. Graphene is considered a significant material for sensing of small molecules because of its properties like quick electron transfer reaction as well as better electrocatalytic activity of graphene-based materials. Enzymes have a wide range of reactions that they can catalyze with high specificity, precision, and selectivity—all these are crucial factors for the designing of sensors. The stability of enzymes can be a problem since they can lose their catalytic activity and function as sensors at higher temperatures. Initially, this problem was solved by utilizing thermophilic enzymes, but now stronger thermophilic enzymes are made or changed via biological engineering (Staiano *et al* 2017). It has been found that factors such as pH, chemical inhibitors, ionic strength, temperature and solvent polarity affect the stability of enzymes. The direct transmission of electrons between electrodes and enzymes is made possible by the graphene-based nanomaterials' structure, which makes them efficient transducers. Moreover, it has been demonstrated that graphene-based materials are good materials for boosting thermal stability, activity of enzyme, and enzyme immobilization (Amine *et al* 2006, Roberts *et al* 2021). Two main principles govern enzyme-based electrochemical biosensors: the first is based on the enzymes' catalytic characteristics (the enzyme converts the target analyte from an electro-inactive form to an electro active one), and the second is on the basis of the inhibition or regulation of enzyme activity (Gaudin 2017).

The graphene-based glucose sensor is one of the most thoroughly researched examples since it is crucial for the diagnosis of diabetes to determine the blood glucose level (Jang *et al* 2012). Typically, enzyme glucose oxidase (GOase) is used to modify the graphene-based electrode to detect glucose. Immobilization or conjugation of enzyme have been done during the casting of the GOase-containing solution on the surface of electrode, which is the last stage in the fabrication process. Lee's group employed CVD method combined with other ways to make gold nanoparticle-tailored nitrogen-doped graphene (AuNP/NG) (Thanh *et al* 2016). A glucose biosensor working on the basis of chemically reduced GO (rGO-GOase) has reported increased amperometric signals for detecting glucose with broad linear range (0.01–10 mM), improved detection limit of 2.0 μM and fast response 20.21 μA mM^{-1} cm^{-2} (Shao *et al* 2010). rGO has higher conductivity than GO because of the presence of higher oxygen content, as a result of which GO-based electrodes are less sensitive than the majority of rGO-based ones. Therefore, regarding sensing performance, rGO-based sensors for glucose are preferred over GO-based ones. A metal nanoparticle-coupled electrode exhibits enhanced catalytic activity for glucose sensing due to peroxide to oxygen conversion, as well as amphiphilicity, solubility, and sensitivity. Along with the glucose oxidase, other enzymes including horseradish peroxidase (HRP) (Zhang *et al* 2011a), alcohol dehydrogenase (ADH) (Prasannakumar *et al* 2012), catalase (Huang *et al* 2011a), and urease (Srivastava *et al* 2012) may be immobilized on the graphene-based electrodes for the development of other enzyme-based sensors.

Other biologically active molecules like antibiotics, vitamins, hormones, metabolic products (Asif *et al* 2019), and toxins (Shukla *et al* 2020) can also be monitored with the help of a fabricated graphene-biosensor. Rapid electron transfer properties make graphene and its derivatives an efficient alternative for the detection of small molecules. Graphene-based biosensors have been used to study dopamine, one of the most important hormones researched in the clinical field. Dopamine assessment is important since changes in its level in the body are directly related to one's state of health. Asif and group developed a biosensor to trace the level of dopamine released inside the neural cell with the help of a n-NiAl LDH/rGO-fabricated electrode (Asif *et al* 2019). Xui and co-workers successfully detected dopamine from a human urine sample with their metal alloy-graphene coated electrode. A developed biosensor showed a 75 nM detection limit and sensitivity of 865.8μA mM^{-1} cm^{-2} (Hui *et al* 2019). Metallic nanoparticle-modified graphene-based electrodes have been utilized for the antibiotic's detection in food and soil samples. For detection of chloramphenicol a combined AuNP/GO electrode was used. The developed electrode successfully detects milk, powdered milk and honey (Karthik *et al* 2016). A multilayered film electrode was fabricated using MWCNTs–ionic liquid and further treated with amine modified GO. The developed aptasensor detects traces of kanamycin between the range of 0.001–100 μM (Karthik *et al* 2016). Shah and group developed an ultrasensitive electrochemical immune-biosensor for detection of penicillin G (Shah *et al* 2022). The adaptability of graphene and its nanocomposites in terms of the chemistry for the recognition of various materials is demonstrated by these sensing platforms. Another group used a carbon paste electrode for sensitive electrochemical detection of Estriol (Charithra and Manjunatha 2019), a graphene paste electrode for Catechol (Manjunatha 2019), a CNT paste electrode for L-tryptophan (Prinith and Manjunatha 2020), Alizarin red (Amrutha *et al* 2019), and formoterol fumarate drug (Hareesha and Manjunatha 2021).

15.6 Conclusion and future outlook

Graphene and materials based on it have been widely used as splendid materials to electrochemically detect a variety of biologically active molecules. Researchers have successfully used graphene for the evaluation due to its status as a rising star in material science due to increased affinity, fouling effect reduction, and improved peak resolution. In-spite of the fact that numerous biosensors have been determined using graphene-modified electrodes, the immediate determination of biomolecules is still in its early stages. Moreover, a small number of biomolecules undergo electrochemical oxidation relatively close to each other. Notwithstanding significant challenges in employing graphene as a material to be deposited onto electrodes, we believe it will be a preferable electrical conductor for its biocompatibility, wide potential range, electrochemical dormancy, abundance, conductivity and other properties, economical affordability, absence of metal contaminants, high conductance, large exterior surface, and improved contact with target materials via π–π interaction. Many facts are related to the electrochemical catalysis and the proteins that the graphene-based electrodes are sensitive to. Two types of interactions exist between biomolecules and structural flaws in oxidized graphene planes:

(i) electrostatic interactions; and (ii) the π–π interaction between graphene nanosheets and annular structures in bio-active molecules.

Graphene has emerged as the best contender to be employed as an electrode-modifying substrate due to its substantial surface area and quick electron transport kinetics. For the effective usage of graphene as an electrode, a few challenges must be addressed. In addition to controlling the production of the graphene material, it is necessary to manage homogeneity and size of the graphene films to get repeatable results, that are critical when constructing a sensing device. Using graphene to modify electrodes is a difficult process. Dropcast electrodes exhibit greater electrochemical catalytic activity and exhibit minimal limit of detection for a variety of biomolecules, however, there are still issues with reproducibility and stability. The flaws in graphene sheets are a significant additional issue. It is important to thoroughly investigate the precise relationship between the deformities of the basal plane and the edges and the electrochemical catalytic activity of graphene.

Author contributions

Conceptualization: SG, MS, DP, PK; writing—original draft preparation: MS, PK, and DP; writing-review and editing: MS, DP and PK; supervision: SG.

Conflict of interest

The authors declare no conflict of interest.

Acknowledgments

The authors are grateful to Science and Engineering Research Board (SERB), New Delhi (Grant No.WEA/2020/000036 and CRG/2020/003014). The authors are grateful for the research endowment under the Intensification of Research in High Priority Area (IRHPA) program from Science and Engineering Research Board (SERB), New Delhi (Grant No. IPA/2020/000069). The authors are grateful for funding provided by Department of Biotechnology (DBT), New Delhi (Grant No. BT/PR34216/AAQ/1/765/2019). DP would like to acknowledge DBT Fellowships (DBT/2021–22/NIAB/1706) provided by the Department of Biotechnology (DBT), New Delhi.

References

Abergel D S L, Apalkov V, Berashevich J, Ziegler K and Chakraborty T 2010 Properties of graphene: a theoretical perspective *Adv. Phys.* **59** 261–482

Allen M J, Tung V C and Kaner R B 2010 Honeycomb carbon: a review of graphene *Chem. Rev.* **110** 132–45

Amine A, Mohammadi H, Bourais I and Palleschi G 2006 Enzyme inhibition-based biosensors for food safety and environmental monitoring *Biosens. Bioelectron.* **21** 1405–23

Amrutha B M, Manjunatha J G, Bhatt A S, Raril C and Pushpanjali P A 2019 Electrochemical sensor for the determination of alizarin red-S at non-ionic surfactant modified carbon nanotube paste electrode *Phys. Chem. Res.* **7** 523–33

Ananda Murthy H C, Gebremedhn Kelele K, Ravikumar C R, Nagaswarupa H P, Tadesse A and Desalegn T 2021 Graphene-supported nanomaterials as electrochemical sensors: a mini review *Results Chem.* **3** 100131

Arduini F, Cinti S, Scognamiglio V and Moscone D 2017 Paper-based electrochemical devices in biomedical field: recent advances and perspectives *Compr. Anal. Chem.* **77** 385–413

Asif M, Aziz A, Wang H, Wang Z, Wang W, Ajmal M, Xiao F, Chen X and Liu H 2019 Superlattice stacking by hybridizing layered double hydroxide nanosheets with layers of reduced graphene oxide for electrochemical simultaneous determination of dopamine, uric acid and ascorbic acid *Microchim. Acta* **186** 1–11

Bacon M, Bradley S J, Nann T, Bacon M, Bradley S J and Nann T 2014 Graphene quantum dots *Part. Part. Syst. Charact.* **31** 415–28

Badillo-Ramírez I, Carreón Y J P, Rodríguez-Almazán C, Medina-Durán C M, Islas S R and Saniger J M 2022 Graphene-based biosensors for molecular chronic inflammatory disease biomarker detection *Biosensors* **12** 244

Bae S *et al* 2010 Roll-to-roll production of 30-inch graphene films for transparent electrodes *Nat. Nanotechnol.* **5** 574–8

Balandin A A, Ghosh S, Bao W, Calizo I, Teweldebrhan D, Miao F and Lau C N 2008 Superior thermal conductivity of single-layer graphene *Nano Lett.* **8** 902–7

Benvidi A, Rajabzadeh N, Mazloum-Ardakani M, Heidari M M and Mulchandani A 2014 Simple and label-free electrochemical impedance amelogenin gene hybridization biosensing based on reduced graphene oxide *Biosens. Bioelectron.* **58** 145–52

Bo Y, Yang H, Hu Y, Yao T and Huang S 2011 A novel electrochemical DNA biosensor based on graphene and polyaniline nanowires *Electrochim. Acta* **56** 2676–81

Bobadilla A D and Seminario J M 2013 Assembly of a Noncovalent DNA junction on graphene sheets and electron transport characteristics *J. Phys. Chem.* C **117** 26441–53

Bolotin K I, Sikes K J, Jiang Z, Klima M, Fudenberg G, Hone J, Kim P and Stormer H L 2008 Ultrahigh electron mobility in suspended graphene *Solid State Commun.* **146** 351–5

Byakodi M, Shrikrishna N S, Sharma R, Bhansali S, Mishra Y, Kaushik A and Gandhi S 2022 Emerging 0D, 1D, 2D, and 3D nanostructures for efficient point-of-care biosensing *Biosens. Bioelectron.* X **12** 100284

Cao A, Liu Z, Chu S, Wu M, Ye Z, Cai Z, Chang Y, Wang S, Gong Q and Liu Y 2010 A facile one-step method to produce graphene–CdS quantum dot nanocomposites as promising optoelectronic materials *Adv. Mater.* **22** 103–6

Cao J, He P, Mohammed M A, Zhao X, Young R J, Derby B, Kinloch I A and Dryfe R A W 2017 Two-step electrochemical intercalation and oxidation of graphite for the mass production of graphene oxide *J. Am. Chem. Soc.* **139** 17446–56

Castro Neto A H, Guinea F, Peres N M R, Novoselov K S and Geim A K 2009 The electronic properties of graphene *Rev. Mod. Phys.* **81** 109–62

Charithra M M and Manjunatha J G 2019 Poly (l-Proline) modified carbon paste electrode as the voltammetric sensor for the detection of estriol and its simultaneous determination with folic and ascorbic acid *Mater. Sci. Energy Technol.* **2** 365–71

Charithra M M, Manjunatha J G G and Raril C 2020 Surfactant modified graphite paste electrode as an electrochemical sensor for the enhanced voltammetric detection of estriol with dopamine and uric acid *Adv. Pharm. Bull.* **10** 247–53

Chauhan N, Maekawa T and Kumar D N S 2017 Graphene based biosensors—accelerating medical diagnostics to new-dimensions *J. Mater. Res.* **32** 2860–82

Chen D, Tang L and Li J 2010 Graphene-based materials in electrochemistry *Chem. Soc. Rev.* **39** 3157–80
Chouhan R S, Jerman I, Heath D, Bohm S, Gandhi S, Sadhu V, Baker S and Horvat M 2021 Emerging tri-s-triazine-based graphitic carbon nitride: a potential signal-transducing nano-structured material for sensor applications *Nano Select* **2** 712–43
Chouhan R S, Shah M, Prakashan D, Ramya P R, Kolhe P and Gandhi S 2023 Emerging trends and recent progress of MXene as a promising 2D material for point of care (POC) diagnostics *Diagnostics* **13** 697
Cosnier S and Holzinger M 2011 Electrosynthesized polymers for biosensing *Chem. Soc. Rev.* **40** 2146–56
Demeritte T, Viraka Nellore B P, Kanchanapally R, Sinha S S, Pramanik A, Chavva S R and Ray P C 2015 Hybrid graphene oxide based plasmonic-magnetic multifunctional nano-platform for selective separation and label-free identification of Alzheimer's disease biomarkers *ACS Appl. Mater. Interfaces* **7** 13693–700
Dey J, Roberts A, Mahari S, Gandhi S and Tripathi P P 2022 Electrochemical detection of Alzheimer's disease biomarker, β-secretase enzyme (BACE1), with one-step synthesized reduced graphene oxide *Front. Bioeng. Biotechnol.* **10** 385
Dong H, Zhu Z, Ju H and Yan F 2012a Triplex signal amplification for electrochemical DNA biosensing by coupling probe-gold nanoparticles–graphene modified electrode with enzyme functionalized carbon sphere as tracer *Biosens. Bioelectron.* **33** 228–32
Dong X C, Xu H, Wang X W, Huang Y X, Chan-Park M B, Zhang H, Wang L H, Huang W and Chen P 2012b 3D graphene-cobalt oxide electrode for high-performance supercapacitor and enzymeless glucose detection *ACS Nano* **6** 3206–13
Dong X, Wang X, Wang L, Song H, Zhang H, Huang W and Chen P 2012c 3D graphene foam as a monolithic and macroporous carbon electrode for electrochemical sensing *ACS Appl. Mater. Interfaces* **4** 3129–33
Du M, Yang T, Li X and Jiao K 2012 Fabrication of DNA/graphene/polyaniline nanocomplex for label-free voltammetric detection of DNA hybridization *Talanta* **88** 439–44
Eivazzadeh-Keihan R *et al* 2022 Applications of carbon-based conductive nanomaterials in biosensors *J. Chem. Eng.* **442** 136183
Elshafey R, Siaj M and Tavares A C 2016 Au nanoparticle decorated graphene nanosheets for electrochemical immunosensing of p53 antibodies for cancer prognosis *Analyst* **141** 2733–40
Falkovsky L A 2008 Optical properties of graphene *J. Phys. Conf. Ser.* **129** 012004
Fernández-Merino M J, Guardia L, Paredes J I, Villar-Rodil S, Solís-Fernández P, Martínez-Alonso A and Tascón J M D 2010 Vitamin C is an ideal substitute for hydrazine in the reduction of graphene oxide suspensions *J. Phys. Chem.* C **114** 6426–32
Gangwar P, Singh S and Khare N 2018 Study of optical properties of graphene oxide and its derivatives using spectroscopic ellipsometry *Appl. Phys.* A **124** 1–8
Gao W, Alemany L B, Ci L and Ajayan P M 2009 New insights into the structure and reduction of graphite oxide *Nat. Chem.* **1** 403–8
Gaudin V 2017 Advances in biosensor development for the screening of antibiotic residues in food products of animal origin—a comprehensive review *Biosens. Bioelectron.* **90** 363–77
Gowtham S, Scheicher R H, Ahuja R, Pandey R and Karna S P 2007 Physisorption of nucleobases on graphene: density-functional calculations *Phys. Rev.* B **76** 033401
Guo J, Gu H, Wei H, Zhang Q, Haldolaarachchige N, Li Y, Young D P, Wei S and Guo Z 2013 Magnetite-polypyrrole metacomposites: dielectric properties and magnetoresistance behavior *J. Phys. Chem.* C **117** 10191–202

Hareesha N and Manjunatha J G 2021 Electro-oxidation of formoterol fumarate on the surface of novel poly(thiazole yellow-G) layered multi-walled carbon nanotube paste electrode *Sci. Rep.* **11** 1–14

Hareesha N, Manjunatha J G, Amrutha B M, Pushpanjali P A, Charithra M M and Prinith Subbaiah N 2021a Electrochemical analysis of indigo carmine in food and water samples using a poly(glutamic acid) layered multi-walled carbon nanotube paste electrode *J. Electron. Mater.* **50** 1230–8

Hareesha N, Manjunatha J G, Amrutha B M, Sreeharsha N, Basheeruddin Asdaq S M and Anwer M K 2021b A fast and selective electrochemical detection of vanillin in food samples on the surface of poly(glutamic acid) functionalized multiwalled carbon nanotubes and graphite composite paste sensor *Colloids Surf.* A **626** 127042

Hou S, Kasner M L, Su S, Patel K and Cuellari R 2010 Highly sensitive and selective dopamine biosensor fabricated with silanized graphene *J. Phys. Chem.* C **114** 14915–21

Hu Y, Li F, Bai X, Li D, Hua S, Wang K and Niu L 2011 Label-free electrochemical impedance sensing of DNA hybridization based on functionalized graphene sheets *Chem. Commun.* **47** 1743–5

Hu Y, Wang K, Zhang Q, Li F, Wu T and Niu L 2012 Decorated graphene sheets for label-free DNA impedance biosensing *Biomaterials* **33** 1097–106

Huang K J, Niu D J, Liu X, Wu Z W, Fan Y, Chang Y F and Wu Y Y 2011a Direct electrochemistry of catalase at amine-functionalized graphene/gold nanoparticles composite film for hydrogen peroxide sensor *Electrochim. Acta* **56** 2947–53

Huang Y, Dong X, Liu Y, Li L J and Chen P 2011b Graphene-based biosensors for detection of bacteria and their metabolic activities *J. Mater. Chem.* **21** 12358–62

Hui X, Xuan X, Kim J and Park J Y 2019 A highly flexible and selective dopamine sensor based on Pt-Au nanoparticle-modified laser-induced graphene *Electrochim. Acta* **328** 135066

Iliut M, Leordean C, Canpean V, Teodorescu C M and Astilean S 2013 A new green, ascorbic acid-assisted method for versatile synthesis of Au–graphene hybrids as efficient surface-enhanced Raman scattering platforms *J. Mater. Chem. C Mater.* **1** 4094–104

Iskandar F, Hikmah U, Stavila E and Aimon A H 2017 Microwave-assisted reduction method under nitrogen atmosphere for synthesis and electrical conductivity improvement of reduced graphene oxide (rGO) *RSC Adv.* **7** 52391–7

Islam S, Shukla S, Bajpai V K, Han Y K, Huh Y S, Ghosh A and Gandhi S 2019a Microfluidic-based graphene field effect transistor for femtomolar detection of chlorpyrifos *Sci. Rep.* **9** 1–7

Islam S, Shukla S, Bajpai V K, Han Y K, Huh Y S, Kumar A, Ghosh A and Gandhi S 2019b A smart nanosensor for the detection of human immunodeficiency virus and associated cardiovascular and arthritis diseases using functionalized graphene-based transistors *Biosens. Bioelectron.* **126** 792–9

Jang H D, Kim S K, Chang H, Roh K M, Choi J W and Huang J 2012 A glucose biosensor based on TiO_2–graphene composite *Biosens. Bioelectron.* **38** 184–8

Jung G S, Qin Z and Buehler M J 2015 Molecular mechanics of polycrystalline graphene with enhanced fracture toughness *Extreme Mech. Lett.* **2** 52–9

Kampouris D K and Banks C E 2010 Exploring the physicoelectrochemical properties of graphene *Chem. Commun.* **46** 8986–8

Karthik R, Govindasamy M, Chen S M, Mani V, Lou B S, Devasenathipathy R, Hou Y S and Elangovan A 2016 Green synthesized gold nanoparticles decorated graphene oxide for sensitive determination of chloramphenicol in milk, powdered milk, honey and eye drops *J. Colloid Interface Sci.* **475** 46–56

Katsnelson M I 2007 Graphene: carbon in two dimensions *Mater. Today* **10** 20–7

Katsnelson M I and Novoselov K S 2007 Graphene: new bridge between condensed matter physics and quantum electrodynamics *Solid State Commun.* **143** 3–13

Kaushik A, Khan R, Solanki P, Gandhi S, Gohel H and Mishra Y K 2021 From nanosystems to a biosensing prototype for an efficient diagnostic: a special issue in honor of professor Bansi D. Malhotra *Biosensors* **11** 359

Kavitha M and Jaiswal M 2016 Graphene: a review of optical properties and photonic applications *Asian J. Phys.* **25** 809–31

Kim Y R, Bong S, Kang Y J, Yang Y, Mahajan R K, Kim J S and Kim H 2010 Electrochemical detection of dopamine in the presence of ascorbic acid using graphene modified electrodes *Biosens. Bioelectron.* **25** 2366–9

Kolhe P, Roberts A and Gandhi S 2023 Fabrication of an ultrasensitive electrochemical immunosensor coupled with biofunctionalized zero-dimensional graphene quantum dots for rapid detection of cephalexin *Food Chem.* **398** 133846

Lee C, Wei X, Kysar J W and Hone J 2008 Measurement of the elastic properties and intrinsic strength of monolayer graphene *Science* **321** 385–8

Li D, Müller M B, Gilje S, Kaner R B and Wallace G G 2008 Processable aqueous dispersions of graphene nanosheets *Nat. Nanotechnol.* **3** 101–5

Li M, Chen T, Gooding J J and Liu J 2019 Review of carbon and graphene quantum dots for sensing *ACS Sens.* **4** 1732–48

Lim C X, Hoh H Y, Ang P K and Loh K P 2010 Direct voltammetric detection of DNA and pH sensing on epitaxial graphene: an insight into the role of oxygenated defects *Anal. Chem.* **82** 7387–93

Liu Y, Xie B, Zhang Z, Zheng Q and Xu Z 2012 Mechanical properties of graphene papers *J. Mech. Phys. Solids* **60** 591–605

Liu Y, Yu D, Zeng C, Miao Z and Dai L 2010 Biocompatible graphene oxide-based glucose biosensors *Langmuir* **26** 6158–60

Lu W, Luo Y, Chang G and Sun X 2011 Synthesis of functional SiO_2-coated graphene oxide nanosheets decorated with Ag nanoparticles for H_2O_2 and glucose detection *Biosens. Bioelectron.* **26** 4791–7

Manjunatha J G, Swamy B E K, Mamatha G P, Gilbert O and Sherigara B S 2011 Poly (maleic acid) modified carbon paste electrode for simultaneous detection of dopamine in the presence of uric acid: a cyclic voltammetric study *Anal. Bioanal. Electrochem.* **3** 146–59

Madhavi G and Reddy V M 2012 Selective determination of dopamine in presence of ascorbic acid by using Triton X-100 poly(safranin) modified carbon paste electrode *Anal. Bioanal. Electrochem.* **4** 595–609

Mahari S and Gandhi S 2022a Electrochemical immunosensor for detection of avian Salmonellosis based on electroactive reduced graphene oxide (rGO) modified electrode *Bioelectrochemistry* **144** 108036

Mahari S and Gandhi S 2022b Recent advances in electrochemical biosensors for the detection of salmonellosis: current prospective and challenges *Biosensors* **12** 365

Mahendra Wijaya I P, Nie T J, Gandhi S, Boro R, Palaniappan A, Hau G W, Rodriguez I, Suri C R and Mhaisalkar S G 2010 Femtomolar detection of 2,4-dichlorophenoxyacetic acid herbicides via competitive immunoassays using microfluidic based carbon nanotube liquid gated transistor *Lab Chip* **10** 634–8

Manjunatha J G 2019 Electrochemical polymerised graphene paste electrode and application to catechol sensing *Open Chem. Eng. J.* **13** 81–7

Manjunatha J G, Kumara Swamy B E, Mamatha G P, Gilbert O, Shreenivas M T and Sherigara B S 2009 Electrocatalytic response of dopamine at mannitol and Triton X-100 modified carbon paste electrode: a cyclic voltammetric study *Int. J. Electrochem. Sci.* **4** 1706–18

Mhaisalkar S G, Tey J N, Gandhi S, Wijaya I P M, Palaniappan A, Wei J, Rodriguez I and Suri C R 2010 Direct detection of heroin metabolites using a competitive immunoassay based on a carbon-nanotube liquid-gated field-effect transistor *Small* **6** 993–8

Min S K, Kim W Y, Cho Y and Kim K S 2011 Fast DNA sequencing with a graphene-based nanochannel device *Nat. Nanotechnol.* **6** 162–5

Monnappa A B, Manjunatha J G G, Bhatt A S and Nagarajappa H 2021 Sensitive and selective electrochemical detection of vanillin at graphene-based poly (methyl orange) modified electrode *J. Sci.: Adv. Mater. Devices* **6** 415–24

Morozov S V, Novoselov K S, Katsnelson M I, Schedin F, Ponomarenko L A, Jiang D and Geim A K 2006 Strong suppression of weak localization in graphene *Phys. Rev. Lett.* **97** 016801

Nandi S, Mondal A, Roberts A and Gandhi S 2020 Biosensor platforms for rapid HIV detection *Adv. Clin. Chem.* **98** 1–34

Narlawar S S and Gandhi S 2021 Fabrication of graphene nanoplatelets embedded 'partition cartridge' for efficient separation of target-bound ssDNA during SELEX *Mater. Today Adv.* **12** 100174

Novoselov K S, Fal'Ko V I, Colombo L, Gellert P R, Schwab M G and Kim K 2012 A roadmap for graphene *Nature* **490** 192–200

Ntsendwana B, Mamba B B, Sampath S and Arotiba O A 2012 Electrochemical detection of bisphenol A using graphene-modified glassy carbon electrode *Int. J. Electrochem. Sci.* **7** 3501–12

Obraztsova E D, Rybin M G and Obraztsov P A 2021 Optical properties of graphene *Graphene: Properties, Preparation, Characterization and Applications* ed V Skakalova *et al* (Elsevier) 2nd edn pp 133–42

Papageorgiou D G, Kinloch I A and Young R J 2017 Mechanical properties of graphene and graphene-based nanocomposites *Prog. Mater. Sci.* **90** 75–127

Park C S, Yoon H and Kwon O S 2016 Graphene-based nanoelectronic biosensors *J. Ind. Eng. Chem.* **38** 13–22

Pattar V P, Atmanand D, Bagoji M, Gokavi N, Manjunatha J G, Kumara Swamy B E and Deraman M 2013 Electrochemical studies of dopamine, ascorbic acid and their simultaneous determination at a poly (rosaniline) modified carbon paste electrode: a cyclic voltammetric study *Anal. Bioanal. Electrochem.* **5** 426–38

Pedersen R, Marchi A N, Majikes J, Nash J A, Estrich N A, Courson D S, Hall C K, Craig S L and LaBean T H 2014 Properties of DNA *Handbook of Nanomaterials Properties* ed B Bhushan *et al* (Springer) pp 1125–57

Pei S and Cheng H M 2012 The reduction of graphene oxide *Carbon* **50** 3210–28

Penmatsa V, Kim T, Beidaghi M, Kawarada H, Gu L, Wang Z and Wang C 2012 Three-dimensional graphene nanosheet encrusted carbon micropillar arrays for electrochemical sensing *Nanoscale* **4** 3673–8

Prasannakumar S, Manjunatha R, Nethravathi C, Suresh G S, Rajamathi M and Venkatesha T V 2012 Graphene-carbon nanotubes modified graphite electrode for the determination of nicotinamide adenine dinucleotide and fabrication of alcohol biosensor *J. Solid State Electrochem.* **16** 3189–99

Prinith N S and Manjunatha J G 2020 Polymethionine modified carbon nanotube sensor for sensitive and selective determination of L-tryptophan *J. Electrochem. Sci. Eng.* **10** 305–15

Pumera M 2011 Graphene in biosensing *Mater. Today* **14** 308–15

Pushpanjali P A, Manjunatha J G, Amrutha B M and Hareesha N 2020 Development of carbon nanotube-based polymer-modified electrochemical sensor for the voltammetric study of curcumin *Mater. Res. Innovations* **25** 412–20

Qaddare S H and Salimi A 2017 Amplified fluorescent sensing of DNA using luminescent carbon dots and AuNPs/GO as a sensing platform: a novel coupling of FRET and DNA hybridization for homogeneous HIV-1 gene detection at femtomolar level *Biosens. Bioelectron.* **89** 773–80

Qiu J D, Huang J and Liang R P 2011 Nanocomposite film based on graphene oxide for high performance flexible glucose biosensor *Sensors Actuators* B **160** 287–94

Raril C, Manjunatha J G and Tigari G 2020 Low-cost voltammetric sensor based on an anionic surfactant modified carbon nanocomposite material for the rapid determination of curcumin in natural food supplement *Instrum. Sci. Technol.* **48** 561–82

Ratinac K R, Yang W, Gooding J J, Thordarson P and Braet F 2011 Graphene and related materials in electrochemical sensing *Electroanalysis* **23** 803–26

Reddy Y V M, Shin J H, Palakollu V N, Sravani B, Choi C H, Park K, Kim S K, Madhavi G, Park J P and Shetti N P 2022 Strategies, advances, and challenges associated with the use of graphene-based nanocomposites for electrochemical biosensors *Adv. Colloid Interface Sci.* **304** 102664

Roberts A, Chauhan N, Islam S, Mahari S, Ghawri B, Gandham R K, Majumdar S S, Ghosh A and Gandhi S 2020 Graphene functionalized field-effect transistors for ultrasensitive detection of Japanese encephalitis and Avian influenza virus *Sci. Rep.* **10** 14546

Roberts A, Kesarwani V, Gupta R and Gandhi S 2022a Electroactive reduced graphene oxide for highly sensitive detection of secretory non-structural 1 protein: a potential diagnostic biomarker for Japanese encephalitis virus *Biosens. Bioelectron.* **198** 113837

Roberts A, Mahari S and Gandhi S 2022b Biological/synthetic receptors (antibody, enzyme, and aptamer) used for biosensors development for virus detection *Advanced Biosensors for Virus Detection: Smart Diagnostics to Combat SARS-CoV-2* ed R Khan *et al* (Academic) pp 113–31

Roberts A, Shahdeo D and Gandhi S 2021 Advancement of immobilization techniques in forensic science *Immobilization Strategies. Gels Horizons: From Science to Smart Materials* ed A Tripathi *et al* (Springer) pp 475–507

Roberts A, Tripathi P P and Gandhi S 2019 Graphene nanosheets as an electric mediator for ultrafast sensing of urokinase plasminogen activator receptor-a biomarker of cancer *Biosens. Bioelectron.* **141** 111398

Shah M, Kolhe P and Gandhi S 2023 Nano-assembly of multiwalled carbon nanotubes for sensitive voltammetric responses for the determination of residual levels of endosulfan *Chemosphere* **321** 138148

Shah M, Kolhe P, Roberts A, Shrikrishna N S and Gandhi S 2022 Ultrasensitive immunosensing of Penicillin G in food samples using reduced graphene oxide (rGO) decorated electrode surface *Colloids Surf.* B **219** 112812

Shahdeo D, Roberts A, Abbineni N and Gandhi S 2020 Graphene based sensors *Comprehensive Analytical Chemistry* (Elsevier) pp 175–99

Shan C, Yang H, Han D, Zhang Q, Ivaska A and Niu L 2010 Graphene/AuNPs/chitosan nanocomposites film for glucose biosensing *Biosens. Bioelectron.* **25** 1070–4

Shan C, Yang H, Song J, Han D, Ivaska A and Niu L 2009 Direct electrochemistry of glucose oxidase and biosensing for glucose based on graphene *Anal. Chem.* **81** 2378–82

Shang N G, Papakonstantinou P, McMullan M, Chu M, Stamboulis A, Potenza A, Dhesi S S and Marchetto H 2008 Catalyst-free efficient growth, orientation and biosensing properties of multilayer graphene nanoflake films with sharp edge planes *Adv. Funct. Mater.* **18** 3506–14

Shao Y, Wang J, Wu H, Liu J, Aksay I A and Lin Y 2010 Graphene based electrochemical sensors and biosensors: a review *Electroanalysis* **22** 1027–36

Shukla S, Haldorai Y, Khan I, Kang S M, Kwak C H, Gandhi S, Bajpai V K, Huh Y S and Han Y K 2020 Bioreceptor-free, sensitive and rapid electrochemical detection of patulin fungal toxin, using a reduced graphene oxide@SnO_2 nanocomposite *Mater. Sci. Eng.* C **113** 110916

Simon J, Flahaut E and Golzio M 2019 Overview of carbon nanotubes for biomedical applications *Materials* **12** 624

Singh P, Pandey S K, Singh J, Srivastava S, Sachan S and Singh S K 2016 Biomedical perspective of electrochemical nanobiosensor *Nanomicro Lett.* **8** 193–203

Sinitskii A, Dimiev A, Kosynkin D V and Tour J M 2010 Graphene nanoribbon devices produced by oxidative unzipping of carbon nanotubes *ACS Nano* **4** 5405–13

Song Y, Luo Y, Zhu C, Li H, Du D and Lin Y 2016 Recent advances in electrochemical biosensors based on graphene two-dimensional nanomaterials *Biosens. Bioelectron.* **76** 195–212

Srivastava R K, Srivastava S, Narayanan T N, Mahlotra B D, Vajtai R, Ajayan P M and Srivastava A 2012 Functionalized multilayered graphene platform for urea sensor *ACS Nano* **6** 168–75

Staiano M, Pennacchio A, Varriale A, Capo A, Majoli A, Capacchione C and D'Auria S 2017 *Enzymes as Sensors Methods in Enzymology* ed R B Thompson *et al* (Academic) vol 589 pp 115–31

Sumitha M S and Xavier T S 2023 Recent advances in electrochemical biosensors—a brief review *Hybrid Adv.* **2** 100023

Sun P Z *et al* 2023 Unexpected catalytic activity of nanorippled graphene *Proc. Natl Acad. Sci.* **120** e2300481120

Thanh T D, Balamurugan J, Lee S H, Kim N H and Lee J H 2016 Effective seed-assisted synthesis of gold nanoparticles anchored nitrogen-doped graphene for electrochemical detection of glucose and dopamine *Biosens. Bioelectron.* **81** 259–67

Uçar A, Aydoğdu Tığ G and Er E 2023 Recent advances in two dimensional nanomaterial-based electrochemical (bio)sensing platforms for trace-level detection of amino acids and pharmaceuticals *TrAC, Trends Anal. Chem.* **162** 117027

Unnikrishnan B, Palanisamy S and Chen S M 2013 A simple electrochemical approach to fabricate a glucose biosensor based on graphene–glucose oxidase biocomposite *Biosens. Bioelectron.* **39** 70–5

Waiwinya W, Putnin T, Pimalai D, Chawjiraphan W, Sathirapongsasuti N and Japrung D 2021 Immobilization-free electrochemical sensor coupled with a graphene-oxide-based aptasensor for glycated albumin detection *Biosensors* **11** 85

Wang C, Wu R, Ling H, Zhao Z, Han W, Shi X, Payne G F and Wang X 2022 Toward scalable fabrication of electrochemical paper sensor without surface functionalization *NPJ Flexible Electron.* **6** 1–8

Wang Z *et al* 2011 Label-free, electrochemical detection of methicillin-resistant staphylococcus aureus DNA with reduced graphene oxide-modified electrodes *Biosens. Bioelectron.* **26** 3881–6

Wu S, Lan X, Huang F, Luo Z, Ju H, Meng C and Duan C 2012 Selective electrochemical detection of cysteine in complex serum by graphene nanoribbon *Biosens. Bioelectron.* **32** 293–6

Wujcik E K, Wei H, Zhang X, Guo J, Yan X, Sutrave N, Wei S and Guo Z 2014 Antibody nanosensors: a detailed review *RSC Adv.* **4** 43725–45

Yasun E, Gandhi S, Choudhury S, Mohammadinejad R, Benyettou F, Gozubenli N and Arami H 2020 Hollow micro and nanostructures for therapeutic and imaging applications *J. Drug Deliv. Sci. Technol.* **60** 102094

Yildiz G, Bolton-Warberg M and Awaja F 2021 Graphene and graphene oxide for bio-sensing: general properties and the effects of graphene ripples *Acta Biomater.* **131** 62–79

Zeng G, Xing Y, Gao J, Wang Z and Zhang X 2010 Unconventional layer-by-layer assembly of graphene multilayer films for enzyme-based glucose and maltose biosensing *Langmuir* **26** 15022–6

Zhan D, Ni Z, Chen W, Sun L, Luo Z, Lai L, Yu T, Wee A T S and Shen Z 2011 Electronic structure of graphite oxide and thermally reduced graphite oxide *Carbon* **49** 1362–6

Zhang P *et al* 2014a Fracture toughness of graphene *Nat. Commun.* **5** 1–7

Zhang Q, Qiao Y, Zhang L, Wu S, Zhou H, Xu J and Song X M 2011a Direct electrochemistry and electrocatalysis of horseradish peroxidase immobilized on water soluble sulfonated graphene film via self-assembly *Electroanalysis* **23** 900–6

Zhang Q, Wu S, Zhang L, Lu J, Verproot F, Liu Y, Xing Z, Li J and Song X M 2011b Fabrication of polymeric ionic liquid/graphene nanocomposite for glucose oxidase immobilization and direct electrochemistry *Biosens. Bioelectron.* **26** 2632–7

Zhang Y, Guo L, Wei S, He Y, Xia H, Chen Q, Sun H B and Xiao F S 2010 Direct imprinting of microcircuits on graphene oxides film by femtosecond laser reduction *Nano Today* **5** 15–20

Zhang Y, Wang Y, Jia J and Wang J 2012 Nonenzymatic glucose sensor based on graphene oxide and electrospun NiO nanofibers *Sensors Actuators* B **171–172** 580–7

Zhang Z, Luo L, Chen G, Ding Y, Deng D and Fan C 2014b Tryptamine functionalized reduced graphene oxide for label-free DNA impedimetric biosensing *Biosens. Bioelectron.* **60** 161–6

Zhen Z and Zhu H 2018 Structure and properties of graphene *Graphene: Fabrication, Characterizations, Properties and Applications* ed H Zhu *et al* (Elsevier) pp 1–12

Zheng Q *et al* 2015 An electrochemical DNA sensor based on polyaniline/graphene: high sensitivity to DNA sequences in a wide range *Analyst* **140** 6660–70

IOP Publishing

Graphene-Based Sensors

Jamballi G Manjunatha

Chapter 16

Stability, sustainability and reliability of graphene-based electrochemical sensors

R Rajkumar, J Maalmarugan, P Karpagavinayagam and C Vedhi

Advanced materials like graphene and the family of two-dimensional crystals are very appealing due to the plethora of applications that could be developed based on their exceptional properties. However, once material development is mature enough for production to be scaled up and products to enter the market, it is critical to place the technology in the context of potential risks to economic well-being, social equity, and environmental harm. The purpose of this review is to highlight the current state of the art in the sustainable development of graphene-related materials, as well as related environmental impact assessment studies using life cycle assessment (LCA). We demonstrate that sustainable development has primarily focused on the use of waste or low-cost materials as precursors. However, the findings of relevant LCA studies reveal the limitations of this approach, which fails to account for the fact that waste recycling is frequently very energy intensive. We present an overview of the life cycle environmental impact assessment performed on different graphene production methods for specific applications ranging from composites to electronics, with a focus on global warming potential and energy demand. Finally, an outlook is provided that focuses on the comparison of the various production routes and the LCA results.

16.1 Introduction

Because of growing environmental concerns, environmental sustainability is becoming an increasingly significant component of manufacturing. Given the complexity and energy requirements of advanced materials' manufacturing and storage routes, environmental impact assessments are critical. Plastics have gained popularity due to their versatile properties, low cost, and ease of production. However, the widespread use of plastics has led to significant environmental issues, particularly

doi:10.1088/978-0-7503-5578-0ch16

in terms of waste generation and limited recycling capabilities. The accumulation of plastic waste in the environment has resulted in the formation of microplastics and nanoplastics, which are tiny particles that can have detrimental effects on ecosystems and human health. For example, plastics is one of the most commonly used materials due to its wide range of tuneable physicochemical properties, low weight, resistance to acids, and low-cost production. Yet, the exponential use of plastics with related increase in their waste, and low value recycling, has led us to face new challenges in material science, associated with generation of microplastics and nanoplastics. There is growing evidence that humans are exposed to such micro- and nanoplastics through inhalation or ingestion and that the small size of these particles may allow for bioaccumulation and retention in the body [1–3]. Hence, it is very important to discuss sustainable development of advanced materials already at the early stages of their commercial exploitation.

Graphene, which is defined as a single layer of graphite, is indeed an advanced material with vast potential in numerous applications. Its unique properties, coupled with the development of various production methods, have paved the way for the creation of a diverse family of graphene-related materials (GRMs) (figure 16.1).

The surface properties of graphene can be easily modified through covalent and non-covalent methods, allowing for fine-tuning to meet specific application requirements. This versatility in producing different grades of graphene is particularly advantageous in developing applications, as each grade can be tailored to suit the needs of a particular application. The 'graphene roadmap' envisions the utilization of GRMs across a broad spectrum of fields, including electronics, photonics,

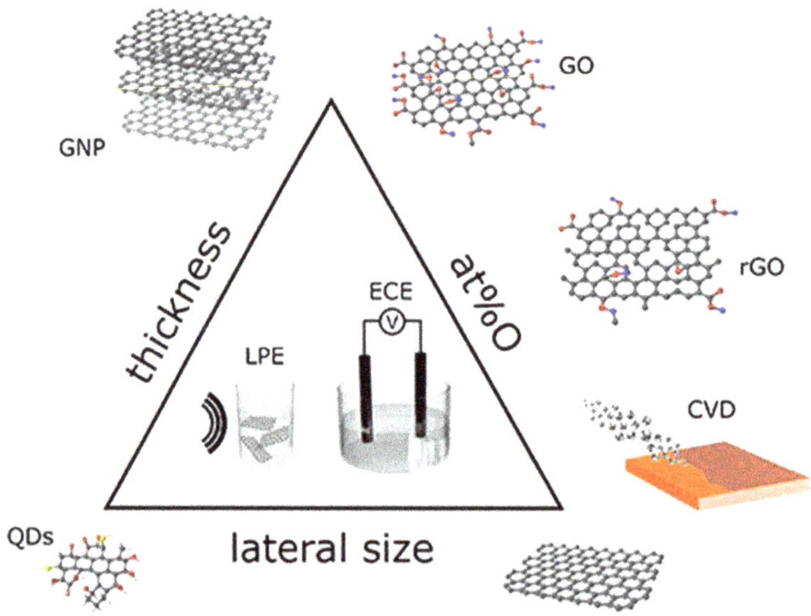

Figure 16.1. Schematic showing the family of GRMs. Reproduced from [85] CC BY 4.0.

sensing, composites, energy storage devices, and bio-applications. While some applications of GRMs are still in the early stages of development, others have already made their way into the market with commercial products available [4–6].

The attempts to synthesize graphene started from 1859 by Brodie, who treated graphite with strong acids to obtain exfoliated graphitic oxide [7]. After that, many methods were developed to separate individual graphene sheets from bulk graphite by overcoming the large interlayer cohesive van der Waals energy of 5.9 kJ mol^{-1} carbon [8] in graphite. These methods are top-down approaches. One of the most popular top-down methods to obtain single-layer graphene sheets is the mechanical exfoliation of highly-oriented pyrolytic graphite (HOPG). However, this method is limited by its low yield [9]. Alternatively, the bottom-up approach is more attractive for the growth of graphene due to its capability to obtain high-quality graphene, such as epitaxial growth on SiC [10], chemical vapor deposition (CVD) on Ni [11], Cu [12] and Ru [13]. Another exceptional bottom-up method, synthesis from organic precursors [14], is different from the aforementioned methods in terms of the growth of large-area graphene, since it can synthesize graphene with well-controlled structure, adjustable solubility and good processability.

However, the solution-chemistry based methods are more practical for high-volume and massive production of graphene at low cost [15]. These methods generally involve the oxidation of graphite-to-graphite oxide, the subsequent exfoliation to graphene oxide (GO) and the final reduction to reduced graphene oxide (rGO, also referred to as graphene in some papers). The oxidation and exfoliation can be achieved by the Hummers method [16] with the aid of strong acids and oxidizers. Many methods have been employed for the reduction of GO, such as chemical reduction using reducing agents (hydrazine) [17–19] dimethylhydrazine [20], hydroquinone [21], vitamin C [22], strong alkaline media [23], bovine serum albumin [24], etc), and electrochemical [25, 26], photochemical [27, 28], thermal [29] and hydrothermal [30] reduction approaches. Although both GO and rGO contain oxygen functional groups, such as epoxy, hydroxyl and carboxyl groups [31], rGO contains less amount of oxygen functional groups after the reduction of GO. These oxygen-containing groups affect their electrochemical performance in terms of electron transfer rate or adsorption/desorption of molecules, and provide anchoring sites to enzymes or other specific species for sensing applications.

Graphene and its derivatives have been widely used in electrochemical studies, particularly in electrochemical sensors, due to their exceptional physical and electrochemical properties and well-developed manufacturing methods. The first section will introduce and discuss the preparation of graphene-based electrochemical electrodes as well as the application of various graphene-based materials for the electrochemical detection of biomolecules and chemicals.

In the subsequent part, we will provide a summary of the works released through the sustainable production and related life cycle assessment (LCA) contest. In the last part of the review, we provide an outlook on the use of waste materials for the manufacture of GRMs by emphasizing the most notable waste resources and methods, taking into account preparation procedures, yield, and final product quality.

16.2 Graphene-based electrochemical sensors

16.2.1 Enzyme biosensors

As mentioned in the Introduction part, graphene possesses outstanding physical and electrochemical properties, such as large surface area, high conductivity, abundant defect sites, superior electrocatalytic activity and fast electron transfer rate, hence various enzyme biosensors based on this attractive material have been fabricated recently. One of the most extensively studied examples is the graphene-based glucose sensor, as the determination of glucose is highly important in diagnosis of diabetes [32]. To detect glucose, the graphene-based electrode is usually modified by glucose oxidase (GOD) in the final step of electrode preparation, i.e. casting of the solution containing GOD onto the electrode.

In addition to GOD, other enzymes may be immobilized onto graphene-based electrodes for construction of various enzyme biosensors, such as horseradish peroxidase (HRP) [33], alcohol dehydrogenase (ADH), catalase [34] and urease [35]. For example, very recently, Prasannakumar *et al* fabricated an alcohol biosensor using graphene-carbon nanotubes (CNTs) modified graphite electrode immobilized with ADH [36]. The as-fabricated electrode showed a linear range of 25–200 μM, a sensitivity of 82.5 nA μM^{-1} cm^{-2} (i.e. 82.5 μA mM^{-1} cm^{-2}) and a detection limit of 10 μM [36]. The sensitivity is a bit higher than that using the ionic liquid-functionalized graphene-based electrode fabricated by Shan *et al* (37.43 μA mM^{-1} cm^{-2}) [37]. The higher sensitivity of graphene-CNT modified electrode could be attributed to the high conductivity of graphene-CNT composite resulted from the interconnections between graphene and CNTs which bridge the gaps among graphene sheets [38].

16.2.2 Nonenzymatic sensors for small biomolecules

Electrochemical sensing of small biomolecules, such as dopamine, ascorbic acid (AA) and uric acid (UA), is another important application of graphene-based electrodes in electrochemistry due to the fast electron transfer kinetics and superior electrocatalytic activity of graphene-based materials. Dopamine is one of the most studied small biomolecules using graphene-based biosensors. Dopamine measurement is crucial since its change in the concentration is closely linked to one's health status [39]. However, it is difficult to be detected selectively and sensitively in the presence of its coexisting species, AA and UA, due to their overlapping voltammetric response [39]. Recently, graphene-based electrodes have shown great performance as electrochemical sensors toward dopamine [40] even in the presence of AA or/and UA due to its excellent physical and electrochemical properties and the feasible electron transfer between graphene and dopamine through the π–π interaction [40, 41].

16.2.3 DNA sensors

With the development of graphene-based electrodes in electrochemical sensors, the application of graphene-based materials has been further explored for detection of

DNA [42–44] and the single-nucleotide polymorphisms (SNPs) without any hybridization or labeling process since the electrode can detect each base in the oligonucleotides quantitatively [45]. Zhou *et al* reported the CrGO-modified glassy carbon electrode (GCE) could be used to detect four free bases of DNA simultaneously, which was impossible to achieve on the GCE or graphite electrode [45]. It could be explained by the abundant edge-plane-like defective sites, antifouling properties and high electron transfer kinetics of CrGO in addition to its unique physical and electrochemical properties [45].

Compared to GCE modified with PTCA functionalized CrGO, the lower detection limit can be attributed to the addition of gold nanoparticles (AuNPs), which can increase the active surface area and conductivity of CrGO and improve the immobilization of probe DNA and magnify DNA hybridization signal. In addition, the abundant imidazolium cations at the outer layer of AuNPs from ionic liquid (IL) could increase the binding of negatively charged probe DNA and hence improve the sensitivity of the electrode [46].

However, the sensor made by graphene quantum dots (GQDs) only showed a linear range of 200–500 nM and a detection limit of 100 nM [47]. The poor sensing performance of GQDs could be attributed to the design of the sensing experiment which used the $[Fe(CN)_6]^{3-/4-}$ as the electrochemical active species for the sensing signal. Theoretically, GQDs with small size would have more edge defects for the interactions with DNA or other analyte molecules, and hence they could produce enhanced signal and exhibit superior sensing performance. Therefore, it is highly expected to construct electrochemical sensors using GQDs based on the well-designed sensing mechanism.

16.2.4 Immunosensors

By immobilizing the antibodies to capture the corresponding antigens on graphene-based electrodes, graphene-based electrochemical immunosensors can be achieved. The high selectivity and sensitivity of the immunosensors can be realized based on the specific interaction between antibodies and antigens. For example, Mao *et al* fabricated an immunosensor using graphene sheet-methylene blue (GS-MB) nanocomposite to detect a model analyte of prostate specific antigen (PSA) [48]. Due to the efficient absorption of antibodies on MB and improved electroactivity of MB in the presence of graphene which possesses a large surface area and excellent conductivity, the sensor showed a linear range of 0.05–5.00 ng ml^{-1} and a detection limit of 13 pg mL^{-1} [48]. The sensitivity can be further improved by using nanomaterials entrapped with antibodies to amplify the electrochemical signals of the immunological reactions, such as quantum dots, CNTs and metal NPs. For example, Yang *et al* constructed a sandwich-type immunosensor using graphene sheets (GSs) and quantum dot functionalized graphene (QD-GS) to immobilize primary antibody (Ab1) and secondary antibody (Ab2), respectively. The as-constructed immunosensor was applied for the detection of PSA and showed a linear range of 0.005–10 ng ml^{-1} and a detection limit of 3 pg ml^{-1} [49]. Lu *et al* built a sandwich-type immunosensor with AuNPs dotted CNTs-graphene composite to detect human chorionic gonadotrophin (hCG) [50]. They particularly emphasized the fixation of AuNPs onto the composite of thionine and mesoporous NPs MCM-41, in which thionine could provide a strong linkage between MCM-41 and AuNPs and enhance the loading of the secondary antibody [50]. In addition, MCM-41 with a high surface-to-volume ratio could also improve the conjugation of biomolecules [50]. In the AuNPs dotted CNTs-graphene composite, AuNPs were used to provide effective electron transport and an ideal microenvironment for the attachment of biomolecules. The coupling between CNTs and graphene was used to improve the electron transfer efficiency and immobilization of the primary antibody [50]. The as-built sensor showed a linear range from 0.005 to 500 mIU ml^{-1} (about 0.0003–30 ng ml^{-1}) and a detection limit of 0.0026 mIU ml^{-1} (about 0.000 156 ng ml^{-1}) [50]. The sensing performance is much superior to the immunosensor fabricated by Li *et al* using nanoporous Au foil and graphene for the detection of hCG with a linear range of 0.5–40.0 ng ml^{-1} and a detection limit of 0.034 ng ml^{-1} [51].

16.3 Environmental impacts of graphene-related materials production processes

As a result of the environmental concerns around the production of nanomaterials, several LCA studies focusing on the environmental impacts of different GRM production processes have emerged in the literature, with the earliest study dating back to 2014 [52]. As can be observed, these studies have looked at few production processes and their variations: chemical oxidation with chemical, biological and thermal reduction using various reduction agents for production of GO and rGO, microwave-assisted exfoliation, liquid phase epitaxy (LPE), electrochemical chloride extraction (ECE), ball milling exfoliation, CVD, and epitaxial growth. All studies included in this section reported direct contributions from the production of GRMs, however, the system boundaries varied from ‘cradle to gate’ [53], ‘cradle to installation’ [54] and ‘cradle to grave’. The functional unit varied across these studies, with the large majority defined in terms of mass, e.g. mg, g or kg of graphene produced. In the particular case of CVD and epitaxial growth, the functional unit has also been defined in terms of surface area in some studies [55], e.g. cm^2 of graphene.

16.3.1 Chemical oxidation with chemical and biological reduction

The studies conducted by Khanam *et al* [56], Cossutta *et al* [57], and Serrano-Luján *et al* [53] assessed the production of GRMs from graphite, and Arvidsson *et al* from hard coal using the chemical oxidation with chemical reduction process. Khanam *et al* [56] also looked at microbial reduction using *Escherichia coli.* All studies reported global warming potential (GWP) and/or energy demand results on a mass basis and are presented in figures 16.2(a) and 16.3(a), respectively.

As it can be observed in figure 16.2(a), there is a large variation in the GWP results reported in these studies. One of the reasons behind this is the different graphite oxidation methods used, e.g. Hummers’ or modified methods (referred to as: Marcano’s, Brodie, Staudenmaier), and the reduction agents employed, e.g. hydrazine, glucose, *E. coli.* In the case of the Hummers’ method using hydrazine as

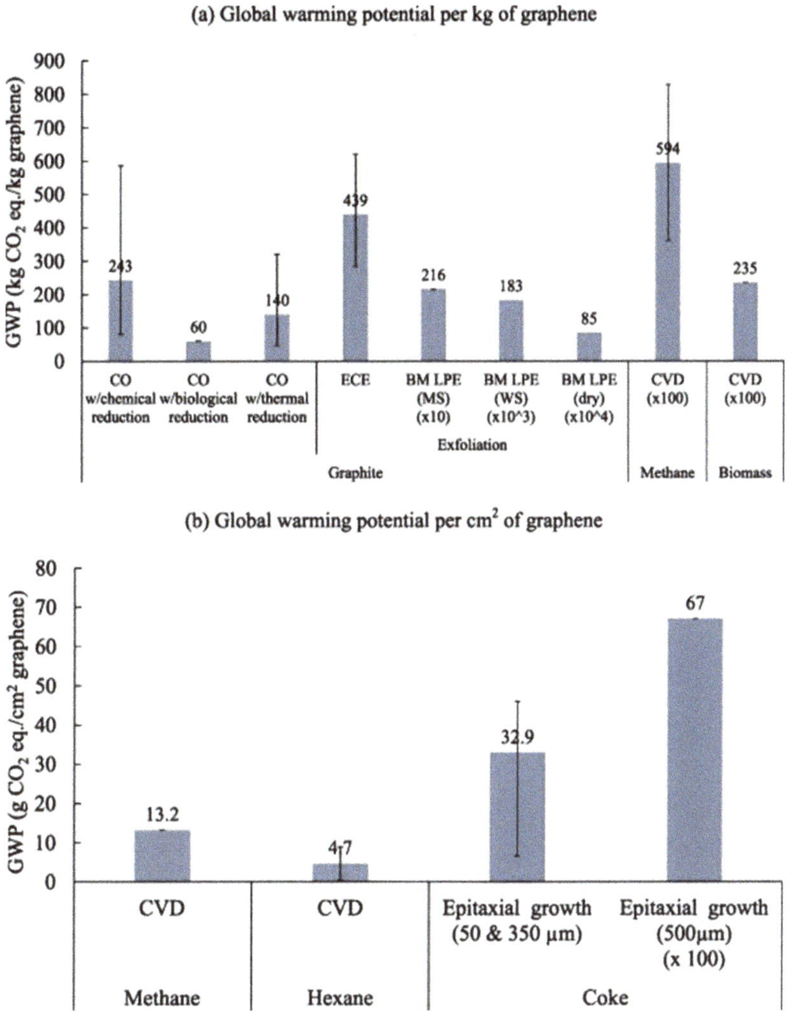

Figure 16.2. GWP of graphene expressed (a) per kg graphene and (b) per cm^2 graphene. Reproduced from [85] CC BY 4.0.

the reducing agent, the GWP is reported at 586 kg CO_2 eq. kg^{-1} graphene, with the production and use of hydrazine and phosphoric acid identified as the main hotspots [53]. When using glucose as the reducing agent, Serrano-Luján *et al* [53] found that the GWP of this process reduces by 40% (343 kg CO_2 eq. kg^{-1} graphene). The GWP associated with some modified versions of the Hummers' method are reported by Cossutta *et al* [57] at 153 kg CO_2 eq. kg^{-1} graphene, 146 kg CO_2 eq. kg^{-1} graphene and 380 kg CO_2 eq. kg^{-1} graphene for the Fugetsu, Bangal and Jeong modifications, respectively.

Besides graphite consumption, these studies have identified the chemical inputs and the acid neutralization step as the main hotspots of this production process, whilst energy consumption was found to be a minor contributor. The oxidative agents used across these methods also contribute to the differences in GWP reported.

Whilst a direct comparison of the results is not possible as the studies employed different methodologies (see figure 16.3(a)), all three studies concluded that the chemical reduction step is energy intensive and was identified as the main hotspot for this process. For example, Serrano-Luján *et al* [53] estimated the cumulative energy demand (CED) of the GRMs and reported values of 33 470 and 38 150 MJ kg^{-1} graphene for the Hummers' method using glucose and hydrazine, respectively, with process electricity consumption contributing with 80%–95% of the total CED (see figure 16.3(a)). Arvidsson *et al* [55] estimated the heat and electricity requirements of the GRM production process using the Hummers' method and the Marcano's modification, and the results are included in figure 16.3(a). The study found that the

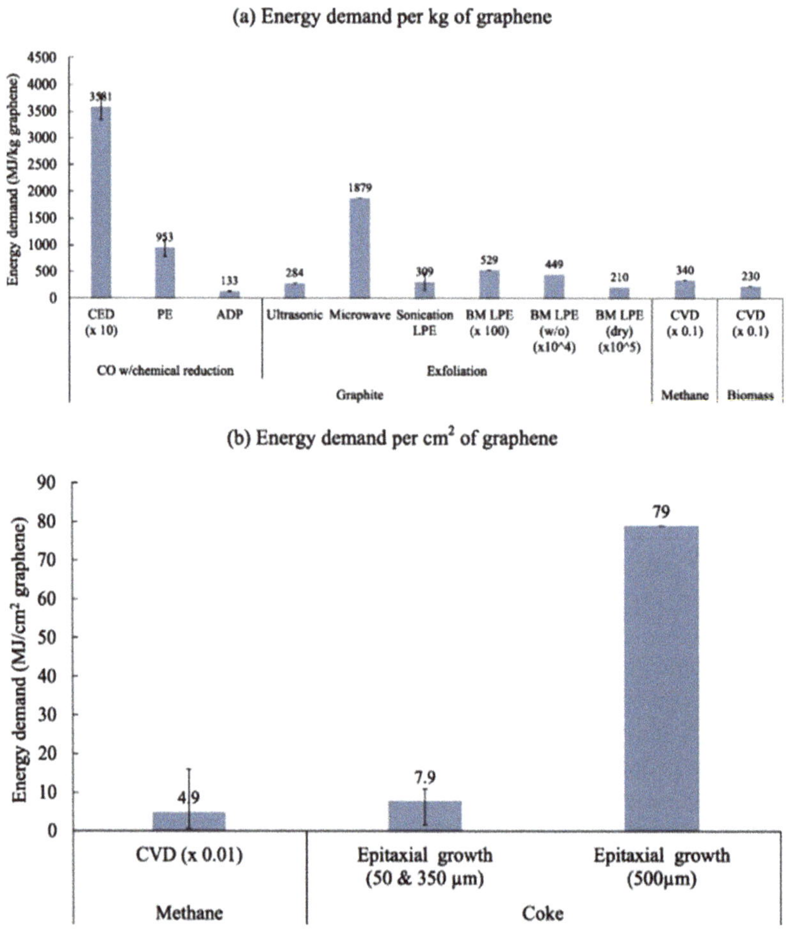

Figure 16.3. Energy demand of graphene produced expressed (a) per kg graphene and (b) per cm^2 graphene. Reproduced from [85] CC BY 4.0.

Hummers' method requires 1100 MJ kg^{-1} graphene, which is around 25% more energy consumption compared to the Marcano's method estimated at 780 and 980 MJ kg^{-1} graphene, depending on the amount of oxidative agent used. The heating requirements of the chemical reduction stage and hydrazine production were identified as the main hotspots for these processes [55]. Khanam *et al* [56] assessed the abiotic depletion potential of fossil energy for the studied GRMs production routes, and the results varied slightly depending on the reduction agent used, 139.38 MJ kg^{-1} graphene for hydrazine and 127.26 MJ kg^{-1} graphene for *E. coli*. The study also identified the heating requirements for the chemical reduction as the main hotspot, and the production of hydrazine accordingly.

16.3.2 Chemical oxidation with thermal reduction

The GWP results for chemical oxidation with thermal reduction presented in figure 16.2(a) correspond to the study by Cossutta *et al* [57]. The study compared five GO production methods from graphite with thermal reduction carried out under the same conditions at laboratory and commercial scale. The chemical oxidation methods included three modifications of the Hummers' method (Fugetsu, Bangal and Jeong) and the Brodie and Staudenmaier methods. The GWP values for this production process range from 86 to 320 kg CO_2 eq. kg^{-1} graphene at laboratory scale, where the Bangal and the Jeong modifications of the Hummers' method have the lowest and highest GWP, respectively. The reported GWP of the Bangal method at the commercial scale (46 kg CO_2 eq. kg^{-1} graphene) is 46% lower compared to the results estimated at laboratory scale due to potential improvements based on experimental insights around energy efficiency and material recovery.

16.3.3 Chemical exfoliation (electrochemical, ultrasonic, microwave-assisted and ball milling)

The studies conducted by Cossutta *et al* [57], Ampah *et al* [58], Pizza *et al* [52], Arvidsson *et al* [55], and Beloin-Saint-Pierre and Hischier [59] assessed the production of GRMs using different exfoliation techniques from graphite. Cossutta *et al* [57] estimated the environmental impacts for the production of graphene sheets via ECE, including GWP results. Ampah *et al* [58] and Pizza *et al* [52] reported energy consumption values for the production of GRMs using both expansion and ultrasonic exfoliation and microwave-assisted exfoliation. In the case of Ampah *et al* [58], the study looked at the production of graphene ink to make coatings for heat management, whilst Pizza *et al* [52] considered the production of epoxy-based nanocomposites using graphite nanoplatelets as fillers. Both studies focused on the energy requirements for the production of GRM components. Ampah *et al* [58], Arvidsson *et al* [55] and Beloin-Saint-Pierre and Hischier [59] looked at variations in ultrasonic-assisted exfoliation for the production of GRMs using different solvents as exfoliation agents. Ampah *et al* [58] and Arvidsson *et al* [55] reported the energy demand of the production of GRMs using sonication and dimethylformamide and dimethyl ether as exfoliating agents, respectively. Beloin-Saint-Pierre and Hischier [59] assessed a ball milling process for the production of

GRMs in water with and without using melamine as the exfoliating agent and reported values for GWP and CED, among other environmental impacts.

16.3.4 Graphene-based electrochemical sensors for heavy metal ions

Graphene-based electrochemical sensors have been developed for environmental analysis for the detection of heavy metal ions (Pb^{2+} and Cd^{2+}) [60, 61]. Li *et al* [60, 61] report that Nafion-graphene composite film based electrochemical sensors not only exhibits improved sensitivity for the metal ion (Pb^{2+} and Cd^{2+}) detections, but also alleviates the interferences due to the synergistic effect of graphene nanosheets and Nafion [60]. The stripping current signal is greatly enhanced on graphene electrodes. The enhanced performance is attributed to the unique properties of the graphene (nanosized graphene sheet, nano-scale thickness of these sheets and high conductivity), which endowed the capability to strongly adsorb target ions, enhanced the surface concentration, improved the sensitivity, and alleviated the fouling effect of surfactants [60, 61].

A variety of electrochemical sensors based on the modified electrode for the detection of UA have been reported in the literature. Therefore, based on nanomaterials, detection methods, detection electrodes, linearity range, limit of detection, and clinical specimens, we performed a detailed comparison of reported techniques of detecting UA. There are various electrochemical methods that can be used to detect UA in clinical specimens, but we can clearly find that each detection method for UA had its own characteristics and advantages. However, the best method to detect UA is to meet the detection requirements of clinical specimens. In contrast, the preparation method of GQDs in this study is simpler, highlighting the advantages in the preparation of modified nanomaterials. In addition, the selectivity, sensitivity, accuracy and linear range of GCE modified with GQDs for UA are very suitable for the detection of UA in clinical specimens. Considering the production cost, the as-prepared electrochemical sensor for detecting UA in this study is more suitable for primary medical institutions, or some medical institutions in poor areas. Therefore, the electrochemical sensing technology was applied to detect UA in serum, which opened up a new option for unconventional detection of UA, provided an experimental basis for the development of portable UA detection technology, and further expanded the application of electrochemical sensing technology in the field of laboratory medicine [62].

16.3.5 Reliability

Graphene has shown excellent performance in direct electrochemistry of enzyme, electrochemical detection of small biomolecules, electroanalysis (electrochemical sensors for bioanalysis and environmental analysis). As a competitor to CNTs, graphene has exhibited superior performance in these applications. However, the development of graphene-based materials/devices is still in its infancy. The future scientific research and application development of graphene-based materials/devices for electroanalysis and electrocatalysis is worthy of discussion here. Novel methods for well-controlled synthesis and processing of graphene should be developed.

As stated in the introduction, graphene has been synthesized with various strategies. However, the economical production approach with high yield is still not widely available. For electrochemical applications, the approach with chemical/thermal reduction of GO looks promising. Graphene sheets from chemical/thermal reduction of GO usually tend to re-stack during the synthesis and the processing. So far, many strategies [63–65] have been suggested in the synthesis and processing of graphene to prevent re-stacking and to improve the dispersion of graphene in solvents. In several recent reports, graphene was produced through the electrochemical reduction of GO, and the electrochemical rGO exhibits much better performance for electrochemical applications than the chemically reduced one [26, 66–68], which indicates a promising strategy in graphene synthesis and processing. Doping graphene with heteroatoms (nitrogen, boron, etc) should be investigated. Heteroatom doping in CNTs has shown to greatly improve the electrocatalytic activity [69, 70]. This strategy has been applied in graphene-based material synthesis [71–74] and theoretical study [75–84], but no reports are available on electrochemical study.

16.4 Outlook

This chapter demonstrates that one of the most commonly used approaches to dealing with environmental harm in graphene production is the use of waste material from various sources. Regardless of all of the various methods examined, it is clear from our review that there is currently no suitable approach for producing high-quality graphene from waste material. The majority of GRMs derived from waste have high levels of defects, contamination, and oxidation, and sometimes no 2D morphology, which may limit their applications in various fields. These limitations are primarily caused by carbon-containing waste.

A fair comparison among the various studies is not feasible due to the different starting points used and various treatment procedures that the unprocessed garbage underwent. In general, graphitic waste products have proven to be the easiest to process and exfoliate, and their characterization frequently allows direct comparisons to commonly prepared GRMs.

From the non-graphitic sources, plant- and animal-derived waste look like the more promising starting materials. This type of waste is abundant, widespread and has low value, often being a cost to be disposed of or recovered through combustion for electricity and heat purposes. Notably, plant-derived wastes contain a high quantity of polysaccharides (e.g. cellulose, lignin) that make these precursors suitable for the graphitization processes. It is also important to underline that the majority of the carbon-containing waste from biological or non-biological sources investigated so far are well 'defined', i.e. not mixed or blended with other wastes or contaminated, causing the yield and quality of the product to worsen. Notably, if the route of producing sustainable GMBs is pursued, the waste needs to be well defined and not mixed with other carbon-free waste.

In conclusion, this review provides an overview on the sustainable routes for production of graphene and related LCA studies, which can be used by the research community to improve the production and the use of GRMs by enabling graphene production to be compatible with the circular economy principles.

References

[1] Campanale C, Massarelli C, Savino I, Locaputo V and Uricchio V F 2020 A detailed review study on potential effects of microplastics and additives of concern on human health *Int. J. Environ. Res. Public Health* **17** 1212
[2] Chen G, Feng Q and Wang J 2020 Mini-review of microplastics in the atmosphere and their risks to humans *Sci. Total Environ.* **703** 135504
[3] Wang W, Gao H, Jin S, Li R and Na G 2019 The ecotoxicological effects of microplastics on aquatic food web, from primary producer to human: a review *Ecotoxicol. Environ. Saf.* **173** 110–7
[4] Novoselov K S, Jiang D, Schedin F, Booth T J, Khotkevich V V, Morozov S V and Geim A K 2005 Two-dimensional atomic crystals *Proc. Natl Acad. Sci. USA* **102** 10451
[5] Novoselov K S, Falko V I, Colombo L, Gellert P R, Schwab M G and Kim K 2012 A roadmap for graphene *Nature* **490** 192–200
[6] Ferrari A C *et al* 2015 Science and technology roadmap for graphene, related two-dimensional crystals, and hybrid systems *Nanoscale* **7** 4598–810
[7] Brodie B C 1859 On the atomic weight of graphite *Philos. Trans. R. Soc. London* **149** 249
[8] Zacharia R, Ulbricht H and Hertel T 2004 Interlayer cohesive energy of graphite from thermal desorption of polyaromatic hydrocarbons *Phys. Rev.* B **69** 155406
[9] Allen M J, Tung V C and Kaner R B 2010 Honeycomb carbon: a review of graphene *Chem. Rev.* **110** 132
[10] Zhang R, Dong Y L, Kong W J, Han W P, Tan P H, Liao Z M, Wu X S and Yu D P 2012 Growth of large domain epitaxial graphene on the C-face of SiC *J. Appl. Phys.* **112** 104307
[11] Weatherup R S, Bayer B C, Blume R, Baehtz C, Kidambi P R, Fouquet M, Wirth C T, Schlogl R and Hofmann S 2012 On the mechanisms of Ni-catalysed graphene chemical vapour deposition *Chem. Phys. Chem.* **13** 2544
[12] Tao L *et al* 2012 Uniform wafer-scale chemical vapor deposition of graphene on evaporated Cu (111) film with quality comparable to exfoliated monolayer *J. Phys. Chem.* C **116** 24068
[13] Sutter P W, Flege J-I and Sutter E A 2008 Epitaxial graphene on ruthenium *Nat. Mater.* **7** 406
[14] Yang X, Dou X, Rouhanipour A, Zhi L, Rader H J and Mullen K 2008 Two-dimensional graphene nanoribbons *J. Am. Chem. Soc.* **130** 4216
[15] Kuila T, Bose S, Mishra A K, Khanra P, Kim N H and Lee J H 2012 Chemical functionalization of graphene and its applications *Prog. Mater Sci.* **57** 1061
[16] Hummers W S and Offeman R E 1958 Preparation of graphitic oxide *J. Am. Chem. Soc.* **80** 1339
[17] Li D, Muller M B, Gilje S, Kaner R B and Wallace G G 2008 Processable aqueous dispersions of graphene nanosheets *Nat. Nanotechnol.* **3** 101
[18] Qi X Y, Pu K Y, Li H, Zhou X Z, Wu S X, Fan Q L, Liu B, Boey F, Huang W and Zhang H 2010 Amphiphilic graphene composites *Angew. Chem. Int. Ed.* **49** 9426
[19] Zhou X Z, Huang X, Qi X Y, Wu S X, Xue C, Boey F Y C, Yan Q Y, Chen P and Zhang H 2009 In situ synthesis of metal nanoparticles on single-layer graphene oxide and reduced graphene oxide surfaces *J. Phys. Chem.* C **113** 10842

[20] Stankovich S, Dikin D A, Dommett G H B, Kohlhaas K M, Zimney E J, Stach E A, Piner R D, Nguyen S T and Ruoff R S 2006 Graphene-based composite materials *Nature* **442** 282
[21] Chen W and Yan L 2011 *In situ* self-assembly of mild chemical reduction graphene for three-dimensional architectures *Nanoscale* **3** 3132
[22] Dua V, Surwade S P, Ammu S, Agnihotra S R, Jain S, Roberts K E, Park S, Ruoff R S and Manohar S K 2010 All-organic vapor sensor using inkjet-printed reduced graphene oxide *Angew. Chem. Int. Ed.* **49** 2154
[23] Yang S, Yue W, Huang D, Chen C, Lin H and Yang X 2012 A facile green strategy for rapid reduction of graphene oxide by metallic zinc *RSC Adv.* **2** 8827
[24] Liu J B, Fu S H, Yuan B, Li Y L and Deng Z X 2010 Toward a universal "adhesive nanosheet" for the assembly of multiple nanoparticles based on a protein-induced reduction/decoration of graphene oxide *J. Am. Chem. Soc.* **132** 7279
[25] Wang Z, Zhou X, Zhang J, Boey F and Zhang H 2009 Direct electrochemical reduction of single-layer graphene oxide and subsequent functionalization with glucose oxidase *J. Phys. Chem.* C **113** 14071
[26] Zhou M, Wang Y L, Zhai Y M, Zhai J F, Ren W, Wang F A and Dong S J 2009 Controlled synthesis of l.arge-area and patterned electrochemically reduced graphene oxide films *Chem. Eur. J.* **15** 6116
[27] Huang X, Zhou X, Wu S, Wei Y, Qi X, Zhang J, Boey F and Zhang H 2010 Reduced graphene oxide-templated photochemical synthesis and in situ assembly of Au nanodots to orderly patterned Au nanodot chains *Small* **6** 513
[28] Williams G, Seger B and Kamat P V 2008 TiO_2-graphene nanocomposites. UV-assisted photocatalytic reduction of graphene oxide *ACS Nano* **2** 1487
[29] Becerril H A, Mao J, Liu Z, Stoltenberg R M, Bao Z and Chen Y 2008 Evaluation of solution-processed reduced graphene oxide films as transparent conductors *ACS Nano* **2** 463
[30] Zhou Y, Bao Q, Tang L A L, Zhong Y and Loh K P 2009 Hydrothermal dehydration for the "green" reduction of exfoliated graphene oxide to graphene and demonstration of tunable optical limiting properties *Chem. Mater.* **21** 2950
[31] Lerf A, He H, Forster M and Klinowski J 1998 Structure of graphite oxide revisited *J. Phys. Chem.* B **102** 4477
[32] Jang H D, Kim S K, Chang H, Roh K M, Choi J W and Huang J 2012 A glucose biosensor based on TiO_2–graphene composite *Biosens. Bioelectron.* **38** 184
[33] Zhang L L, Cheng H H, Zhang H M and Qu L G 2012 Direct electrochemistry and electrocatalysis of horseradish peroxidase immobilized in graphene oxide–Nafion nanocomposite film *Electrochim. Acta* **65** 122
[34] Huang K J, Niu D J, Liu X, Wu Z W, Fan Y, Chang Y F and Wu Y Y 2011 Direct electrochemistry of catalase at amine-functionalized graphene/gold nanoparticles composite film for hydrogen peroxide sensor *Electrochim. Acta* **56** 2947
[35] Srivastava R K, Srivastava S, Narayanan T N, Mahlotra B D, Vajtai R, Ajayan P M and Srivastava A 2011 Functionalized multilayered graphene platform for urea sensor *ACS Nano* **6** 168
[36] Prasannakumar S, Manjunatha R, Nethravathi C, Suresh G S, Rajamathi M and Venkatesha T V 2012 Graphene-carbon nanotubes modified graphite electrode for the determination of nicotinamide adenine dinucleotide and fabrication of alcohol biosensor *J. Solid State Electrochem.* **16** 3189

[37] Shan C S, Yang H F, Han D X, Zhang Q X, Ivaska A and Niu L 2010 Electrochemical determination of NADH and ethanol based on ionic liquid-functionalized graphene *Biosens. Bioelectron.* **25** 1504
[38] Tung V C, Chen L M, Allen M J, Wassei J K, Nelson K, Kaner R B and Yang Y 2009 Low-temperature solution processing of graphene–carbon nanotube hybrid materials for high-performance transparent conductors *Nano Lett.* **9** 1949
[39] Shang N G, Papakonstantinou P, McMullan M, Chu M, Stamboulis A, Potenza A, Dhesi S S and Marchetto H 2008 Catalyst-free efficient growth, orientation and biosensing properties of multilayer graphene nanoflake films with sharp edge planes *Adv. Funct. Mater.* **18** 3506
[40] Ma X Y, Chao M Y and Wang Z X 2012 Electrochemical detection of dopamine in the presence of epinephrine, uric acid and ascorbic acid using a graphene-modified electrode *Anal. Methods* **4** 1687
[41] Sheng Z H, Zheng X Q, Xu J Y, Bao W J, Wang F B and Xia X H 2012 Electrochemical sensor based on nitrogen doped graphene: Simultaneous determination of ascorbic acid, dopamine and uric acid *Biosens. Bioelectron.* **34** 125
[42] Feng L Y, Chen Y, Ren J S and Qu X G 2011 A graphene functionalized electrochemical aptasensor for selective label-free detection of cancer cells *Biomaterials* **32** 2930
[43] Giovanni M, Bonanni A and Pumera M 2012 Detection of DNA hybridization on chemically modified graphene platforms *Analyst* **137** 580
[44] Lin L, Liu Y, Tang L H and Li J H 2011 Electrochemical DNA sensor by the assembly of graphene and DNA-conjugated gold nanoparticles with silver enhancement strategy *Analyst* **136** 4732
[45] Zhou M, Zhai Y and Dong S 2009 Electrochemical sensing and biosensing platform based on chemically reduced graphene oxide *Anal. Chem.* **81** 5603
[46] Hu Y W, Hua S C, Li F H, Jiang Y Y, Bai X X, Li D and Niu L 2011 Green-synthesized gold nanoparticles decorated graphene sheets for label-free electrochemical impedance DNA hybridization biosensing *Biosens. Bioelectron.* **26** 4355
[47] Zhao J, Chen G F, Zhu L and Li G X 2011 Graphene quantum dots-based platform for the fabrication of electrochemical biosensors *Electrochem. Commun.* **13** 31
[48] Mao K X, Wu D, Li Y, Ma H M, Ni Z Z, Yu H Q, Luo C N, Wei Q and Du B 2012 Label-free electrochemical immunosensor based on graphene/methylene blue nanocomposite *Anal. Biochem.* **422** 22
[49] Yang M H, Javadi A and Gong S Q 2011 Sensitive electrochemical immunosensor for the detection of cancer biomarker using quantum dot functionalized graphene sheets as labels *Sensors Actuators* B **155** 357
[50] Lu J J, Liu S Q, Ge S G, Yan M, Yu J H and Hu X T 2012 Ultrasensitive electrochemical immunosensor based on Au nanoparticles dotted carbon nanotube–graphene composite and functionalized mesoporous materials *Biosens. Bioelectron.* **33** 29
[51] Li R, Wu D, Li H, Xu C X, Wang H, Zhao Y F, Cai Y Y, Wei Q and Du B 2011 *Anal. Biochem.* **414** 196
[52] Pizza A, Metz R, Hassanzadeh M and Bantignies J-L 2014 Life cycle assessment of nanocomposites made of thermally conductive graphite nanoplatelets *Int. J. Life Cycle Assess.* **19** 1226–37
[53] Serrano-Luján L, Víctor-Román S, Toledo C, Sanahuja-Parejo O, Mansour A E, Abad J, Amassian A, Benito A M, Maser W K and Urbina A 2019 Environmental impact of the production of graphene oxide and reduced graphene oxide SN *Appl. Sci.* **1** 179

[54] Scott R P and Cullen A C 2016 Reducing the life cycle environmental impacts of kesterite solar photovoltaics: comparing carbon and molybdenum back contact options *Int. J. Life Cycle Assess.* **21** 29–43

[55] Arvidsson R and Molander S 2017 Prospective life cycle assessment of epitaxial graphene production at different manufacturing scales and maturity *J. Ind. Ecol.* **21** 1153–64

[56] Khanam P N, Popelka A, Alejji M and AlMaadeed M A 2017 Biotechnological production process and life cycle assessment of graphene *J. Nanomater.* **2017** 5671584

[57] Cossutta M, McKechnie J and Pickering S J 2017 A comparative LCA of different graphene production routes *Green Chem.* **19** 5874–84

[58] Ampah A D, Pagone E and Salonitis K 2019 *Life Cycle Assessment of Graphene as Heating Element* (Springer) pp 283–97

[59] Beloin-Saint-Pierre D and Hischier R 2021 Towards a more environmentally sustainable production of graphene-based materials *Int. J. Life Cycle Assess.* **26** 327–43

[60] Li J, Guo S J, Zhai Y M and Wang E K 2009 High-sensitivity determination of lead and cadmium based on the Nafion-graphene composite film *Anal. Chim. Acta* **649** 196

[61] Li J, Guo S J, Zhai Y M and Wang E K 2009 Nafion–graphene nanocomposite film as enhanced sensing platform for ultrasensitive determination of cadmium *Electrochem. Commun.* **11** 1085

[62] Shan C S, Yang H F, Han D X, Zhang Q X, Ivaska A and Niu L 2009 Water-soluble graphene covalently functionalized by biocompatible poly-L-lysine *Langmuir* **25** 12030

[63] Bai H, Xu Y X, Zhao L, Li C and Shi G Q 2009 Non-covalent functionalization of graphene sheets by sulfonated polyaniline *Chem. Commun.* **2009** 1667–9

[64] Si Y and Samulski E T 2008 Synthesis of water soluble graphene *Nano Lett.* **8** 1679

[65] Xu Y X, Bai H, Lu G W, Li C and Shi G Q 2008 Flexible graphene films via the filtration of water-soluble noncovalent functionalized graphene sheets *J. Am. Chem. Soc.* **130** 5856

[66] Guo H L, Wang X F, Qian Q Y, Wang F B and Xia X H 2009 A green approach to the synthesis of graphene nanosheets *ACS Nano* **3** 2653

[67] Ramesha G K and Sampath S 2009 Electrochemical reduction of oriented graphene oxide films: an in situ Raman spectroelectrochemical study *J. Phys. Chem.* C **113** 7985

[68] Shao Y Y, Wang J, Engelhard M H, Wang C M and Lin Y H 2009 Facile and controllable electrochemical reduction of graphene oxide and its applications *J. Mater. Chem.* **20** 743

[69] Dong J P, Qu X M, Wang L J, Zhao C J and Xu J Q 2008 Electrochemistry of nitrogen-doped carbon nanotubes (CN_x) with different nitrogen content and its application in simultaneous determination of dihydroxybenzene isomers *Electroanalysis* **20** 1981

[70] Deng S Y, Jian G Q, Lei J P, Hu Z and Ju H X 2009 A glucose biosensor based on direct electrochemistry of glucose oxidase immobilized on nitrogen-doped carbon nanotubes *Biosens. Bioelectron.* **25** 373

[71] Wei D C, Liu Y Q, Wang Y, Zhang H L, Huang L P and Yu G 2009 Synthesis of n-doped graphene by chemical vapor deposition and its electrical properties *Nano Lett.* **9** 1752

[72] Subrahmanyam K S, Panchakarla L S, Govindaraj A and Rao C N R 2009 Simple method of preparing graphene flakes by an arc-discharge method *J. Phys. Chem.* C **113** 4257

[73] Wang X R, Li X L, Zhang L, Yoon Y, Weber P K, Wang H L, Guo J and Dai H J 2009 N-doping of graphene through electrothermal reactions with ammonia *Science* **324** 768

[74] Li X L, Wang H L, Robinson J T, Sanchez H, Diankov G and Dai H J 2009 Simultaneous nitrogen doping and reduction of graphene oxide *J. Am. Chem. Soc.* **131** 15939

[75] Groves M N, Chan A S W, Malardier-Jugroot C and Jugroot M 2009 Improving platinum catalyst binding energy to graphene through nitrogen doping *Chem. Phys. Lett.* **481** 214
[76] Nambudumada S P and Manjunatha J G 2023 Voltammetric analysis of hazardous azo-dye indigo carmine at simple and cost-effective carbon paste sensor modified with L-phenylalanine *Sens. Technol.* 2218877
[77] Pushpanjali P A, Manjunatha J G, Tigari G and Fattepur S 2020 Poly(niacin) based carbon nanotube sensor for the sensitive and selective voltammetric detection of vanillin with caffeine *Anal. Bioanal. Electrochem.* **12** 553–68
[78] Manjunatha J G 2019 Electrochemical polymerised graphene paste electrode and application to catechol sensing *Open Chem. Eng. J.* **13** 81–7
[79] Prinith N S, Manjunatha J G and Raril C 2019 Electrocatalytic analysis of dopamine, uric acid and ascorbic acid at poly (adenine) modified carbon nanotube paste electrode: a cyclic voltammetric study *Anal. Bioanal. Electrochem.* **11** 742–56
[80] Prinith N S and Manjunatha J G 2020 Polymethionine modified carbon nanotube sensor for sensitive and selective determination of L-tryptophan *J. Electrochem. Sci. Eng.* **10** 305–15
[81] Raril C and Manjunatha J G 2018 Sensitive electrochemical analysis of resorcinol using polymer modified carbon paste electrode: a cyclic voltammetric study *Anal. Bioanal Electrochem.* **10** 488–98
[82] Pushpanjali P A, Manjunatha J G and Srinivas M T 2020 Highly sensitive platform utilizing poly (l-methionine) layered carbon nanotube paste sensor for the determination of voltaren *Flat Chem.* **24** 100207
[83] Hareesha N, Manjunatha J G, Amrutha B M, Pushpanjali P A, Charithra M M and Prinith Subbaiah N 2021 Electrochemical analysis of indigo carmine in food and water samples using a poly (glutamic acid) layered multi-walled carbon nanotube paste electrode *J. Electron. Mater.* **50** 1230–8
[84] Hareesha N, Manjunatha J G, Amrutha B M, Sreeharsha N, Asdaq S B and Anwer M K 2021 A fast and selective electrochemical detection of vanillin in food samples on the surface of poly (glutamic acid) functionalized multiwalled carbon nanotubes and graphite composite paste sensor *Colloids Surf.,* A **626** 127042
[85] Munuera J, Britnell L, Santoro C, Cuéllar-Franca R and Casiraghi C 2021 A review on sustainable production of graphene and related life cycle assessment *2D Mater.* **9** 012002

www.ingramcontent.com/pod-product-compliance
Ingram Content Group UK Ltd.
Pitfield, Milton Keynes, MK11 3LW, UK
UKHW051904100926
8556IPUK00005B/759

9 780750 355766